# The Book of Tempeh

## Praise for *The Book of Tempeh*

**Frances Moore Lappé**—*"The Book of Tempeh,* prepared with such care, should be read by all who would like their eating pleasures to match their political understanding."

**Soyfoods**—"Stunningly impressive in its thoroughness and vastly wholistic in its vision. A milestone in the accelerating soyfoods renaissance. An unparalleled success."

**Library Journal**—"Unique in focus and truly excellent treatment . . . highly recommended."

**Whole Foods**—"The research is thorough and the recipes are mouth-watering."

**East West Journal**—". . . an invaluable handbook . . . the sole practical guide for making tempeh at home or on a larger scale."

**Food & Wine**—"A trailblazing work on a little-known but much-loved food in Indonesia . . . the definitive word on a food well worth knowing."

**Vegetarian Times**—"Carries on the tradition of excellence established by *The Book of Tofu*. If you have used that modern classic or *The Book of Miso* you have no doubt come to expect nothing but the best from these authors: an historical and cultural context, thorough research, complete nutritional analysis, delectable recipes, and profuse illustrations. Add to that a compelling analysis of the stark picture of world hunger and the revolutionary role of soyfoods in meeting critical protein shortages, and you have the formula for Soyfoods Center books. This latest is a bright and lively account of what will surely be one of the most important foods of the future—tempeh."

### BOOKS BY
### William Shurtleff & Akiko Aoyagi

The Book of Tofu

Tofu & Soymilk Production

The Book of Miso

Miso Production

The Book of Tempeh

Tempeh Production

History of Tempeh

Soyfoods Industry and Market:
    Directory and Databook (annual)

The Book of Kudzu

# THE BOOK OF
# TEMPEH

# William Shurtleff
# Akiko Aoyagi

## Illustrations by AKIKO AOYAGI

SECOND EDITION

HARPER COLOPHON BOOKS
Harper & Row, Publishers
New York, Cambridge, Philadelphia, San Francisco
London, Mexico City, São Paulo, Singapore, Sydney

*A Soyfoods Center Book*

# Hardcover Professional Edition of The Book of Tempeh

An enlarged professional edition of *The Book of Tempeh,* for those who wish to explore the subject in more detail, is available from The Soyfoods Center. In addition to all of the information contained in this paperback edition, it contains over 88 pages of additional detailed material. The Glossary is greatly enlarged.

Appendix A: A Brief History of Tempeh East and West
Appendix B: Tempeh in Indonesia
Appendix C: The Varieties of Tempeh
Appendix D: Soybean Production and Traditional Soyfoods in Indonesia
Appendix E: The Microbiology and Chemistry of Tempeh Fermentation
Appendix F: People and Institutions Connected with Tempeh
Appendix G: Tempeh Shops in the West
Appendix H: Onchom or Ontjom
Weights, Measures, and Equivalents
Glossary, Bibliography, Index

---

Grateful acknowledgment is made for permission to reprint excerpts from *The Book of Tofu* by William Shurtleff and Akiko Aoyagi, Copyright © 1976 by William Shurtleff and Akiko Aoyagi; and from *The Book of Miso* by William Shurtleff and Akiko Aoyagi, copyright © 1977 by William Shurtleff and Akiko Aoyagi. Reprinted by permission of Autumn Press.

THE BOOK OF TEMPEH, Second Edition, Revised and Updated.

*Book design by Akiko Aoyagi*

*Typography by Stephanie Winkler*

Library of Congress Cataloging in Publication Data

Shurtleff, William, 1941
    The book of tempeh.
    "A Soyfoods Center book."
    Bibliography: p.
    Includes index.
    1. Soybeans. 2. Protein. 3. Tempeh.
    4. Onchom or ontjom. 5. Indonesia
    6. Vegetarian cookery. I. Aoyagi,
Akiko, 1950- joint author. II. Title. III. Title: Tempeh.
TX814.5.B4S54  1985   641.6'5'655   84-48621
ISBN 0-06-091265-0 (pbk.)
"Cover illustration by Akiko Aoyagi"                    85 86 87 88 10 9 8 7 6 5 4 3 2 1

DEDICATION . . . .

Within this lifetime, to awaken to our true nature, original enlightenment,
    and perfect oneness with all beings, things, and events
To manifest this realization in our daily life as love
To vow and endeavor with our whole body and mind
    to help all beings cross over to the other shore of liberation before doing so ourselves

THE BODHISATTVA'S VOW

He will never go to heaven
    who is content to go alone.

BOETHIUS, A.D. 450

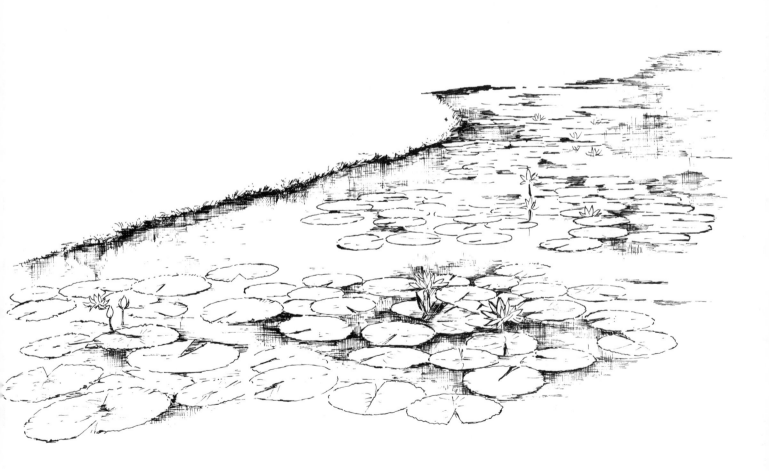

# Acknowledgments

Part of the satisfaction and adventure of writing a book lies in meeting people from around the world who share a common interest and in learning from them what is often the knowledge or skill acquired during a lifetime of devoted study and practice. Without these teachers and friends, this book could not have been written. For us the name of each person here is surrounded with vivid memories. We wish to give deepest thanks . . .

To American tempeh researchers Clifford W. Hesseltine, Keith H. Steinkraus, Earl W. Swain, and Hwa L. Wang.

To North America's first generation of fine tempeh producers Cynthia and Albert Bates, John and Charlotte Gabriel, Lewis Headrick, Alexander Lyon, and Don Wilson (all from The Farm), Joan Harrison, Henoch Khoe, Rudy Kohler, Gale Randall, Earl Lepper, Russ Pals, Benjamin Hills, Robert Walker, and the tempeh collective at Wolfmoon Bakery.

To Indonesian tempeh researchers Purwo Arbianto, Sutarjo Brotonegoro, J. K. Dewipadma, D. Dwidjoseputra, Indrawati Gandjár, Edi Guhardja, Sri Hartadi, Mr. Hermana, Nasruddin Ilyas, Ir. Jutono, R. B. Kasmidjo, M. Khumaidi, Ko Swan Djien, Mas Loegito, Dedi Muchadi, Djiteng Roedjito, Mr. Roestamsjah, Susono Saono, M. S. A. Sastroamijojo, Ibrahim Sastroamihardja, Joedoro Soedarsono, Slamet Sudarmadji, Mr. Sudigdo, Ignatius Sueharto, Ir. Suhardjo, Lindayati Tanuwidjaya, and F. G. Winarno.

To producers of fine tempeh in Indonesia Manan B. Samun, Oeben Sjarim, Pedro Sudjono, and many others too numerous to mention here.

To fine Indonesian tempeh cooks and nutritionists, Mmes. Mariyati Sukarni and Amini Nasution of IPB in Bogor, Ms. Uken Sakaeni of the Nutrition Research and Development Institute in Bogor, Ms. Li Emma of Ruma Sakit Mata in Yogyakarta, Mr. and Ms. Muhammad Mustam of Tokyo, and Ms. Marianka de Nijs of Los Angeles.

To tempeh researchers Robert Djurtoft of Denmark, Kiku Murata of Japan, Napavarn Noparatnaraporn of Thailand, Natividad Dionkno-Palo of the Philippines, and Yeo Quee Lan of Malaysia.

To Bob Hamaker, onchom research scholar from The Netherlands; Ho Coy Choke, geneticist and onchom research scholar from Malaysia; John J. Ellis, Paul Curtis, Ralph Emerson, David D. Perkins, and Barbara Turner, all mycologists or geneticists from the United States; and Michio Kozaki, mycologist from Japan. Ted Hymowitz and Noel Vietmeier, plant and agricultural research scholars from the United States; Yohanni Johns, renowned Southeast Asian cookbook author.

The majority of these people have kindly reviewed portions of our manuscript and offered helpful suggestions. Special thanks go to Dr. Keith H. Steinkraus, Dr. Clifford W. Hesseltine, and Dr. Hermana, who have given most generously of their knowledge and time from the inception of this project until its completion.

Finally we would like to thank Buz Wyeth and our editors, Nahum Waxman and Margaret Cheney.

Doubtless this book has its share of errors in fact or judgment. These, of course, are the responsibility of the authors alone.

# Contents

# What Is Tempeh?

## THE SOYFOOD WITH CULTURE

Tempeh (pronounced TEM-pay) is a popular Indonesian fermented (cultured) food consisting of tender-cooked soybeans (or occasionally other legumes, seeds, cereal grains, or even coconut) bound together by a dense cottony mycelium of *Rhizopus* mold into compact, ¾-inch-thick white cakes or patties. Sold fresh, refrigerated, or frozen, these are usually sliced and fried until their surface is crisp and golden brown. Served this way, their flavor and texture have been described variously as reminiscent of "southern fried chicken," "savory veal cutlets," or "seafood fillets"; their pleasant, subtly sweet fresh aroma as "nutty," "cheesy," "mushroomy" or "yeasty like freshly baked bread."

Tempeh owes much of its good flavor, sliceable meatlike texture, and excellent nutritional properties to the process of fermentation, the same process that gives us cheese (made with molds and bacteria), yogurt (made with lactic-acid bacteria), bread (made with yeasts), and a number of our other favorite natural foods.

Full-bodied and satisfying, tempeh is unique in its ability to serve as a delicious main course at meals in place of meat, chicken, or fish, and to become the protein backbone of a vegetarian diet. Convenient in that it requires only a few minutes of cooking (or it may be sold in ready-to-eat form), tempeh is also remarkably versatile and can be served in hundreds of different Western-style recipes. You can use it to make Tempeh Burgers or Tempeh, Lettuce & Tomato Sandwiches; topped with applesauce it becomes Tempeh Chops; pureed it makes creamy dressings or spreads; crispy slices or cubes are wonderful used like croutons in salads or soups, or added to pizza toppings, stir-fried rice, casseroles, sauces, or tacos. Among true connoisseurs, the favorite (and simplest) way of serving it is shallow-fried or deep-fried as Seasoned Crisp Tempeh or Coriander & Garlic Crisp Tempeh.

Fresh soy tempeh contains an average of 19.5 percent protein, which compares very favorably with chicken (21%), beef (20%), hamburger (13%), eggs (13%), and milk (3%). Moreover, the quality of this protein (the percent of the total protein that can actually be utilized by the body) is only slightly less than that of chicken or beef. Unlike tofu, East Asia's other primary source of high-quality soy protein, tempeh is a whole food from which we derive the benefits of that unanalyzable essence of wholeness plus natural dietary fiber and the full abundance of soy's fine nutrients. And tempeh is the world's richest known vegetarian source of vitamin $B_{12}$, one of the ingredients most often lacking in vegetarian diets. It is also highly digestible; during tempeh's fermentation, the molds produce enzymes that partially break down or digest the soy proteins and oils, making them more easily assimilated by the body. Simultaneously the beans are tenderized and the necessary cooking time greatly reduced to yield a handy, quick-cooking food. The result is a product of better nutritional value and flavor than cooked soybeans. One of the few fermented vegetarian foods made without salt, it is also an excellent diet food, containing only 157 calories per 100-gram (3½-ounce) serving, which costs only about 26 cents. Finally, it is very low in saturated fats and contains absolutely no cholesterol—in short, an ideal food for every health-minded person.

Although soy tempeh is by far the most popular type, it is by no means the only one. Some of our other favorites are peanut (or peanut & soy) tempeh, millet (or

8

millet & soy) tempeh, rice & soy tempeh, wheat & soy tempeh, whole-wheat-noodle tempeh, and okara tempeh; the latter is made from the soy pulp left over after the preparation of tofu or soymilk.

The more than thirty varieties of tempeh can all be grouped into five basic types: (1) *legume tempehs,* including traditional Indonesian varieties made from soy, velvet, winged, leucaena, and mung beans, plus a host of modern developments including lupin, cowpea, broadbean, and garbanzo tempeh; (2) *grain* (or *seed*) *& soy tempehs,* such as wheat & soy or rice & soy varieties; (3) *grain tempehs,* featuring rice, barley, millet, wheat, oats, or rye; (4) *presscake tempehs,* which transform into nutritious delicacies ingredients that might otherwise become wastes: okara, presscakes (residues after extracting oils) of peanuts, coconut, and mung beans, defatted soy meal, and various mixtures of these. Coconut-presscake tempeh *(tempé bongkrek),* it should be noted, occasionally becomes highly toxic and should *never* be prepared or consumed; and (5) *nonleguminous seed* or *seed & legume tempehs,* such as rubber-seed or sesame & soy tempeh.

In Indonesia, various extenders (or adulterants) are often used in tempeh to lower the price and prevent waste of nutrients. Extenders include, in order of popularity, okara, cassava, mung-bean presscake, soybean hulls, sweet potatoes, coconut or coconut presscake, and papaya.

Finally, tempeh may be sold in any of four stages of fermentation. *Premature tempeh,* removed from its incubator 4 to 6 hours early, has a relatively firm texture; *mature tempeh* is the basic common type; *slightly overripe tempeh,* used 2 or 3 days after it reaches maturity, has a subtle Camembert aroma; and *overripe tempeh* (tempe busuk), used 3 to 5 days after it reaches maturity, has a pronounced aroma and soft consistency quite like those of a fine Camembert cheese. Each of the many types of tempeh is described in detail in the professional edition of this book.

Making your own tempeh at home is surprisingly easy, especially now that top-quality starter is readily available in the United States. Using storebought soybeans (35 cents per pound) and utensils found in a typical kitchen, you can prepare tempeh for about 34 cents a pound; the actual preparation times takes about 1 to 1½ hours followed by a 24-hour incubation. Whole soybeans are washed, brought just to a boil and soaked overnight, then dehulled by being rubbed between the palms of your hands. Cooked for 45 minutes, they are drained, cooled to body temperature, and mixed with a little of the spore culture starter. The mixture is then packed into perforated sandwich bags or bread pans and incubated (as for yogurt) at about 88°F. (31°C.) for 22 to 26 hours. It is now ready to be cooked and served. Any remaining portions may be refrigerated and will stay fresh for 3 to 5 days.

Tempeh is an ideal food for use in developing countries as a source of tasty and inexpensive high-quality protein:

1. Its commercial production requires only the simplest, decentralized, low-level technology; no machines are necessary, the process is labor intensive, and production costs are low. The only ingredients are soybeans, water, and a little home-grown starter.

2. The warm or tropical climates characteristic of so many developing countries greatly facilitate the tempeh fermentation, allowing year-round incubation at room temperature without consumption of energy for heating.

3. The fermentation is unusually simple and short (24 to 48 hours) as compared with several months for many other fermented foods.

4. Tempeh makes it possible to utilize what might otherwise become waste products from food processing operations; okara and the various presscakes from peanuts, mung beans, and cassava are transformed into tasty and nutritious foods.

5. Tempeh has a flavor and texture, appearance and aroma that are well suited to use in local cuisines throughout the world: it is an ideal meat substitute and extremely versatile.

Tempeh originated hundreds of years ago in Central and East Java, and is now Indonesia's most popular soy-protein food, making use of over 64 percent of the country's total soybean production. Tempeh production in Indonesia is still a household art, and most of the 41,000 shops that make fresh tempeh daily are family run. They employ a total of 128,000 workers, who produce each year 169,000 tons of tempeh having a retail value of $85.5 million (U.S.). The tempeh is prepared in much the same way as described above except that the precooked soybeans are dehulled underfoot in bamboo baskets, allowed to undergo a prefermentation in the soak water, inoculated with a starter grown on soybeans sandwiched between hibiscus leaves, and wrapped in small packets with banana leaves or packed into perforated plastic bags that are incubated for about 48 hours at balmy air temperature (77° to 80° F., 25° to 27° C.). The tempeh is then taken to the market and sold fresh in the wrapper for about 23 to 30 cents (U.S.) a pound (1977). It is a key protein source (and protein booster) in the diet of Indonesia's rice-loving millions and is also popular in the Netherlands, Surinam, Malaysia, Singapore, and other areas where Indonesians have settled.

And now tempeh has come to North America. Although it first came to be known and studied here only as recently as the late 1950s, it experienced a sudden rise in popularity starting about 1975, together with the interest in natural foods, low-cost protein, meatless or vegetarian diets, ecology, and simpler, more satisfying life styles. Yet, in spite of its short history here, tempeh has already received a warm welcome and remarkably strong reviews. The first came in 1971, when "Foods of Tomorrow" section of *Food Processing* magazine did a study of "specialty fermented foods" to test their potential acceptability in the American market. It concluded:

*Of all fermented foods, tempeh with its high ratings in taste, nutritional benefits, and simple, low-cost pro-*

cessing techniques, appears to be the most likely candidate for Americanization and may be one of the next to appear in the U.S. marketplace.

In 1976, *Organic Gardening and Farming* magazine, too, began to get excited about tempeh. The staff sent tempeh kits and a questionnaire to sixty readers across the country and asked for feedback. Again the results were most promising, and led to numerous follow-up articles:

*People who've gotten the starter and made their own tempeh have enjoyed the experience and loved the food they made themselves. "Very good eating," reported Mark Kane. "The whole family liked it immensely," said Marilyn Emerson. "I couldn't believe soybeans could taste that good (though we eat and enjoy tofu)," said Robert Shea, "It has a mild flavor like veal."*

In June, 1977, Robert Rodale, editor of *Prevention* magazine and *Organic Gardening,* after considerable experimentation with tempeh cookery and homemade preparation, did a feature article followed by a visit to the first tempeh shop to be started by an American. His enthusiasm added greatly to the growing interest throughout America:

*Tempeh is on its way up. Before long it will be eaten widely and lovingly across this land of ours. . . . It has the potential to be a real main dish that could lead soybeans to the forefront of the health food advance. . . . I'm convinced that tempeh is just sitting there waiting for lightning to strike, the way it hit yogurt. All the signs point to that happening, and soon. There's a tremendous need for new main-dish foods that aren't meat or milk products. . . . Yogurt has become the modern, healthful, "with-it" snack and lunch food for people on the move. Tempeh, in the same stylish fashion, can become the main dish that makes sense for our ecological and nutrition-conscious future. Not only is it inexpensive, but it's much lower in cholesterol than fat-rich meat and eggs. That point about cholesterol could be the clincher for tempeh's future popularity.*

*Ultimately tempeh could get into the mainstream of the American diet. Now, it's a unique and good health food. With careful handling and promotion, it could eventually challenge the fast-food hamburger. If you've ever kicked yourself for missing a chance to get in on the ground floor of a new industry when it was just starting to grow, now's your chance to make amends. We could well be on the verge of an "age of tempeh," and those who start working in that field now could find the rewards great.*

The Farm, a spiritual community of 1,400 people living on 1,700 acres in Tennessee, has played a leading role in introducing tempeh to America. In September, 1977, in a major article in *Mother Earth News* magazine, they wrote:

*Tempeh was a hit here on The Farm the very first time we tried it. We've been making and eating tempeh*

for four years now and we still think it's the greatest food since yogurt . . . We like this healthful taste treat because it can be served in so many delightful ways that we never get tired of it. . . . Children come back for seconds and thirds.

On September 21, 1977, Michio Kushi, one of America's best-known teachers of natural foods and healing, speaking in Washington, D.C., to President Carter's executive committee on food policy, recommended the use of traditional, naturally fermented soyfoods products such as soy sauce, miso, and tempeh.

Since the publication of *The Book of Tempeh* in July 1979 and of *Tempeh Production* in March 1980 (both by Shurtleff and Aoyagi), tempeh has caught on quickly in the West. In July 1979 there were only 19 tempeh companies in the Western world, but by early 1984 the number had jumped to 82, including 53 in the U.S.A., 18 in Western Europe, 5 in Australia, and 4 in Canada. The addresses of all of these are given in Appendix B. Most of these businesses are run by non-Indonesian Westerners and the industry is thriving. In America from 1981 to late 1983 tempeh production grew at the remarkable rate of 33 percent a year!

Thus, for the first time in history, tempeh is being made commercially and enjoyed in the West. And thousands of people are also making their own tempeh at home and helping it to find its new home in America's evolving cuisine. Tempeh is truly a food for the people—people everywhere. Its unique combination of virtues promises to make it a food of worldwide interest during the decades to come . . . a key part of the soyfoods revolution.

*Cuts of fresh tempeh*

# Preface

Having lived in Japan for the past seven years, writing books about tofu (soybean curd) and miso (fermented soy paste), Akiko and I have developed a deep respect for the remarkable soybean, which is known throughout East Asia as "the meat of the fields" and which a growing number of experts feel will be a key protein source for the future on planet earth.

We were first introduced to tempeh in March, 1975, when friends at The Farm, a large spiritual community in Tennessee, sent us their newly published vegetarian cookbook. We read the section on tempeh with great interest.

That week, on an introduction from the Indonesian Embassy, we visited Mr. Muhammad Mustam, a former tempeh maker living in Tokyo. He and his wife showed us how to make tempeh, and we were surprised at how quick and easy the process was. Two days later, they invited us back to their home and prepared a feast of their favorite Indonesian tempeh recipes. Such appealing textures and savory flavors we had rarely tasted before. We were so impressed that we included the Mustams' tempeh-making method and five recipes in our *Book of Tofu* that was just going to press.

We then wrote the United States Department of Agriculture's regional research center in Peoria, Illinois, requesting some tempeh starter and all available tempeh literature, and soon, at our home in Tokyo, we were successfully preparing tempeh and experimenting with using it in Western-style recipes.

### Studying Tempeh in America

Upon returning to America in July, 1977, we found that *Organic Gardening* magazine, as part of a reader research project, had sent us (and fifty-nine other interested people throughout America) a tempeh kit, complete with instructions and recipes. We used it with good results and felt their method was the best and easiest we had yet seen.

At a branch of the Tennessee farm, located in San Rafael, California, we studied the way their community tempeh maker prepared it on a scale of about twenty-two pounds a day. In their natural-foods delicatessen we enjoyed delectable Tempeh Burgers, Deep-Fried Tempeh Cutlets, Tempeh with Creamy Tofu Topping, and a host of other specialties on the menu.

In the fall of 1976, we began a four-month speaking tour around America. Trying to do for soybeans what Johnny Appleseed did for apples and George Washington Carver did for peanuts, we gave over seventy programs. On the trip, we were surprised to see how many people already knew of tempeh and were interested in learning more about it. We met with Drs. Hesseltine, Wang, and Swain, tempeh research scholars from the United States Department of Agriculture Northern Regional Research Center in Illinois, who explained that as a result of an article on tempeh we had written for *Mother Earth News,* they had received hundreds of requests for tempeh starter. We also met with Dr. Keith H. Steinkraus, another leading American tempeh research scholar, who talked at length of his interest in seeing tempeh become known and enjoyed throughout the world, and showed us color slides he had taken of tempeh in Java, which whetted our desire to go there and study this fine food firsthand. On our trip we observed tempeh production at seven commercial shops in America. We visited The Farm in Tennessee, where tempeh is made on a community scale, and later their sister group in Louisiana, which operates on a smaller scale. In Nebraska we met with Mr.

Gale Randall, the first Caucasian American to open a commercial tempeh shop, and at the trip's end we visited three Indonesian tempeh shops in the Los Angeles area.

Soon we began to collect and study the extensive body of scientific literature that has been published about tempeh, mostly by Western Scholars, since 1960. To our surprise we found that tempeh is better known in the scientific community and has had more written about it than any other traditional soybean food, yet among the public at large it is not nearly as well known or widely used as tofu, miso, or shoyu (natural Japanese soy sauce).

### Indonesia in a Nutshell

By March of 1977 we had become so interested in and fond of tempeh that we decided to write a full-fledged book on the subject. To do this, we realized we would have to go to Indonesia to see firsthand what we had read so much about. We sent numerous letters to Indonesian tempeh researchers, who all kindly encouraged us to visit them. We began to study the language intensively and read everything we could get our hands on.

We learned that the Republic of Indonesia, formerly the Netherlands East Indies, is a vast collection of more than 13,000 islands; only half are large enough to have names and less than 1,000 are inhabited. From the air, this lush equatorial archipelago looks like a beautiful jade-green necklace against a background of turquoise, sapphire, and aquamarine, strung between Malaysia and the top of Australia, a span of some 3,200 miles, or more than the distance from California to New York. The fifth-most-populous nation on earth (135 million people in 1978), it is also the world's biggest collection of islands, the names of which are as beautiful as the landscape itself: Java (spelled and pronounced Jawa in Indonesia), Sumatra, Kalimantan (Borneo), Sulawesi (Celebes), the Moluccas, Timor, and a host of other outer islands. Java, the main island and home of the capital, Jakarta, lies about 7 degrees of latitude south of the equator. It has a land area about the size of the state of New York, inhabited in 1975 by 84 million people, or some 62 percent of Indonesia's total population; this makes it the most densely populated region of its size on earth, with a population density of over 1,600 people per square mile (over 600 per square kilometer). Perhaps no other fact has such an important bearing on the daily life of the people or distinguishes it so clearly from that of North America.

The sprawling, loosely knit nation is divided into twenty-six provinces; each province is divided into districts (kabupaten; also called regencies) and each district into subdistricts (kecamatan); the smallest unit is the village (desa). Java has five provinces: West, Central, and East Java, Jakarta, and the Special District of Yogyakarta. The latter is ruled by a sultan, whose authority is said to derive from the prophet Muhammed. West Java or

Sunda is inhabited by the Sundanese people, whereas Central and East Java are inhabited mostly by people of Javanese cutlture from which the island took its name (East Java has a substantial population of Madurese from the island of Madura, which is part of the province of East Java). Formerly antagonistic, the Sundanese first joined with the Javanese and Madurese in 1928 to help fight Dutch colonialism.

Indonesia has a rich and ancient cultural history; it had huge and magnificent temples centuries before Europe's great Gothic cathedrals were dreamed of, and its

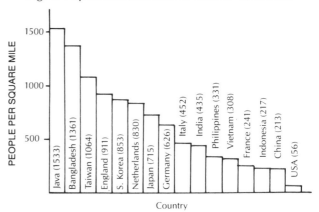

Fig. 1. Population densities of various countries.

famed arts—exquisite batik, shadow puppet theater, gamelan music, and Javanese dance—are as alive today as they were a millennium ago. Yet, for 1,500 years, until she gained independence in 1945, Indonesia was under foreign domination. The first outside influence came from India with the arrival of Hinduism in A.D. 400. The islands remained Hindu for about a thousand years and prospered greatly. During the 1200s, Marco Polo wrote: "Java is a very rich island, producing all the precious spices that can be found in the world. It is visited by great numbers of ships and merchants who buy a great range of merchandise, reaping handsome profits and rich returns. The quantity of treasure in the island is beyond all computation."

The Arabs, looking for spices and converts, arrived in about 1400 and by 1475 had become an important social force, propagating Islam, living as traders in the commercial centers, and selling their spices to Europe for fantastic prices. When Columbus set sail in the late 1400s, he was looking for these East Indies in hopes of getting control of the spice trade. Soon the search was on and the Portuguese were the first to find the way. By 1511 they had gained control of the crucial Strait of Malacca, but the Dutch arrived soon and by 1668 were masters of the Indonesian islands. For three hundred years thereafter they systematically exploited the foreign bounty to finance the prosperity of modern Holland. After World

War I the independence movement gained momentum and, after a long and bloody struggle, the Dutch were driven out. The country gained independence on August 17, 1945, and became a republic in 1949. During the struggle against colonialism people separated by great cultural differences and geographical distances developed a sense of nationhood and began to shape a common language out of the ancient Malay tongue.

On August 17, 1972, the Indonesian language, which is written phonetically, was modernized, in part because of a Malaysian-Indonesian effort to make the two similar languages even more similar. There were four basic spelling changes, *dj* becoming *j* (*Djakarta* is now *Jakarta*), *j* becoming *y* (*Jogjakarta* is now *Yogyakarta*), *oe* becoming *u* (*oelek* is now *ulek*), and *tj* becoming *c* (*ketjap* and *ontjom* are now *spelled kecap* and *oncom* but *pronounced kechap* and *onchom*). Finally, the letter *v* (as in *Java*) has always been pronounced *w* (*Jawa*). Throughout this book we have followed the new spelling conventions with two exceptions, both aimed at avoiding mispronunciation of key terms in the West: for basic foodstuffs or place names written without italics in the text, and for recipe titles, we write the new *c* as *ch* (*kechap, onchom*), and a terminal sounded *e* as *eh* (*tempeh, tapeh*) in the text but as *é* in recipe titles (*Saté, Taugé*). Other terms in parentheses follow the new spelling exactly.

### *Studying Tempeh in Indonesia*

On May 10 our plane touched down in Bali to start what was to be, for us, an adventure, a fascinating research trip, and a honeymoon all rolled into one. Bali, just east of Java, may be the loveliest island anchored in any ocean. It is unquestionably better than even the legends about it! A lush picturebook landscape teeming with sarong-clad brown-skinned people and their gentle life style, it is the epitome of fertile, tropic opulence, blessed with an abundance of rainfall, rich soil, and a cornucopia of luscious fruits that boggle the imagination and delight

the palate. (Still, a large percentage of the population is very poor.) The balmy climate, ideal for making tempeh, has an average year-round temperature of 79°F. (26°C.) at sea level and a range of only 68° to 86°F. (20° to 30°C.). In the mornings we visited tempeh shops or spent hours in the markets learning the Indonesian names of all the new foods; in the evenings we stayed in a bungalow ($2.90 a night for two) at Kutha beach—simply the most breathtaking wild spot we have ever seen—12 miles of unbroken white beach and turquoise surf as pristine as it was on the sun's birthday.

Flying from Bali to Jakarta, a city that some say is best avoided and others find fascinating, we immediately caught a typical jam-packed bus that trundled and careened on busy, narrow West Javanese roads the 40 miles south to Bogor, one of three main centers of tempeh research in Indonesia. Founded 495 years ago, Bogor is built around a huge and verdant botanical garden, surely one of the most magnificent in the world. Hundreds of deer roam the vast lawns and over 15,000 species of exotic palms, orchids, water lilies, and the like lend grace and beauty to the surrounding city. The climate is pleasant, with May mornings crystal clear and evenings capped with torrential rains and ear-splitting thunder. Traveling in bechaks, Bogor's famous brightly painted, three-wheeled pedicabs, I spent most of each day talking with research scholars, visiting tempeh shops and taking hundreds of pictures, while Akiko worked with a number of very fine professional cooks and nutritionists watching and learning how to cook Indonesian tempeh recipes. I always met Akiko for lunch at the place she was studying, so I could sample every recipe that had been prepared.

We quickly realized that Indonesia has one of the world's great cuisines, and one which has not yet been given the attention it deserves. Doing magic with coconut milk, exquisite spices and herbs, and creative culinary techniques, our host cooks would daily summon forth a pageant of magnificent foods with flavors, textures, and aromas that we had not known existed. The carefully orchestrated contrasts and counterpoints caught our palates unaware: tingling ginger sautés, rich and creamy peanut sauces or gentle coconut-milk sauces, smooth and steamy leaf-wrapped hors d'oeuvres, spice-laden and supercharged chili sambal toppings, soft and crunchy fried-rice dishes, followed by desserts of fruits delicately crisp and brimming with juice. With tempeh always used as the basic motif, our cooks developed what seemed to be unending original themes, using lemongrass and laos, tamarind and turmeric, plus many other seasonings you will discover in the recipes to follow; yet rarely, to our surprise, did they use the very spices that gave these "Spice Islands" their name: nutmeg, mace, cloves, and pepper.

Reading Indonesian cookbooks helped us to realize that the forces that had shaped the flavors we enjoyed had originated at the four corners of the world, as successive waves of immigrants, traders and adventurers, pirates and chefs, drawn by the lure and fragrance of these islands at the crossroads of the ancient world, had cross-pollinated them with the finest of their local cooking secrets. From India came curries; from China, the wok and stir-frying; from Arabia, Middle Eastern culinary techniques and ingredients; from Holland, the heritage of Western European cuisine. Each tradition is distinguishable to this day, yet always creatively blended and married with the gastronomic arts of ancient Indonesia. The result offers a challenge to all who enjoy fine food to greatly expand their culinary horizons.

From Bogor, we traveled by bus to Bandung, the second main center of tempeh research. The capital of West Java and the fourth-largest city in the republic, Bandung is situated in a mountain range 2,000 feet above sea level; its cool evenings, wide tree-lined avenues, and spacious parks make it a popular vacation area. Here we studied the largest tempeh shops in Indonesia and were impressed by the advanced state of the work being done by top Indonesian research scholars. We also studied onchom (ontjom), a food which consists of cakes of either peanut presscake (left over after peanut oil extraction) or okara (the pulp left after soymilk extraction) cov-

ered with a brilliant orange mycelium of *Neurospora* mold and served in many of the same ways as tempeh.

From Bandung we took a pufferbelly train some 300 miles to Yogyakarta (formerly spelled Jogjakarta), the smallest province in the nation, located in the south of Central Java. The trip gave us a chance to see more of the countryside. Terraced rice paddies, chiseled into the steep mountainsides like stairways for a race of giants, have carved the intensively cultivated landscape into living sculpture, while the lowlands, planted intensively in rice, bananas, coconut, cassava, and soy, present a richly embroidered study in emerald. We found Yogya (as the local people call it) to be Java's most charming and livable city. The capital of the country from 1945 to 1949, it is still governed by the ancient line of sultans. Well known as the "main door to traditional Javanese culture," the city attracts thousands of young artists from around the world. Nearby is the great Buddhist temple Borobudur and the stately Hindu temple Prambanan, built in the eighth and tenth centuries respectively, reminders of the antiquity and magnificence of Javanese civilizations. Masters of their own culture and guardians of their cuisine,

*Terraced rice paddies in Java*

the people of Yogya also have a great university and a traditional market that exceeds in size and color even the fabled markets of West Africa. Here was tempeh in the greatest abundance and variety we had seen, each small cake wrapped in shiny green pieces of banana leaf. Each morning we went to the marketplace to bargain (in Indonesian) for our breakfast fruits—fresh papayas, mangosteens, salaks, durian, and tiny bananas. The laughing women in their colorful batik sarongs, the delightful mélange of earthy aromas, the great crimson mounds of fresh chilies, and the grains and beans in bulging handwoven baskets were a feast for the senses.

*Woman selling tempeh in Bali market*

Our days in Yogya were very full, talking with researchers, visiting tempeh shops that used modern machinery or the traditional banana-wrapping techniques, and enjoying the sweeter flavors that characterize the local cuisine. By now we were able to carry on fairly coherent conversations in Indonesian (a little language goes a long way!), which proved to be indispensable to our work. And each evening, as the heat of the day mellowed into evening, we went to the favorite international hangout for artists and enjoyed a tall cold glass of "white mango juice," an ambrosial puree of soursop and ice.

Our four weeks in Indonesia seemed to pass in an instant, yet were most productive: we learned and prepared over 70 Indonesian tempeh recipes, studied in depth the production process in ten shops, talked with virtually all of the country's top tempeh research scholars, and took more than 300 color slides.

Yet, beneath the surface of Indonesia's natural beauty and the charm of her people, there is another world of which we were constantly aware and which, in fact, was a major reason for our interest in Indonesia and in tempeh. Here lurks the same complex of increasingly serious problems found in so many less-developed countries: overpopulation, environmental destruction, growing concentration of wealth, power, and land in the hands of a small elite accompanied by the impoverishment of the landless masses, technological disruption, and serious malnutrition. Java is a chilling example of what well may be in store for the rest of the Third World, a sobering reminder that time is running out.

Population is at the root of Java's problems. Indonesia's birth control efforts have been as effective as any, but that only means that if present trends continue, the country's total population of 135 million will double in the next 35 years instead of the next 30 years. Actually, however, the country's farmland and other ecosystems could never begin to support that many people; massive starvation and environmental breakdown would probably intervene to limit population growth. Sumitro Djojohadikusumo, a leading Indonesian population analyst, predicts that even if the fertility drops by one-fourth by the year 2000 Java's population will still increase by 74 percent to 146 million and Indonesia's total population will increase by 92 percent to 250 million. This would give Java an average density of almost 2,800 people per square mile (1,100 per square kilometer), a figure greater than the present density of most populated and urbanized centers in Western Europe. Today, Java's population is growing by over 2 million people per year. With 60 percent of Java's land already cultivated and with the burgeoning population overflowing onto what is now farmland, where will all these extra people get their food? In 1977, Indonesia had a staggering one-third of the world's rice imports, 2.9 million tons or 49 pounds per capita. Yet some analysts are surprised that the total food imports are still so low.

Many leading Indonesian and Western observers fear that Java is rushing inexorably toward an ecological crisis that will undercut food production before any theoretical agronomic limits can be reached. Under the enormous pressure of population expansion, deforestation is progressing at a frightening pace, largely to clear land to grow more food. Simultaneously Indonesia, like so many developing countries, is experiencing a massive energy crisis: the dwindling supply of firewood, caused in part by the rise in petroleum and therefore kerosene prices. The search for food and fuel is causing unprecedented erosion, which has begun to silt up Java's reservoir-and-irrigation system, one of the most extensive, intricate, and delicate in the world, and the lifeline of the country's rice cultivation. The pernicious cycle of floods and droughts also takes its toll on the low ricelands, causing further food shortages.

The so-called Green Revolution, with its new strains of rice that are responsive to high-technology irrigation systems and fertilizers, was supposed to be the answer to feeding the burgeoning population. Indeed, between 1968 and 1978 rice production jumped 50 percent and rice yields are still 36 percent higher than the average of all other rice producers in southern Asia. In spite of population growth, average per capita rice consumption has increased by 25 percent from 200 to 250 pounds a year. Yet the figures are misleading, for, as a growing number of authoritative rural studies by development experts such as William Collier, Benjamin White, Ingrid Palmer, and others clearly show, the general welfare and per capita food consumption *among the poor* has steadily *decreased,* and the gap between the rich and the poor has continued to widen. The modern agricultural technologies have benefited primarily those few with the money or credit to afford them and the skills to use them. The accompanying mechanization has left large numbers of poor farmers unemployed. Each generation the size of family farms, already small, grows ever smaller as it is divided among the male heirs. Plots too small to farm are sold (often to the rich) and more and more families become landless. Present studies show that about *half* of Java's population is landless, and another quarter nearly so — shocking figures for a predominantly rural farming society. In some areas export crops take precedence over subsistence crops; in East Java the government uses as much as 30 percent of the land to grow sugarcane, a heritage of the colonial era and a reminder that there is still land in reserve. Traditional village welfare and harvest institutions once ensured that the poor would have at least a minimum to eat, but modernizing influences, and technology, population pressures, and socio-economic injustices, are rapidly rendering these important institutions ineffective.

Indonesia's per capita GNP in 1978 was a mere $240 per year, ranking the country in the bottom 20 percent of the world's 172 nations. Inflation is now running at 10 to 20 percent. Fifty percent of the rural population is still illiterate. The four poorest areas, with the poorest listed first, are Yogyakarta, Central Java, East Java, and West Nusa Tenggara; the four best off, in order of "affluence," are South Sumatra, North Sumatra, West Java, and Bali.

A United Nations FAO nutrition survey of 36 developed and developing countries conducted in 1971 showed that Indonesia ranked at the *very bottom* of the scale on daily per capita consumption of *all three* essential nutrients sampled: calories (1,760 vs. 2,100 recommended), protein (38.4 grams,) and fat (22.8 grams). It must be emphasized that these statistics are averages for the total population and that the nutrient intake of the poor is considerably below these averages. Moreover, tempeh is a protein lifeline for a large proportion of the entire population. Many rural Javanese families eat only one or two rice-centered meals a day. Samusudin, in 1971, reported that 30 to 50 percent of all Indonesian children between the ages of one and three years suffer from protein-calorie malnutrition, and Prawinaregara, in 1974, reported other serious nutritional problems, including vitamin A deficiency, which causes serious eye problems and even blindness; iodine deficiency, causing endemic goiter; and lack of iron resulting in anemia. Partially owing to the country's poor nutritional status, 35 percent of all children now die before the age of five.

In recent years the term "developing countries" has come to be widely used. Yet in the case of Indonesia, as with many other poor countries, the term may no longer be applicable. For most impartial observers now agree that the lot of the majority of the people, and especially of the rural poor, is going from bad to worse.

### Bringing It All Back Home

Back in Japan, while continuing our daily practice of meditation, we set to work full time on our book. Every other day, we prepared a large batch of homemade tempeh, which Akiko used to develop a variety of tasty recipes. I carried on an extensive correspondence with tempeh researchers around the world, delved into a study of the microbiology and chemistry of tempeh fermentation, and was invited to give a presentation on tempeh in Bangkok at a United Nations-sponsored Symposium on Indigenous Fermented Foods and the Global Impacts of Applied Microbiology; there a total of seventeen papers on tempeh were presented, showing growing interest in the subject.

During our more than three years of studying, making, and eating tempeh, we have developed a deep respect for this unique food, which we feel has a vital contribution to make to people in both developed and developing countries. We believe it may someday be recognized as one of Indonesia's most important gifts to the world.

*Masked Indonesian figure*

# 1

# Soybeans: Protein Source of the Future

During the 1970s, some 15 million people died each year of starvation and malnutrition-caused diseases, and 75 percent of those who died were children. This is 41,000 deaths each day (a large stadiumful) and over 1,700 each hour. In the 15 minutes it will take you to read this chapter, an additional 428 people will have died of starvation. Moreover, estimates by the United Nations Food and Agriculture Organization (FAO) and the World Bank, respectively, suggest that from 450 million to 1,000 million people (from 11 to 25% of the world's population), most of them living in the developing countries, do not receive sufficient food. Even 20 million Americans living in slums and backward areas are malnourished. And in some societies 40 to 50 percent of the children die before the age of five, mostly from nutrition-related causes. The survivors often suffer such severe and chronic malnutrition that their physical, mental, and social development are permanently retarded. Malnutrition and infectious diseases commonly occur in the same child and each magnifies the other in a vicious circle; infection can precipitate malnutrition and malnutrition increases susceptibility to infection. The circle soon becomes a downward spiral as malnutrition leads to poor health, education, and work, which in turn generate more poverty and malnutrition. If people eat poorly, they *do* poorly. This is the cycle of suffering which is already the dominant reality of daily life for many poor people throughout the world.

Malnutrition and hunger are manifestations of extreme poverty, which in turn are rooted in political and economic structures that engender deprivation. One of the most significant historical developments of recent times is the splitting of the world into two megacultures — the rich and the poor. So great is the gap between the two that the total disappearance of the poor countries — two-thirds of the world's population — would decrease total world consumption of all resources, food, and energy by only 10 to 20 percent. In 1977 the World Bank estimated that 23 percent of the human population lives in extreme poverty on incomes of less than $75 a year! Thirty-six of the poorest nations, with a total population of 1.2 billion people, have a per capita gross national product of less than $200, and in roughly half of these countries, per capita food production has de-

Fig. 1.1   The Changing Pattern of World Grain Trade[1]

| Region | 1934-38 | 1948-52 | 1960 | 1970 | 1976[2] |
|---|---|---|---|---|---|
| | | (Million Metric Tons) | | | |
| North America | + 5 | +23 | +39 | +56 | +94 |
| Latin America | + 9 | + 1 | 0 | + 4 | − 3 |
| Western Europe | −24 | −22 | −25 | −30 | −17 |
| E. Europe & USSR | + 5 | − | 0 | 0 | −27 |
| Africa | + 1 | 0 | − 2 | − 5 | −10 |
| Asia | + 2 | − 6 | −17 | −37 | −47 |
| Australia & N.Z. | + 3 | + 3 | + 6 | +12 | + 8 |

[1] Plus sign indicates net exports; minus sign, net imports.
[2] Preliminary estimates of fiscal year data.

SOURCE: Brown, *The Twenty-Ninth Day* (1978). Derived from FAO and USDA data and the author's estimates.

17

creased since 1960. By contrast, countries such as Sweden, Switzerland, and the United States have per capita GNPs of over $7,500, or some 37 *times* the $200 figure. Moreover, the gap between the rich minority and the poor majority *within* the developing countries is often greater than the gap between the rich and poor nations. And both of these gaps are rapidly widening.

As but one example of the increasingly serious picture, we note that, prior to World War II, the developing areas of Asia, Africa, and Latin America were all able to produce enough grain to feed their people, with a small excess for export. (Fig. 1.1). The situation in each of these areas — and especially in Asia — has steadily deteriorated until today all are major importers, dependent primarily on North America, which has a virtual monopoly on the world's most precious resource — food. Most countries are rapidly losing the ability to feed themselves.

### The Causes of Hunger and Starvation

Of the various analyses concerning the causes of world hunger, there are two that we find especially worthy of careful consideration. The first, which represents the view of much of the international development community, is well expressed in *The Twenty-Ninth Day* by Lester R. Brown, president of the Worldwatch Institute. Brown's multifaceted analysis sees the primary cause of the problem as rapid population growth and the increased stresses it puts on the planet's four major biological systems: croplands, oceanic fisheries, grasslands, and forests. In each area, as demand exceeds the sustainable yield, populations begin to eat away at the biological base that sustains them; in economic terms, they con-

Fig. 1.2. Projected population densities in various regions

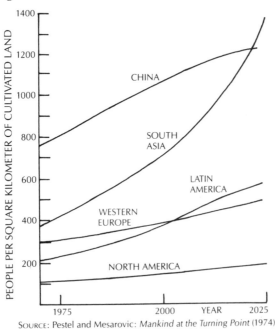

SOURCE: Pestel and Mesarovic: *Mankind at the Turning Point* (1974)

sume principal as well as interest. Population pressures coupled with unsound land-use patterns lead to erosion, deforestation, flooding, overgrazing and other disturbances in the ecosystem, which in turn affect food production. Secondary causes of hunger include the increasing use of basic food crops as livestock fodder in affluent diets.

**Population:** At the present growth rate of 1.64 percent, the population of the planet will double every 43 years. Many of the poorer nations, with growth rates of 3 percent, double every 23 years and (theoretically) increase a staggering nineteenfold every century! During the seventies the rate of increase began to slow, perhaps for the first time in history — a promising sign. Yet each day 178,000 new faces appear at the breakfast table and 64 million more passengers are added to spaceship earth each year. At the same time, there has been very little net increase in the world's agricultural acreage since 1955. According to the United Nations median projections, the world's population is expected to rise from the present 4.1 billion figure to 6 billion shortly before the year 2000, with about 90 percent of the additional 2 billion people being born in developing countries. Brown states the urgent need to attack the problem of population growth from all directions at once: making family planning services universally available; liberating women from traditional roles; meeting basic social needs, such as adequate health care, nutrition, and literacy, that are usually associated with reduced fertility; educating the people about the consequences of rapid population growth; and reshaping national economic and social policies to encourage small families. Countries such as China and Singapore that have moved on all five fronts at once have been spectacularly successful in reducing their birth rates. In Indonesia, a well-organized family planning program (which has not been accompanied by extensive social restructuring) has led to a decline in birth rate only slightly less than that in China. Each of us can make a vital contribution to population control by limiting our family size to two children or less.

**Affluence:** As people (including the upper classes in developing countries) become more affluent, they tend to increase their consumption of meat, poultry, dairy products, and eggs. In 1975, to produce these products, roughly 33 percent of the world's grain harvest (78 percent of the U.S. grain harvest) was fed to livestock, yet could have been used for human food. This represents a total of 450 million tons, or a whopping 248 pounds for every person on the planet. Thus, according to Brown, while population growth accounts for about two-thirds of the annual growth in world food demand, this affluent pattern of feeding livestock grain (the feedlot system which was developed in the U.S. in the 1950s as a way of reducing grain surpluses) accounts for the rest. Since more than 14 pounds of feedlot fodder protein are required to produce 1 pound of beef protein, a small increase in the demand for meat leads to an enormous decrease in the amount of grain and soy proteins available for human consumption and simultaneously tends to raise prices. Thus hungry people throughout the world are now in direct competition with livestock for basic foods. Because of their meat-heavy diet based on the feedlot system, Americans consume about *five times* as much basic foods (and *twenty times* as much total resources) per capita as people in the developing countries. This wasteful pattern should be phased out as soon as possible. Diverting grain away from livestock and into exports would shore up our balance of payments, which has been adversely affected by oil imports, while helping to feed hungry people abroad (see Fig. 1.1). Increasing our own consumption of grain (and thereby decreasing our consumption of animal fats) would lower food costs and promote better health. Continuing the present pattern will only aggravate the present world food crisis.

Brown sees hunger as an extremely complex problem with no single cause and requiring a combination of approaches. Population growth, patterns of land ownership, environmental deterioration, affluent diets, miserable social conditions, inappropriate technology, and hunger all form a vicious circle which must be attacked at every point along its circumference.

The second and newest analysis of the causes of world hunger is the pioneering work *Food First: Beyond the Myth of Scarcity* by Frances Moore Lappé and Joseph Collins. Called ''One of the most stimulating books in years'' by the *New York Times Book Review,* it has been highly praised by the emerging American hunger movement and by Third World leaders alike as offering a penetrating new perspective and dispelling traditional myths that prevent us from grasping how hunger is generated. Predictably, it has evoked considerable controversy within the U.S. international development community. With an impressive array of documentation, the authors challenge the basic assumption that the main causes of world hunger are too many people and not enough food. They show that this standard diagnosis simply does not fit the facts: there is now more than enough food in the world for everyone (so much that we can wastefully feed a third of all grains to livestock), and virtually every country has the resources for people to free themselves from hunger. The outrage and tragedy is that hunger exists in the face of plenty. Thus the new analysis emphasizes that hunger and poverty, more than being technical problems of producing more food and fewer children, are primarily a result of economic, political, and social structures that block development and prevent or discourage people from producing their own food. At the core of the problem is a worldwide concentration of wealth and power in the hands of a small minority in each country. Inequality and undemocratic control over productive resources are the primary restraints on food production and its equitable distribution. The critical issue is who controls the agricultural resources and therefore benefits from them. The overpopulation-scarcity analysis distracts us from understanding the core of the problem and working to conceive and build more just economic structures. Moreover, it allows our continued support of foreign elites and their part-

ners, multinational corporations, who increasingly represent primary causes of the problem. Thus economic and political causes of hunger have persistently been overlooked, partly because they are so complex, deeply rooted, and difficult to remedy, and partly because it benefits the power structure in developed countries to do so. To those who argue that this approach is as narrow and one-sided as its opposite, the authors point out that attempting — as we so often do — to alleviate hunger without restructuring the basic systems that promote it not only leads to misdirected and ineffective action, it frequently makes the situation even worse. For example:

**Population:** We are accustomed to seeing overpopulation and rapid population growth as primary causes of hunger. Yet a deeper look reveals that both hunger and rapid population growth are symptoms of the same deeper disease — the insecurity and poverty of the majority resulting from the control over national productive resources by a few. High birth rates are symptomatic of the failures of a social system, reflecting people's need to have many children in an attempt to provide laborers to increase meager family income, to provide old-age food security, and to compensate for the high infant death rate, the result of inadequate nutrition and health care. Parents in this tragic double bind — acting quite rationally in having large families yet knowing full well they may not be able to feed them — often have no interest in birth-control programs. Indeed as Lappé and Collins point out: ''To attack high birth rates without attacking the causes of poverty that make large families the only survival option is not only fruitless, it is a tragic diversion our planet simply cannot afford.'' Birth rates do not decline significantly until social and economic changes or restructuring make it a rational and feasible option for people to have fewer children. In case after case where fundamental socio-economic reforms have been made in the areas of health care, nutrition, education, and land reform, creating a basic environment of food security and increasing the likelihood of child survival, birth rates have dropped and the drop can be accelerated by family planning programs. Yet, Lappé and Collins argue, if these basic social changes are not made, birth-control programs, being symptomatic cures, have little or no effect.

If ''too many people'' caused hunger, we would expect to find the most hunger in countries having the most people per acre of cultivated land. Yet no such pattern exists. China, for example, where starvation was eradicated in only 25 years, has *twice* as many people per acre of cropped land as India (and 7.6 times as many as North America), while well-nourished Taiwan has twice as many as Bangladesh. Or take Brazil, which has more cultivated acreage than the United States, yet in recent years the percent of the people undernourished has increased from 45 to 72 percent. In each case, the difference lies in the economic and political systems.

One often hears that environmental destruction from deforestation, erosion, overgrazing, and the like is a result of population pressure and the need to expand food production. Yet, in many cases, Lappé and Collins show that these problems have deeper socio-economic causes: farmers are forced to cultivate steep hillsides, causing erosion, since the fertile lowlands are used to grow luxury crops for export. Thus safeguarding the world's argricultural environment and people freeing themselves from hunger are complementary goals.

Lappé and Collins clearly do *not* attempt to discount the long-term consequences of population growth. They *do* maintain that overpopulation is not now a primary cause of hunger and that even halving the world's numbers would not affect the gross inequalities in control over resources that generate hunger.

**Narrow Focus on Increasing Productivity:** The standard analysis that hunger is a result of food scarcity has inevitably led to programs to increase food production, generally through technological modernization. Yet, when new agricultural technology enters a system based on severe economic and power disparities, it selectively benefits the rich, those who have some combination of land, money, credit worthiness, and political influence: the rich get richer and the poor poorer. A major study by the International Labor Organization (confirmed in a similar study by the United Nations Research Institute for Social Development) documents that in seven South Asian countries comprising 70 percent of the rural population of the nonsocialist world, while there has been a rise in per capita grain *production,* the food *consumption* of the rural poor is less than it was ten to twenty years ago: more people are more hungry and poorer than ever before. In many cases, the so-called Green Revolution, expected to abolish hunger, has actually made it worse. The high cost of the ''inputs'' necessary to grow the special hybrid seeds — inputs such as irrigation, chemical fertilizers and pesticides, farming equipment, and the seeds themselves — have favored the larger, wealthier farmers, allowing them to expand their acreage, often by buying out the poor. Farm mechanization has left large numbers of farm workers without jobs, causing mass migration to city slums. Not only are the poor left without a source of food or income, the large landowners often move the land out of staple food production for local people and start to grow export crops such as grapes and flowers, which greatly increase per-acre profits. Third World agriculture is an increasingly popular way for a small elite to get rich. In 1973 two-thirds of Colombia's Green Revolution rice went into livestock feeding or beer production. Thus hunger is often made worse when approached as a technical problem with inappropriate technology. Hunger can only be overcome by the transformation of social relationships in which the majority directly participate in building a democratic economic system.

**International Food Exploitation:** In recent years, large agribusiness corporations from the industrialized countries have begun to buy up huge quantities of food from developing countries (where land and labor costs are often as low as 10 percent of those in the U.S.) for sale in supermarkets at home. They have turned the planet into a Global Farm to supply a Global Supermarket, where the poor must compete with affluent foreigners for food grown in their own countries. In 1976, for example, the United States imported $13.6 billion in agricultural products, or about $65 per capita. In this same year well over 50 percent of the winter and early spring vegetables sold in United States supermarkets were grown in Latin America on land formerly used to feed local people. Moreover, while plundering scarce food supplies from low-income nations, we simultaneously put farm workers out of work in our own country and allow a rapidly increasing proportion of our farmland to be concentrated in the hands of large corporations. In the United States there were only half as many farmers in 1975 as in 1950.

**Land Monopolization and Misuse:** In many developing countries a small and wealthy percentage of the population owns a major portion of the land. Based on a study of 83 developing countries, slightly more than 3 percent of all landowners, those with 114 acres or more, controlled almost 80 percent of all farmland. Nowhere are land and wealth distributed less equitably than in Latin America; the great bulk of the population consists of *campesinos* who are abysmally poor. According to a study by the Inter-American Committee for Agricultural Development, the largest 1.5 percent of Latin American farms cover 47 percent of the land, while 33 percent of the farms occupy only 1.3 percent of the land area. Wealthy landowners, with their great political power, prevent land reform and redistribution. Yet, where redistribution has taken place, farmers who own their land are found to produce consistently 2½ to 3½ times as much food per acre as tenant farmers and sharecroppers, who have no incentive to make long-term agricultural improvements or to farm the land to full capacity. Government policies and basic economic systems also have a major effect on establishing incentive systems that lead to increased or decreased food production. Large landholders often utilize only a small portion of their potential land; a 1960 study of Colombia, for example, found that the largest holders controlled 70 percent of the country's land, of which they planted only 6 percent. And in all countries small farms have higher per-acre yields than large farms. A World Bank study of six Latin American countries shows that the value of output on small farms is from three to fourteen times as high per acre as that on large farms.

Thus elite-dominated governments, whose control is often reinforced by our own government's military, economic, and political policies, extract whatever and as much as they can from the land and the labor of the peasant majority. Not scarcity or resources but gross underutilization characterizes every society where wealth is controlled by a few and where those who work the land do not have direct control over it.

**Cash Crop System of Export Agriculture:** The finest land in developing countries is now widely used to grow export cash crops (coffee, sugar, cocoa, bananas, etc.) instead of food for the people. And what once were basic foods, such as fruits and vegetables, have recently also come to be exported. This system generally benefits the upper classes, who live an imported life style while the poor go hungry. During the severe famines of the early 1970s, many of the hardest-hit countries were exporting basic foods to Europe, America, and Japan. Even in such disasters the allocation of food resources is based first on profits. In Central America and the Caribbean, where as many as 70 percent of the children are undernourished, approximately half of the agricultural land, invariably the best, is made to produce crops and cattle for export instead of basic food for local people. Clearly agriculture must become, first and foremost, a way for people to produce the food they need and only secondarily a source of foreign exchange.

*Food First* is a book that every person seriously interested in world hunger should read and read again. Because it succeeds so effectively in exposing the traditional myths about the causes of world hunger, it is destined to have an enormous impact on public thinking concerning this problem for years to come. Above all, the crucial questions it raises will have to be dealt with; they will not go away.

## Protein Source of the Future

The two basic yardsticks for measuring the quality of diets are protein and calories; protein measures the diet qualitatively, calories quantitatively. Protein-calorie malnutrition is the term now commonly used to describe the undernutrition of the poor. Protein is by far the more expensive of these two key nutrients, and the one in shortest supply. In their widely acclaimed *Mankind at the Turning Point: The Second Report to the Club of Rome,* Mesarovic and Pestel state that "Protein deficiency is already the most serious aspect of the present food supply situation. In more than half the world, the protein content of the average diet is about two thirds of the daily need." They go on to show, using carefully developed computer projections, that even under the most optimistic assumptions, especially in South and Southeast Asia, the protein gap is expected to increase dramatically in the years to come. At the same time, people living in the affluent nations generally have excess protein consumption, as reported in a study of thirty-six nations conducted in 1971 by the United Nations Food and Agricultural Organization (FAO). Figure 1.3 shows the per capita daily protein consumption for the ten countries with the highest and the ten with the lowest values. The average consumption for the former is 96 grams per person per day, and for the latter, 46.1 grams, or less than half. Note that the lowest consumption (Indonesia's) is only 35 percent of the highest (New Zealand's). Roughly 65 grams total protein or 43.1 grams usable protein per day is considered the recommended daily allowance for typical Western adults; the allowance for smaller-statured Southeast Asians is somewhat less. Again it must be stressed that the protein intake of the poor in each country is generally far below the national average.

Protein is particularly essential in the diets of children, who need about twice as much per pound of body weight as adults; it provides the building blocks which allow for full development of the body and brain. Severe protein deficiency can lead to kwashiorkor, a dreadful condition characterized by bloated bellies, apa-

Fig. 1.3. Per capita protein consumption in rich and poor countries (Source: FAO, 1975)

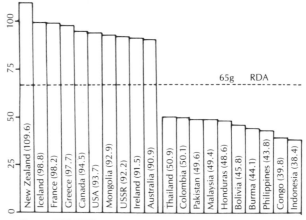

Source: FAO (1975)

thy, and wasted muscles, which leaves its victims looking like little more than bags of skin and bones. If treated soon enough, all degrees of protein malnutrition can be cured by a well-balanced, protein-rich diet.

In the search to find ways to bridge the growing protein gap with reliable sources of low-cost high-quality protein, a growing number of scientists, research scholars, agricultural economists, and nutritionists are predicting that soybeans will be not only a key protein source for the future on planet earth, but the number one alternative protein source with greatest potential for expansion in human diets. Today, most of the world's high-quality protein comes from three sources: fish, beef, and soybeans. Fish catches have dropped precipitously and prices have risen since 1970, owing largely to overfishing; with ocean pollution steadily increasing, contamination will remain a major concern. Beef, which has never been a significant protein source for most of the world's poor, may decrease in importance in developed countries because of skyrocketing prices and concern with the health dangers of excess animal fat and cholesterol consumption. Soy protein, which is much more abundant and less expensive than that from fish or beef, holds great promise to meet great future needs. Here are ten reasons why:

**1. Optimum land utilization:** A given area of land planted in soybeans can produce much more usable protein than the same land put to use for any other conventional farm crop — from 33 to 360 percent more protein than second-place rice and *twenty times* as much as if the land were used to raise beef cattle or grow their fodder. Granted, futuristic high-yielding sources such as single-celled proteins (SCP, including yeasts, bacteria, and molds), microalgae (such as chlorella, spirulina, and scenedesmus), leaf proteins, or synthetics may eventually prove to be safe, economically feasible, and acceptable

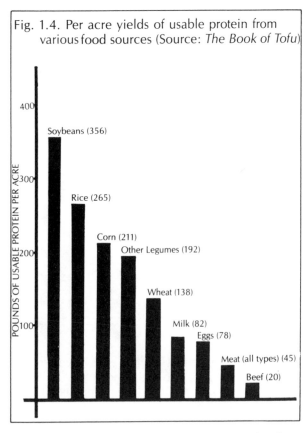

Fig. 1.4. Per acre yields of usable protein from various food sources (Source: *The Book of Tofu*)

to the consumer, but delicious low-cost soyfoods are available now. And as farmland grows increasingly scarce and costly, the soybean's ability to maximize protein output will become more and more important.

**2. Lowest cost:** Soybeans are presently the least-expensive source of protein in virtually every country of the world. And at a time when there is a steady increase in the number of poor and protein-deficient people as well as the number of people trying to live more simply and economically, the soybean's low cost will unquestionably be a prime factor favoring its increasing use as a protein source.

**3. High nutritional value:** Soybeans are an excellent source of key nutrients. The whole beans contain 35 to 38 percent protein, more than any other traditional unprocessed food. And soybeans have the highest-quality protein of any legume, being the only one containing all the eight (or ten) essential amino acids needed by the body. In widely used soyfoods such as tofu, tempeh, and miso, the protein quality is approximately equal to that found in chicken or beef. And, because soy-protein foods contain an abundance of lysine, the amino acid deficient in most cereal grains, they are an outstanding protein booster; served together with grains or a grain-based diet, soy can increase the amount of usable protein by up to 42 percent at no extra cost. Low in saturated fats and free of cholesterol, soybeans are also low in calories; and they have one of the lowest-known ratios of calories to protein. Rich in minerals (iron and calcium), soyfoods are also virtually free of the chemical toxins and pesticide residues found concentrated in meat, fish, and

dairy products. In the Orient, they are widely viewed as a food that promotes good health and long life.

**4. Time tested:** For over two thousand years, soy-protein foods have served as the protein backbone of the diet for more than one-fourth of the world's population, the people of East Asia.

**5. Remarkably versatile:** Perhaps no other plant serves as the source of so many different foods as the soybean. In countries like China, Japan, Indonesia, and Korea countless generations of farmers, craftspeople, and cooks have engaged in a vast experiment to find the best ways of transforming soybeans into delicious high-protein foods and savory seasonings. Today we can enjoy traditional fermented foods such as tempeh, miso, shoyu and soy sauce, wine-fermented tofu, natto, hamanatto, and soymilk yogurt. Nonfermented foods include tofu (the most popular of all soy-protein foods), fresh green soybeans, roasted soybeans, soynuts, natural (full-fat) soy flour, *kinako* (roasted full-fat soy flour), soy sprouts, and whole dry soybeans. Modern Western technology has brought us defatted soy flour and grits, textured soy protein (TVP), soy protein concentrates, isolates, spun fibers, and a host of other products. The fact that the soybean can be used for both protein and oil is the secret of its success in the United States. Soy-oil products are the most widely used edible oils and fats in America, accounting for 65 percent of the total market and giving a per capita usage of 30.7 pounds in 1975; 44 percent of this was used in cooking and salad oils, 32 percent as vegetable shortenings, and 24 percent as margarine. Most of the lecithin used in the United States is derived from soybeans.

**6. Appropriate technology:** All of the traditional soyfoods can be produced in cottage industries using village-level or intermediate small-is-beautiful technology. It is precisely this inexpensive, decentralized technology that is most relevant to the majority of people in developing countries who are most in need of protein as well as to the growing number of people in post-industrial societies practicing voluntary simplicity or seeking meaningful work.

**7. New dairylike products:** The food industry (and particularly the dairy industry) in the United States and other industrialized countries is now spending large sums of research money on the development of dairylike products such as soymilk, soy cheese (including a type that melts), soy yogurt, soymilk ice cream, sour cream, whipped cream, cream cheese, and mayonnaise, and creamy tofu or soymilk dressings. These low-cost, tasty, cholesterol-free foods are expected gradually to take over a major portion of the market from their increasingly costly dairy counterparts in much the same way that margarine has already captured 74 percent of the butter market. In recent years, the total number of dairy cattle in America has begun a steady decline. And already over 150 *million* bottles and cartons of soymilk are sold each year in Hong Kong alone, with Singapore, Malaysia, Thailand, and Japan close behind.

**8. Hardy and adaptive:** Soybeans can be grown on a variety of soils under a remarkably wide range of climatic conditions, from equatorial Brazil to Japan's snowy northernmost island of Hokkaido, both major soybean-producing areas. They are relatively resistant to diseases and pests and are widely grown without fertilizers or irrigation. There is now a rapid increase in soybean acreage in Third World countries as new strains are developed.

**9. Free nitrogen fertilizer:** Soybeans are a legume that works together with nature to fertilize and enrich the soil. *Rhizobium* bacteria in the plant's root nodules extract nitrogen from the air and fix it in the soil, where it stimulates the growth of both the soybeans themselves and of other crops planted later in the same field or intercropped. With the prices of chemical nitrogen fertilizers skyrocketing (they more than tripled between 1973 and 1976), their ecological hazards causing concern (they cause eutrophication in waterways, and preliminary research indicates that their heavy use leads to the release of nitrous oxide, which depletes the ozone layer), their production being highly energy intensive, and the fossil-fuel supply from which they are made being limited and nonrenewable, soybeans are expected to serve as an increasingly important nitrogen source for the planet's emerging organic agriculture. While humans now add some 40 million tons of nitrogen to the soil in the form of nitrogen fertilizers, nature annually fixes an estimated three times that amount via legumes, soil microflora, and other means. Using soybeans in the traditional pattern of crop rotation and mixed cropping not only enriches the soil with nitrogen, the rotation itself helps to control plant diseases and weeds, aids the growth of beneficial soil organisms, saves the precious mantle of topsoil from wind and water erosion, and maintains a better plant nutrient balance — all of which reduce the need to use expensive and harmful chemical pesticides and herbicides. And plowing the soybean plants under as ''green manure'' also increases the essential organic matter in the soil.

**10. Energy and resource efficient:** In an increasingly energy-short world, soybeans are the most efficient known converters of fossil-fuel energy into protein, having a ratio of about 2 to 1. Other dry beans are 3.7 to 1, corn is 3.6 to 1 and other plants lag far behind. Forage- (or grass-) fed beef, the most efficient animal protein source, has a ratio of 10 to 1, and feedlot beef, the least efficient animal source, has a ratio of 27 to 1. When processing and transportation costs are added in, one unit of energy input can produce 90 times as much soy protein (delivered to the consumer) as meat protein. Moreover, it takes from 10 to 100 times as much water and 3 to 6 times as much space to produce livestock protein as it does plant protein.

All of these ten factors work together synergistically, reinforcing one another, to give added weight to the prediction that soybeans will be a key protein source for the future on planet earth.

Rarely if ever has an agricultural crop enjoyed such rapid and prolonged expansion as the soybean. Between 1950 and 1975, the world soybean harvest nearly quadrupled, rising from 16 to 61 million metric tons; in the decade between 1965 and 1975, the harvest more than

doubled, reflecting an enormous and accelerating world-wide appetite for high-quality protein. In 1978 the top six producers with their percentage shares of the world market shown in parentheses were: United States

Fig. 1.5. World soybean production (Source: *The Soybean Digest Blue Book*, 1978)

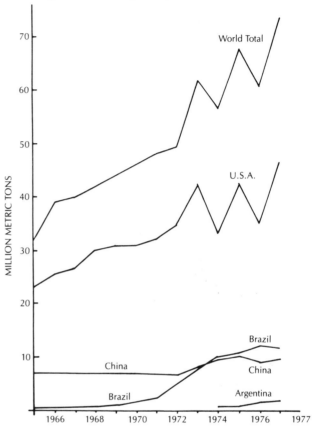

(61.2%), China (17.3%), Brazil (15.6%), Argentina (1.8%), Indonesia (1.0%), and USSR (0.8%).

Enough soybeans are now produced each year to provide 42 pounds for every person on the planet. (The United States alone produces about 488 pounds for every American.) If all this soy protein were used directly as food, it could fulfill an estimated 36 to 46 percent of the yearly protein requirements of every person in the world, and if it were served together with cereal grains (as it usually would be), the above figures, owing to protein complementarity, could increase to 50 to 60 percent, not including the protein supplied by the grains alone. Thus we see that soybeans are one of the planet's great renewable resources.

During the final quarter of this century, population growth and rising affluence are projected to double world protein demand as they did during the third quarter. Can soybeans continue to meet this demand? In spite of intensive efforts by agricultural scientists over the two decades from 1956 to 1976, soybean yields have increased only 33 percent; during the same period corn yields nearly tripled. The failure to achieve the long-awaited breakthrough may be due to the incontrovertible fact that, as a legume, soybeans fix their own nitrogen and respond only modestly to applications of nitrogen fertilizer. Therefore, the main way for farmers to increase production is to expand acreage, as they have steadily for the past 25 years. Yet, as Lester Brown comments, "With most of the formerly idled U.S. farmland now active, and with one hectare (or acre) in every six on U.S. farms now planted in soybeans, it is unclear how much more land farmers can shift to soybean production and still statisfy the expanding demand for other crops, especially cereals." It is expected that the greatest acreage increases will take place in countries like Brazil and Argentina. During the 1960s, U.S. farmers sold their soybeans on the world market for an average of $2.63 a bushel (4.4 cents a pound or $97 a ton). Between 1969 and 1972 the price increased two and one-half times and, though slightly lower by 1979, continued to maintain a high rate, averaging $5.98 a bushel. This soaring price reflects the inability of agricultural scientists to raise yields significantly, a worldwide scarcity of land on which to grow soybeans, the deterioration of oceanic fisheries (livestock once fed anchovies are now fed soybeans), and considerable price manipulation by the U.S. government. Lester Brown concludes: "Given the difficulties in expanding the world supply of high-quality protein, strong upward pressure on protein prices seems likely to continue unabated. If the price of soybeans (perhaps the best single indicator of the tightening world protein supply) continues to rise, reducing protein hunger may become even more difficult than it already is." Yet the ways in which soy protein is used will determine, in large part, its availability and price.

### Present Patterns of Soy-Protein Utilization

We have seen above that the world presently has an abundance of soy protein. What happens to all this protein? An estimated 40 percent of the total is used directly as food, primarily in East Asia. Yet, in spite of serious protein shortages, some 55 percent of the total is used as livestock fodder and the remaining 5 percent finds industrial applications, primarily in Western countries which have not yet learned to produce and enjoy fine soy protein foods. Worldwide, there are five basic patterns of direct and indirect soybean food use: (1) as unprocessed whole dry or fresh green soybeans; (2) as soy flour; (3) as traditional East Asian low-technology foods; (4) as modern high-technology processed foods (concentrates, isolates, TVP, etc.), and (5) as fodder for livestock which convert them into meat, eggs, and dairy products. The third and fifth patterns are particularly relevant to our discussion here.

The East Asian pattern developed thousands of years ago in cultures that have long had to live with realities of high population density and therefore obtain most of their proteins directly from plants. By using soyfoods in balanced combination with rice and other grains to create free complementary protein and by developing savory high-protein soy-based seasonings (such as miso and shoyu) having meatlike flavors, the people have created a diet that makes optimum use of the earth's ability to provide them with tasty natural foods which are nutritious and low in cost. Even in the most modern and technologically advanced parts of Japan and other East Asian countries, these foods are still prepared using fairly labor-intensive cottage industry or intermediate technologies requiring low energy input and creating little or no pollution. The work is often seen as a craft, found to be deeply satisfying — its own reward. In some cases it is raised to the level of an art or even a form of spiritual practice, like the work of the potter or the sword maker. In recent years, the principles of East Asia's traditional soy-protein technologies have been skillfully applied to both medium and large-scale production, resulting in lower costs and, in some cases, improved uniform quality. This traditional pattern may well be the pattern of the future in developing countries and is presently attracting a great deal of interest in North America and Europe, especially among those interested in fine natural foods, vegetarian diets, right livelihood, appropriate "soft path" technology, self-sufficiency, simpler life styles, and the problems of providing enough food for everyone on the planet. Dr. Keith Steinkraus of Cornell University's New York State Agricultural Experiment Station believes that the development of these traditional foods will one day be regarded as one of East Asia's most important contributions to food-processing technology.

The affluent Western pattern of soy utilization, which developed in the United States during the 1920s, involves extracting the high-quality soy oil from the beans (using hexane solvent), feeding the remaining defatted soy meal (which contains all the protein) to livestock, and then eating the livestock. By using soybeans indirectly in this way, we make extremely *inefficient* use of the earth's ability to provide us with protein. The process of running soybeans (and/or cereal grains) through animals and then eating the animals is so wasteful, uneconomical, and energy expensive as to be virtually inexcusable, especially at a time when, worldwide, protein is in such short supply. On the average it takes 7 to 10 pounds of soy or grain protein to make 1 pound of livestock protein, but in the case of the feedlot steer, the least efficient converter, it takes from 15 to 21 pounds. In other words, more than 93 percent of the protein fed to a steer is lost to human consumption. The low "return on investment" from other livestock products is shown in Figure 1.6.

The scale of waste generated by the feedlot system is immense. In America, for example, an astonishing 50 percent of all our farmland is used to grow crops that are fed to animals; we also feed them a full 78 percent of our finest cereal grains, including 90 percent of our corn, 80 percent of our oats, and 24 percent of our wheat. Soybeans are one of America's largest and most important farm crops, second only to corn (and ahead of wheat) in dollar value ($7.5 billion to the farmer in 1976), and third in total production and acreage (46.7 million metric tons and 58.3 million acres). We saw above that America is by far the world's largest producer of soybeans, with about 61 percent of the total; about half of this production is exported, making soybeans the number one U.S. farm export — one way we pay for our immense petroleum imports. All but about 3 percent of the remaining nonexported soy protein is fed to livestock or used for nonfood industrial uses. And the soy protein exported to Western and Eastern European countries, the Soviet Union, Iran, and other increasingly affluent areas is generally used in the American pattern of fodder and oil.

Fig. 1.6. Protein consumed vs. protein returned (Source: *The Book of Miso*)

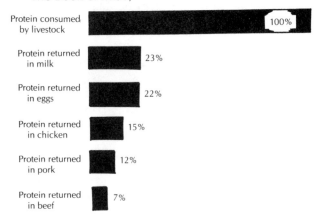

| Protein consumed by livestock | 100% |
| Protein returned in milk | 23% |
| Protein returned in eggs | 22% |
| Protein returned in chicken | 15% |
| Protein returned in pork | 12% |
| Protein returned in beef | 7% |

Seen from this point of view, there is no shortage of protein on the planet, only gross misuse.

Each of us can begin today to stop or decrease our support of this system by moving in the direction of meatless or vegetarian diets and learning to use our abundant soy protein directly in fine, low-cost foods — such as tempeh. This can be an important step, although hopefully only a first one, toward a deepened personal commitment to end hunger and starvation on our planet. We must not forget, however, that until widespread economic reforms are enacted based on a new economic order, the grain "saved" by our direct consumption either will never get produced, or at least probably won't wind up in the bellies of the poor, who are generally too poor to pay U.S. farmers to grow their food. Ultimately, elimination of hunger will come not through international redistribution of food but through redistribution of control over food-producing resources. In such a democratic economic system, soyfoods can serve as a key source of low-cost protein for people everywhere.

## New Developments

In recent years, together with the rapidly growing interest in soy-protein foods from countries around the world, there have been some particularly interesting and promising developments in both North and South America.

Latin American nations, realizing the huge economic and nutritional potential of soybeans, have begun to farm them on a massive scale. In 1975 Brazil (with extensive technical aid from Japan) passed China to become the world's second-largest producer. Between 1968 and 1981, Brazil boosted its production from a mere 654,000 tonnes (metric tons) to 15.5 million tonnes, for an almost 24-fold increase in only 13 years.

With export value totaling almost $1 billion (U.S.), soybeans have surpassed coffee as Brazil's number one export crop, and the government is actively encouraging farmers to turn their coffee plantations into soybean farms. Other Latin American countries are now actively following Brazil's lead; top producers for 1981 are Argentina (3,500,000 tonnes), Mexico (680,000), and Paraguay (600,000).

Unfortunately, in parts of Brazil, farmers planted their lucrative soybeans on land that had been used to grow the traditional black beans, the staple protein source of low-income families. Per capita consumption of these table beans in Brazil fell by one-third between 1971 and 1974, while the price tripled. Moreover, most of the soybeans were exported for livestock fodder to Japan, Europe, and the USSR. Thus, while the Brazilian GNP, foreign reserves, and soybean farmers' income was bolstered, malnutrition increased among the poor, whose meager protein reserves had, in effect, been sold to wealthy foreigners.

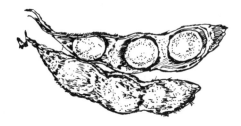

*Soybeans in the pod*

Fortunately, however, in 1975, representatives from the Latin America soybean-growing countries met in Mexico with experts from around the world to discuss ways of using soybeans directly as protein foods for the people. The first conference of its kind, this historic event is already showing important results in Latin America, with Mexico taking the lead in introducing soy protein (primarily in the form of soy flour) into traditional, widely accepted foods such as tortillas, breads, and pasta; the conference is expected to have important implications for people in all developing countries. Additional work is now needed to expand soybean acreage, use a greater percentage of the soy protein in local diets, develop local soyfood production, and create recipes using these foods in ways suited to indigenous cultures and tastes.

In North America, starting in the early 1960s, traditional soy-protein foods suddenly began to play an important role in the diet, especially among young people and those moving away from the "affluent diet" toward natural foods and vegetarian diets. At present there are over ten million vegetarians in America, and the number is rapidly increasing as people realize that a meatless diet makes it possible for the earth to feed about *seven* times as many people as the present American meat-centered diet. Meatless diets are healthier, more economical, more humane, and less fattening. Medical research has shown that those who avoid meat have lower rates of heart disease, bowel and lung cancer, and gastrointestinal disorders, plus lower blood pressure and serum cholesterol, triglyceride, uric acid, and pesticide levels. They live longer, have more stamina, and approach their ideal weight more closely than their meat-eating counterparts. This new awareness, coupled with increased concern over dietary cholesterol and saturated fats plus a strong new interest in health, nutrition and fitness, led to major declines in per capita red meat consumption from the mid-1970s into the 1980s—and an increasingly positive image for balanced meatless or vegetarian diets.

Reflecting the growing interest in appropriate technology, Americans are using traditional or middle-level equipment to produce their own soyfoods. As of early 1984 some 191 U.S. companies were making tofu, 53 made tempeh, 29 made soymilk, 21 made soy sauce, 21 made soy sprouts, 12 made miso, and 10 made soynuts. In several innovative companies a number of these foods are produced under one roof, a pattern unprecedented in East Asia. The Farm, the spiritual community of 1,100 people living on 1,750 acres in Tennessee, has taken the historic step of growing soybeans (6,000 bushels on 200 acres in 1978) for use primarily as a protein-rich staple in its pure vegetarian diet; most of the rest of the crop has been shared with people in famine-stricken areas abroad. The Farm is also training people from many countries in soyfood production at its Tennessee headquarters, and has worked with villagers in Guatemala to start a tofu shop and soy dairy there.

In America soyfoods enjoy an increasingly positive image among all classes of people and are quickly being adapted to American culture and cuisine in new and creative ways. Tofu, for example (which is now sold in most supermarkets on the West Coast), is used to make creamy salad dressings, tasty sandwiches, and protein-rich casseroles; deep-fried tempeh and tofu are made into vegetarian tempeh or tofu burgers, which are sold at delicatessens. These foods are especially popular at America's 6,000 natural food stores. In 1978 the Soyfoods Association of North America was founded; its nationwide magazine, *Soyfoods,* is linking together people interested in all aspects of these new and ancient foods. A company in California called Bean Machines is specializing in developing, importing, and selling tofu and soymilk equipment. At least 50 soyfoods cookbooks have been published since 1975, and attractive kits for home preparation of tofu, tempeh, and miso are now sold commercially. Tempeh burgers and tofu burgers are all the rage, second only to tofu ice creams! At the consumer level, this new generation of soyfoods has upstaged the modern, high-tech soy protein products such as meat extenders made from textured soy flour or spun soy protein fibers which, in the early 1970s, were seen as the wave of the future. These latter products, however, are selling fairly well to institutions and food manufacturers, and soy flour is used domestically in baked and canned goods, and sent to developing countries in various food supplements.

Since the publication of our *Book of Tofu* and *Book of Miso,* we have received over 400 invitations to do speaking engagements and television programs concerning soyfoods. What is perhaps most significant in this small but growing trend is that many young Americans, often considered the most affluent generation on the planet, are turning to lives of voluntary simplicity and trying to live lightly on the planet, eating in ways that ensure there will be enough for everyone. While the people of most other countries, as they become more affluent, tend to increase their consumption of meat and dairy products, thereby aggravating the world food crisis, many Americans are now working to reverse this trend. And soybeans are playing a promising role in this development.

In February of 1977 a Gallup poll in America showed a remarkable shift in the public awareness of soyfoods. The sampling of 1,543 adults across the country found that 33 percent believe that soybeans will be the most important source of protein in the future—ahead of fish at 24 percent and meat at 21 percent. Fifty-five percent believe "soy products have a nutritional value equal or superior to meat," and 54 percent said they "had eaten foods containing soy protein as a prime ingredient within the past 12 months." Younger age groups living in large cities and those with college educations were the most favorable to soy protein, indicating that support for soyfoods is likely to grow in the future.

*Dry soybeans*

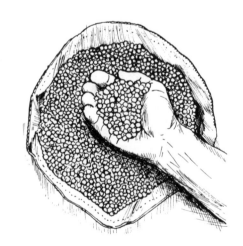

## An Idea Whose Time Has Come

Most futurologists and serious students of history and international problems agree that the world stands on the threshold of a basic social transformation and is now entering a period of rapid, sustained, and perhaps cataclysmic change. "Like earlier 'revolutions,'" points out Lester Brown, "this one could raise us to a higher level of humanity. But unlike others, it must be reckoned with in advance. Whether the impending transformation will be orderly or convulsive depends on our foresight and will."

The period from 1940 to 1970 was one of unrestrained optimism concerning the ability of technology to solve mankind's most difficult problems. Nuclear technology would provide us with abundant energy, genetics and new food and agricultural technologies would make food shortages a thing of the past, and people throughout the world would enter a new era of prosperity. Yet, since 1970, environmental deterioration coupled with food and energy shortages have brought scientists, technologists, and the public at large to a sobering realization of the limits of technology and its promises. Few people still look for "quick technical fixes." A new and much less optimistic view of the future has begun to emerge.

Sophisticated studies and computerized projections of past and present trends modified to include various future options and scenarios have appeared in respected and highly influential books, such as *The Limits of Growth* and its successor *Mankind at the Turning Point*. They point out that we are engaged in a process of cancerous, undifferentiated growth, are using up irreplacable resources at a rate that simply cannot be sustained, and are putting unbearable pressure on each of our life-support systems. Increasingly estranged from nature, we are rapidly reducing biological diversity, thereby greatly reducing the stability of nature's delicate biological systems and increasing the probability of their future collapse. Six major and overlapping crises—population, food, energy, nuclear war, pollution, and raw materials—will soon begin to have severe planetary repercussions, first in the low-income countries and especially in South and Southeast Asia. In 1977 an independent study by the Asian Development Bank, a close relative of the World Bank, predicted massive and virtually unavoidable famines in South Asia by 1985.

There can be no doubt that we still possess the funds, the technology, and the potential to reverse these disturbing trends. But do the people of the world have the individual and political will, the wisdom and creativity, to act while we still have time? Today the nations of the world spend $400 *billion* a year (a sum that exceeds the total yearly income of the poorest half of mankind) and employ the talents of one-fourth of the world's scientists to develop means of military destruction; $228 billion of this is spent by the United States and the Soviet Union. Most countries spend more on military security than they do on educating their youth, despite the growing recognition that national security can no longer be

defined primarily in military terms. Moreover, since 1960, the commitment of almost all developed countries (except Sweden) to foreign aid has shown a steady decline.

Fig. 1.7    Where the World's Money Goes

**Yearly Global Expenditures**

| | Billions of Dollars |
|---|---|
| Military arms and armaments | $400 |
| Education | 320 |
| Health care | 207 |
| Alcoholic beverages | 100 |
| Foreign aid | 20 |

**Yearly U.S. expenditures**

| | |
|---|---|
| Alcoholic beverages | $24.7 |
| Tobacco products | 14.8 |
| Radios, televisions, records, etc. | 14.6 |
| Jewelry | 6.3 |
| Barber shop, beauty parlor services | 4.3 |
| Foreign aid | 4.0 |

SOURCE: *Overseas Development Council, Washington, D.C.*

The present momentum of the world and all its forces, traveling on a course which many projections show will almost certainly lead to widespread suffering, is great indeed. A powerful and revolutionary force will be needed to swerve its course, to turn it around. We believe that force will come primarily from one source: *deeply committed individuals,* people who are not afraid to work hard and selflessly, to study deeply, to serve, and to persist. People who realize that global change does not begin on a global level, but begins with single individuals then expands to ever larger groups. People who know that "All the forces in the world are not so powerful as an idea whose time has come."

The time has now come for each of us, regardless of occupation or outlook, to solemnly resolve to put an end to hunger and starvation on this planet. Even now there is more than enough food for everyone in the world. And starvation is no more an inevitable part of the human situation than smallpox, which we have now totally eradicated. For the first time in human history, hunger appears to be a solvable problem. Yet the challenge is great, so great, in fact, that we probably cannot overcome it without transforming both the old world order and ourselves. At this node in history, perhaps at the beginning of what history will subsequently view as a new age, the events and forces throughout the planet and in each of our lives proclaim more loudly and clearly each day that the old and confining perspectives of enmity and greed, selfishness and nationalism keep us separate and in bondage, and no longer work. This time of crisis thus offers us all a great opportunity—if not a mandate—to liberate ourselves and others.

Already in North America and Western Europe, the prevailing ethic of unlimited growth has begun to be supplanted by three new and closely interrelated ethics, which promise to help accommodate human numbers, ways of living, and consumption to basic human needs and the earth's limited resources. *Voluntary simplicity,* based on a frugal, aesthetic, and joyous life style that is outwardly simple and inwardly rich, in which waste and unnecessary consumption have no place, embodies local self-reliance, decentralization, right livelihood, appropriate technology, and sharing with those in need, offers hope in dealing with the complexities of the modern world, which have already begun to exceed our analytical capacity, and represents a realistic alternative to centrally enforced austerity. *Ecological awareness,* based on an understanding of the deep interdependence and fundamental oneness of all things, feels the beauty of fragile ecosystems, opposes the drift toward monoculture by encouraging diversity among humans and nonhumans alike, realizes that one's own good cannot be separated from the good of all, and thus works to build a just world community and a new economic order. *Self-realization,* as we acknowledge that materialistic scientism, the doctrine of inevitable "progress," and the security offered by a technocratic society have proved unable to satisfy the deepest needs of the human spirit, provides an alternative and more satisfying reason for and way of living, based on increased awareness, service and love, and new forms of consciousness.

Today we seem to stand on the brink of a new synthesis and paradigm that as yet has no name. There is an awakening planetary consciousness that borrows from East and West and links us with the most creative energies of our ancient past; in transforming our images of reality and self, it transforms our hearts.

Out of this new consciousness and these new ways of living, a vital concern for world hunger has begun to emerge. Many feel that food will be the overriding social, political, economic, medical, ecological, and moral issue of the next decade. Around the world people are working to democratize control over food-producing resources. We are learning to recognize the forces (including forces such as the Global Supermarket, the feedlot system, and some foreign aid that may operate with our unwitting support or in our name) that keep people hungry and in bondage, keep them from producing the food they need. Because of a renewed sense of the sacredness of all life, an aversion to gross injustice, and an innermost yet deepest desire for no less than the liberation of all beings, we find ourselves compelled to join, each in his or her own way, in the struggle with people everywhere who find themselves trapped in suffering. This is an ancient path of liberation, likewise, of ourselves. This is a new practice, a vision, and a calling. And this book has only one central purpose: the furtherance of the immense work, to bring an end to hunger and starvation on our planet.

And, in this work, watch how soybeans join with us all to help feed a hungry world.

*Selling beans and grains in a Javanese market*

# 2
# Tempeh as a Food

Although soybeans are a rich source of high-quality protein and other nutrients, only a portion of these are available to the body when the whole beans are served in their baked, boiled, or roasted forms. However, during the short tempeh fermentation, *Rhizopus oligosporus* brings about a total transformation, unfolding a panorama of delicious new flavors and aromas, creating a unique texture and appearance, while simultaneously enhancing the nutritional value and digestibility. Figure 2.2 shows the composition of nutrients for various types of tempeh.

### Rich in High-Quality Protein

Tempeh is prized above all as a source of tasty, high-quality protein. The protein value of any food depends on two basic factors: the *quantity* of protein in the food and the *quality* of that protein. Quantity is usually expressed as a simple percentage of weight. By comparing the figures in Fig. 2.1, it can be seen that tempeh and other soy-protein foods rank high on the list—ahead of some meats and dairy products.

The protein *quality* of a food depends largely on its digestibility and on the degree to which the configuration of the eight (or ten) essential amino acids that make up the protein matches the pattern required by the body. There are five standard measures of protein quality: Protein Efficiency Ratio (PER), Biological Value (BV), Net Protein Utilization (NPU), Chemical Score, and Amino Acid Score (formerly called the Protein Score). To determine the first three requires the use of animals (usually rats), while the last two can be quickly calculated once the amino-acid composition of the food is known.

NPU, defined as the Biological Value times the true

Fig. 2.1 Percentage of Protein in Various Foods

| Food | Percent of Protein by Weight |
| --- | --- |
| Soy flour, defatted | 51 |
| Tempeh, sun-dried soy | 43 |
| Soy flour, whole | 40 |
| Soybeans, dry | 35 |
| Cheeses | 30 |
| Fish | 22 |
| Chicken | 21 |
| Beef (steak) | 20 |
| Tempeh, fresh soy | 20 |
| Hamburger, raw | 18 |
| Eggs | 13 |
| Wheat | 12 |
| Tofu, regular | 8 |
| Brown rice, uncooked | 8 |
| Milk, whole dairy | 3 |

Source: Standard Tables of Food Composition (Indonesia and U.S.A.) and Lappé, *Diet for a Small Planet* (1975).

digestibility (or coefficient of digestibility) of the food, is a widely used measure of the protein in a food that can actually be utilized by the body; it represents the ratio of nitrogen retained by the body to nitrogen intake. It is a common misconception that the protein found in animal foods is somehow basically different from (and superior to) plant protein. In fact, however, there is no basic difference; it is simply a matter of degree. As Figure 2.3 shows, although animal foods tend to have the highest NPU ratings, a number of plant protein sources—including tempeh—rank quite high on the scale.

29

Fig. 2.2  Composition of Nutrients in 100 Grams of Tempeh

| Type of Tempeh | Food Energy Calories | Moisture Percent | Protein Percent | Fat Percent | Carbo-hydrates (incl. fiber) Percent | Fiber Percent | Ash Percent | Calcium Mg | Phos-phorus Mg | Iron Mg | Thiamine Vit. B$_1$ Mg | Riboflavin Vit. B$_2$ Mg | Niacin (Nicotinic Acid) Mg | Reference |
|---|---|---|---|---|---|---|---|---|---|---|---|---|---|---|
| Soy tempeh, fresh | 157 | 60.4 | 19.5 | 7.5 | 9.9 | 1.4 | 1.3 | 142 | 240 | 5.0 | 0.28 | 0.65 | 2.52 | See note |
| Soy tempeh, sun-dried | – | 8.9 | 43.1 | 18.0 | 26.2 | 2.8 | 3.8 | – | – | – | – | – | – | Murata, 1967 |
| Soy tempeh, freeze-dried | – | 1.9 | 46.2 | 23.4 | 25.8 | 2.7 | 2.7 | – | – | – | – | – | – | Murata, 1967 |
| Soy tempeh, dry basis | – | 0 | 54.6 | 14.1 | 27.9 | 3.1 | 3.5 | – | – | – | – | – | – | Iljas, 1969 |
| Wheat & soy tempeh, dry weight | – | 0 | 33.1 | 12.1 | 49.6 | 2.7 | 2.6 | – | – | – | – | – | – | Wang, 1968 |
| Okara tempeh, fresh | – | 84.9 | 4.0 | 2.1 | 8.4 | 3.9 | 0.7 | 226 | – | 1.4 | 0.1 | – | – | Gandjar, 1977 |
| Wheat tempeh, dry weight | – | 0 | 18.2 | 2.0 | 74.9 | 3.1 | 1.8 | – | – | – | 0.3 | 0.32 | 13.5 | Wang, 1966, 1968 |
| Winged-bean tempeh, fresh | 212 | 58.2 | 16.0 | 9.0 | 15.4 | 1.9 | 1.4 | 186 | 177 | 2.2 | 0.2 | – | – | Gandjar, 1977 |
| Velvet-bean tempeh, fresh | 141 | 64.0 | 10.2 | 1.3 | 23.2 | 3.0 | 1.3 | 42 | 15 | 2.6 | 0.09 | – | – | ITFC, 1967 |
| Leucaena tempeh, fresh | 142 | 64.0 | 11.0 | 2.5 | 20.4 | 2.6 | 2.1 | 42 | 15 | 2.6 | 0.19 | – | – | ITFC, 1967 |
| Coconut-presscake tempeh, fresh | 119 | 72.5 | 4.4 | 3.5 | 18.3 | 2.4 | 1.3 | 27 | 100 | 2.6 | 0.08 | – | – | ITFC, 1967 |
| Soy tempeh, deep-fried | – | 50.0 | 23.0 | 18.0 | 8.0 | 2.0 | 1.0 | – | – | – | – | – | – | Amara, 1976 |
| Peanut-presscake onchom | 187 | 57.0 | 13.0 | 6.0 | 22.6 | 2.9 | 1.4 | 96 | 115 | 27.0 | 0.09 | – | – | ITFC, 1967 |

Note: These nutritional data should be viewed as approximate rather than absolute because of variations in production methods, raw materials, moisture content, etc. The values for fresh soy tempeh are an average from three sources: Indonesian Tables of Food Composition (ITFC), Hermana (1972), and an unpublished study of tempeh sampled from four areas in Jakarta. Dashes indicate figures not yet available. Data on vitamin B$_{12}$ are given below. Nicotinic acid and Niacin are synonyms.

Fig. 2.3 Protein Quality (NPU) of Various Foods

| Food | NPU (percent) |
| --- | --- |
| Eggs | 94 |
| Milk, whole cow's | 82 |
| Fish | 80 |
| Tempeh, wheat & soy | 76 |
| Cottage cheese | 75 |
| Soybeans, fresh green | 72 |
| Brown rice | 70 |
| Cheeses | 70 |
| Wheat germ | 67 |
| Beef and hamburger | 67 |
| Oatmeal | 66 |
| Tofu | 65 |
| Chicken | 65 |
| Soybeans (dry) and soy flour | 61 |
| Tempeh, soy | 56 |
| Soy sprouts | 56 |
| Peanuts | 43 |
| Lentils | 30 |

SOURCES: Non-tempeh data from Lappé, *Diet for a Small Planet* (1975). Tempeh NPU figure is an average of Bai et al., 1975 (58.7), Hermana, 1972 (57.0), and Djurtoft and Jensen, 1977 (52.4).

By combining the figures for protein quantity and quality (NPU), we can compare the true protein value of various foods. Soy tempeh, for example, contains 19.5 percent protein, 56 percent of which is actually usable by the body. Thus, a 100-gram (3½-ounce) serving can supply us with 100 × 0.195 × 0.56 or 10.9 grams of *usable* protein. This is a full 25.3 percent of the daily adult male requirement of 43.1 grams. The same amount of usable protein could be supplied by 3.1 ounces of steak (at a much higher cost), 5.2 ounces of hamburger, 1.6 cups of milk, 1.9 eggs, or 1.9 ounces of firm cheese. Yet each of these animal or dairy products contains a relatively high level of cholesterol, saturated fats, and chemical toxins; and their production via the feedlot system involves the large-scale waste of basic foods.

Built into the NPU figures are figures for coefficient of digestibility, which for tempeh average a relatively high 86.1 percent (Djurtoft and Jensen, 1977; Hackler et al., 1964; Van Veen, 1950).

Protein Efficiency Ratio (PER), defined as the gain in weight of a growing animal divided by its protein intake, measures the efficiency with which the body can use protein for growth. It is the most widely used technique today for biological *in vivo* evaluation of protein quality. The standard reference PER is that of casein, milk protein, which is usually taken as 2.5 but may vary. Various studies (Hackler et al., 1964; Smith et al., 1964; Wang et al., 1968) have shown an average PER for soy tempeh of 2.43 with an average PER for the casein control of 2.81. This value is higher than all other soyfoods. Moreover, deep-frying the tempeh for a typical three minutes at 385° F. (196° C.) decreased the PER by only 2 percent (Hackler et al., 1964). Even more impressive, Wang et al., (1968) found that wheat & soy tempeh has a PER of 2.79,

higher than that of eggs, all meats, and every dairy product except milk. Likewise, soy & peanut tempeh has a PER of 2.61 (Bai et al., 1975), and soy tempeh fortified with only 0.3 percent of the amino acid methionine attains the remarkably high PER of 3.09 (Smith et al., 1964). All of the above PER figures are based on experiments done using rats. Recently, however, a number of key experiments done at MIT by Scrimshaw and Young, two of the world's top nutritionists, have shown that the PER for soy-protein foods consumed by humans is higher than for those consumed by rats. They conclude that "soy can replace beef in a typical balanced diet without altering the nutritional value of the diet."

Amino-acid score, defined as the ratio of the limiting amino acid to the corresponding amino acid of the reference pattern multiplied by 100, is widely used by the United Nations. Figure 2.4 below shows that the amino-acid score of soy tempeh is 78.

Biological Value (BV), a measure based on the amount of nitrogen actually used by the body, is defined as the ratio of nitrogen retained (not excreted) by the body to nitrogen absorbed (digested). The BV of cooked soybeans is 75, above most other vegetable proteins but below meat at 80 and eggs at 96. Djurtoft (1977) found that soy tempeh has a BV of 58.7 and a Chemical Score (the greatest percentage deficit in an essential amino acid compared with a reference protein) of 63.9. Corresponding figures plus amino-acid analyses for broadbean tempeh, cowpea tempeh, and broad-bean & wheat tempeh are given in the same publication.

### High Protein Complementarity

Both tempeh and soybeans are "complete proteins," containing all of the eight (or ten including cystine and tyrosine) essential amino acids, which are the building blocks of protein and which must be supplied to the body since it cannot synthesize them itself. Perhaps even more important is the fact that soy protein has an extremely high lysine content which can be used to complement the typically lysine-deficient cereal grains, the primary foods in most traditional cultures. In fact, in many parts of the world today, the simple and inexpensive practice of serving soy protein together with the indigenous cereal grains could be a major step in improving the nutritional status of the population.

Every protein food has a "limiting" or "first limiting" amino acid, the one that is in shortest supply relative to a reference pattern; like the weakest link in a chain, it determines the protein quality of that food. As shown in Figure 2.4, tempeh's limiting amino acid is the methionine-cystine pair; it is well endowed with the relatively scarce amino acid lysine:

The principle of protein complementarity involves combining two foods that have opposite strengths and weaknesses in their amino-acid profiles. Cereal grains are short on lysine but well endowed with methionine-cystine. Thus soy and rice, for example, have amino-acid

Fig. 2.4 The Amino-Acid Composition of Tempeh Compared with the FAO/WHO Reference Pattern (expressed in milligrams per gram of nitrogen)

| Amino Acids | FAO/WHO Pattern[1] | Tempeh[2] (Soybean) | Tempeh as Percent of FAO/WHO Pattern | Soybeans[2] | Eggs[2] |
|---|---|---|---|---|---|
| Methionine-cystine | 220 | 171 | 78 | 165 | 342 |
| Threonine | 250 | 267 | 107 | 247 | 302 |
| Valine | 310 | 349 | 113 | 291 | 437 |
| Lysine | 340 | 404 | 119 | 391 | 417 |
| Leucine | 440 | 538 | 122 | 494 | 547 |
| Phenylalanine-tyrosine | 380 | 475 | 125 | 506 | 588 |
| Isoleucine | 250 | 340 | 136 | 290 | 378 |
| Tryptophan | 60 | 84 | 140 | 76 | 106 |
| Total | 2,250 | 2,627 | | 2,460 | 3,117 |
| Methionine | | (71) | | (84) | (192) |
| Cystine | | (101) | | (81) | (150) |
| Tyrosine | | (170) | | (165) | (240) |
| Phenylalanine | | (305) | | (341) | (348) |

Note: Amino acids in shortest supply in tempeh are listed first. Methionine and cystine are called the "sulfur-containing amino acids"; phenylalanine and tyrosine the "aromatic amino acids."

SOURCES: (1) Provisional Amino Acid Scoring Pattern, Technical Report Series 522 of WHO, 1973; (2) Food Composition Table for Use in East Asia, 1972; two studies by Murata, 1967; and one by Djurtoft, 1977. Figures in parenthesis have no corresponding FAO/WHO referent.

patterns that complement each other as shown in Figure 2.5. By serving these two foods at the same meal in the proper proportions (about 4 parts rice to 1 part tempeh, dry weight) we are able to fortify or strengthen the limiting amino acid of each food and thus substantially increase the protein quality of the resultant combination; in effect, we have "created" new usable protein at no extra cost.

Fig. 2.5. Limiting amino acids in rice and tempeh

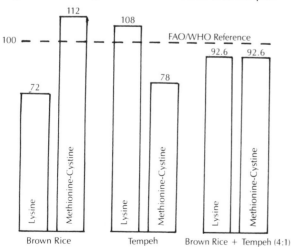

Most traditional societies have practiced protein complementarity as a fundamental (if only intuitive) principle in their daily diets. Indonesians have their tempeh and rice; Chinese, Japanese, and Koreans their tofu (plus other soyfoods) and rice; American Indians their corn and beans (or succotash); Latin Americans their corn or wheat tortillas and frijoles; Middle Easterners their chickpeas (garbanzos) and bulgur; and the people of India their chapaties or rice and lentils or dhal. Interestingly enough, the use of only small amounts of tempeh (or other soy-protein foods) combined with grains and other foods can produce large increases in usable protein. Thus, tempeh is the perfect complement to Indonesia's rice-centered diet, for by serving only 1¾ ounces of soy tempeh with 1¼ cups (uncooked) brown rice, we obtain 32 percent more usable protein than if we served these two foods separately. Hence, tempeh's unique amino-acid composition makes it not only a basic protein *source,* but also a truly remarkable protein booster.

Each of the following combinations provides exactly 50 percent of the daily adult requirement of usable protein, or the equivalent of that found in 4½ ounces of (uncooked) steak. (All quantities of grains refer to the raw, uncooked product.)

Fig. 2.6 Combining Foods to Increase Protein

| Combination | Percent Increase |
|---|---|
| 1 cup whole-wheat flour or 6 slices bread, 1½ tablespoons sesame butter, 1½ ounces soy tempeh | 42 |
| 1 cup whole-wheat flour or 6 slices bread, 2¼ ounces soy tempeh | 32 |
| 1¼ cups brown rice, 1¾ ounces soy tempeh | 32 |
| 3 tablespoons each peanut and sesame butter, 2 ounces soy tempeh | 25 |
| ⅜ cup each whole-wheat flour and brown rice, 3½ ounces soy tempeh | 24 |
| ½ cup each whole-wheat flour and brown rice, 1¼ tablespoons peanut butter, 2¾ ounces soy tempeh | 15 |

SOURCE: Adapted from Lappé, *Diet for a Small Planet* (1975).

The protein-rich combinations of tempeh with whole-wheat flour suggest a variety of tempeh (and nut butter) sandwiches or burgers, noodle, pasta, or bulgur-wheat dishes, and even taco, chapati, or pizza dishes. Note also that wheat and soy tempeh is, itself, an example of protein complementarity, which explains its high NPU. In Indonesia, special high-protein foods used to combat infant and preschool malnutrition are prepared by combining tempeh with one or more grains and other low-cost nutrients; one of the better-known developments is tempeh-fish-rice (TFR), which contains 23.7 percent protein and has an NPU of 67. In other areas, nutritionists are experimenting with the use of tempeh (often fortified with methionine) as an inexpensive substitute for powdered milk in infant diets.

At this point we would like to add a word of clarification and caution. Like a growing number of nutritionists, we believe that most people typically require less protein than official U.S. figures advocate, and that a moderate protein diet with the majority of the protein obtained directly from plant sources (such as grains and soy) is the most healthful. The typical American overconsumption of protein (twice as much as recommended by FAO), especially animal protein, easily overloads the system with saturated fats and with uric acid, which has been reported to cause both kidney damage and cancer.

### World's Richest-Known Vegetarian Source of Essential Vitamin B₁₂

In 1977, Liem, Steinkraus, and Cronk at Cornell University's New York State Agricultural Experiment Station announced that tempeh is the first vegetarian food shown to contain nutritionally significant quantities of vitamin $B_{12}$ necessary to fulfill the U.S. recommended daily allowance (RDA).

Vitamin $B_{12}$ (cyanocobalamin) is considered an essential human nutrient necessary for the proper formation of red blood cells and the prevention of pernicious anemia. Yet it is required in extremely small amounts and the minimum has not yet been determined. An adult RDA of 3 micrograms (mcg) was established by the Committee on Dietary Allowances of the National Academy of Sciences in 1979; the corresponding figure set by the United Nations Food and Agricultural Organization is 2 micrograms.

Vitamin $B_{12}$ is synthesized by bacteria, which are its ultimate source in the human diet. Most people get their daily quota from dairy products, eggs, and meat. Ruminants are furnished with $B_{12}$ by bacteria in their intestinal tracts but in humans it is synthesized by bacteria only in the lower bowel, where it cannot be assimilated. It has previously been known that $B_{12}$ is present in several non-animal products such as fermented soyfoods, sea vegetables, and microalgae (in each case the source is bacteria), but the amounts were considered insignificant. Consequently the growing number of Westerners practicing complete vegetarian (vegan) diets have been concerned about obtaining sufficient vitamin $B_{12}$ from natural sources and some have, often begrudgingly, felt they had to resort to vitamin-pill supplements.

But now the situation has changed. The researchers found that the vitamin $B_{12}$ content of tempeh presently made and sold in North America ranges from 1.5 to 6.3 micrograms per 3½-ounce (100-gram) dried portion, enough by itself to meet the typical RDA. Moreover, Dr. Paul Curtis, biology professor at Eisenhower College, New York, has identified one producer of this important vitamin as the bacterium *Klebsiella*. When *Klebsiella* is deliberately added to the starter used to make tempeh, the $B_{12}$ content of the finished fresh product can rise as high as 14.8 micrograms per 100 grams. At this level a typical 100 gram (3½-ounce) serving of tempeh will provide

14.8 micrograms of vitamin $B_{12}$, or 493 percent of the adult RDA. The only known food which is a more concentrated source of vitamin $B_{12}$ is beef liver; a 100-gram serving provides 80.0 micrograms of vitamin $B_{12}$, or 2670 percent of the adult RDA. The next two best sources, canned tunafish and eggs, provide only 73 percent and 67 percent respectively of the adult RDA per 100-gram serving.

This breakthrough discovery with tempeh offers the possibility of producing a commercial tempeh starter containing *Klebsiella* that can be used at home and in tempeh shops to prepare tempeh rich in both protein and vitamin $B_{12}$. Tempeh made from pure-culture *Rhizopus* starter is relatively low in vitamin $B_{12}$; a 100-gram portion contains only 0.047 micrograms or less than 2 percent of the RDA. However, since *Klebsiella* is common throughout our environment, it can enter at various stages of the tempeh-making process and may steadily increase if one batch is used to inoculate the next. Vitamin $B_{12}$ is not destroyed by cooking but can be decreased by direct sunlight as during sun drying; the body can also store large amounts of it for later use.

Fig. 2.7. Grams of dietary fiber in 100 grams of various foods

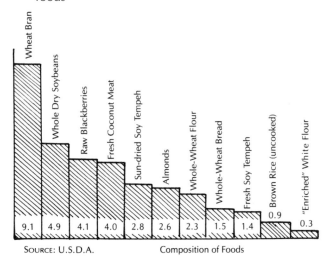

| | |
|---|---|
| Wheat Bran | 9.1 |
| Whole Dry Soybeans | 4.9 |
| Raw Blackberries | 4.1 |
| Fresh Coconut Meat | 4.0 |
| Sun-dried Soy Tempeh | 2.8 |
| Almonds | 2.6 |
| Whole-Wheat Flour | 2.3 |
| Whole-Wheat Bread | 1.5 |
| Fresh Soy Tempeh | 1.4 |
| Brown Rice (uncooked) | 0.9 |
| "Enriched" White Flour | 0.3 |

SOURCE: U.S.D.A.    Composition of Foods

### A Whole Food

Unlike tofu, East Asia's other most widely used source of soy protein, tempeh is a whole food containing plenty of natural fiber, as shown in Figure 2.7. Note that the crude fiber shown represents only about one-third to one-fourth of the actual dietary fiber. According to modern nutritional theory, dietary fiber or roughage promotes good health and regularity by stimulating and cleaning the digestive tract, especially the large and small intestines. High-fiber diets also lower serum cholesterol. Moreover, tempeh allows us to enjoy the unanalyzable essence of wholeness which cannot be evaluated with numbers yet which is the way that Mother Nature brings us our foods.

### An Excellent Diet Food;
### Low in Saturated Fats

Tempeh is an excellent diet food containing only 157 calories per 100-gram serving, or 356 calories per 8-ounce portion—good news for that 67 percent of all Americans who report they are trying to lose weight. Like all soyfoods, it is low in saturated fats and contains an abundance of lecithin plus essential polyunsaturates, such as linoleic and linolenic acids, as shown below; these perform the vital functions of emulsifying, dispersing, and eliminating deposits of cholesterol and other fatty acids that may have accumulated in the vital organs and bloodstream.

Fig. 2.8   Fatty Acids in Soy Tempeh

Fatty Acids

| | |
|---|---|
| Saturated, total | 0.595 grams (14.5% of total) |
| 16.0 (palmitic) | 0.420 g |
| 18.0 (stearic) | 0.175 g |
| Monounsaturated, total | 0.713 grams (17.3% of total) |
| 18.1 (oleic) | 0.713 g |
| Polyunsaturated, total | 2.803 grams (62.2% of total) |
| 18.2 (linoleic) | 2.510 g |
| 18.3 (linolenic) | 0.293 g |

SOURCE: Wagenknecht (1961).

Tempeh is also low in carbohydrates (9.9%), with starch being entirely absent or found in only very small amounts. This makes tempeh (as well as most other soy-protein foods) ideal for use in diabetic diets.

### Free of Cholesterol,
### A Cholesterol Reducer

Tempeh contains no cholesterol and it makes a delicious and inexpensive replacement for main-course dishes, like meat and eggs, high in saturated fats. In short, it allows you to avoid animal fat and cholesterol without losing protein and flavor. And, perhaps even more important, scientists have recently discovered that tempeh and other soy-protein foods actually have the ability to lower cholesterol rates. C. R. Sirtori and five colleagues working at medical research centers in Milan, Italy, fed soy-protein foods to people with very high cholesterol levels *who were unable to lower them by eating low-fat diets.* However, when these same people ate soyfoods, their cholesterol level dropped into the normal range. Similar experiments in Japan using dried-frozen tofu yielded identical results. Formerly, the only option for people with high cholesterol was to stop eating or cut down on foods that cause cholesterol levels to rise. But now the soybean is thought to be far more potent against cholesterol than anything else we've known about. Perhaps this is one reason why the Japanese, who use soybeans in the forms of tofu, miso, and shoyu as regular parts of

their daily diet, have one of the world's lowest rates of mortality from coronary heart diseases. As Robert Rodale, editor of America's best-selling health food magazine, *Prevention,* said recently: "When that knowledge about soybeans seeps into the public consciousness, as it is sure to do, there's going to be the most powerful push toward eating soy foods we've ever seen."

### Highly Digestible

Many foods made from beans are considered difficult to digest, partially owing to problems of intestinal gas. But not tempeh. Dr. Doris Calloway of the University of California Department of Nutritional Sciences, in a study of the gas-forming properties of 16 legumes and legume foods, found that tempeh was "essentially nonflatulent" and produced no more intestinal gas than a typical nonlegume diet. (The second-most-readily-digestible food tested was tofu.) Other researchers have found that the tempeh-making processes of soaking, cooking, and fermenting the soybeans greatly reduce their oligosaccharides (complex sugars such as raffinose and stachyose) believed to cause flatulence.

A widely used measure of digestibility is the digestibility coefficient. The value for tempeh is 86.1 percent, which, while lower than that of tofu (95%), is still very good. The process of fermentation makes the soybeans softer and more tender, since enzymes produced by the mold actually predigest a large portion of the basic nutrients, breaking them down into soluble solids and soluble nitrogen. The active protease enzymes, produced in abundance, break down more than 50 percent of the original protein into amino acids and other water-soluble products that are readily assimilated by the body. Likewise, by the end of the 24-hour fermentation, the abundance of strong lipase enzymes break down (hydrolyze) about one-third of the neutral fats into readily assimilated fatty acids. Whole soybeans ordinarily require lengthy cooking before they become tender; however, natural enzymes from the *Rhizopus* mold serve as almost magic tenderizers, making it possible for the final cooking time (and fuel) to be greatly reduced. The tempeh-making process whisks away the subtle "beany" soy qualities and replaces them with a highly pleasing flavor and aroma. Cooking the soybeans once or twice before inoculation and then cooking the tempeh before it is served inactivates trypsin inhibitors which, in uncooked beans, would limit the full assimilation of soy protein by the body. The result of these many biochemical transformations is a food that is so gentle on the digestive system that in Indonesia it is used to wean babies and nourish elderly adults.

### A Good Source of Vitamins and Minerals

In addition to vitamin $B_{12}$ discussed earlier, tempeh is also a good source of numerous other vitamins (especially B vitamins, which the tempeh molds can synthesize) and minerals, most of which are quite heat stable.

Fig. 2.9    Vitamins and Minerals in Soy Tempeh

| Nutrient | Amount per 100 grams fresh tempeh | | U.S. Recommended Daily Allowance | | 100 grams tempeh as a percent of RDA |
|---|---|---|---|---|---|
| Vitamin A | 42 | I.U. | 5000 | I.U. | 1% |
| Thiamine (B₁) | 0.28 | mg | 1.5 | mg | 19% |
| Riboflavin (B₂) | 0.65 | mg | 1.7 | mg | 28% |
| Niacin | 2.52 | mg | 20 | mg | 13% |
| Pantothenic Acid | 0.52 | mg | 10.0 | mg | 7% |
| Pyridoxine (B₆) | 830 | mcg | 2000 | mcg | 42% |
| Folacin (Folic Acid) | 100 | mcg | 400 | mcg | 25% |
| Cyanocobalamin (B₁₂) | 3.9 | mcg | 3.0 | mcg | 130% |
| Biotin | 53 | mcg | 300 | mcg | 18% |
| Calcium | 142 | mg | 1000 | mg | 14% |
| Phosphorus | 240 | mg | 1000 | mg | 24% |
| Iron | 5 | mg | 18 | mg | 28% |

SOURCES: Values have been obtained by averaging data from Steinkraus et al. (1961), Roelofson and Thalens (1964), Murata (1965, et al, 1970), Noparatnaraporn et al. (1977), and Liem et al. (1977).

Of the world's various vitamin and mineral deficiencies, by far the most widespread is anemia, a condition resulting from inadequate intake of iron or other vitamins as well as from iron losses to blood parasites, like hookworm. Anemia saps one's energy needed to do work and raises susceptibility to disease. Tempeh (and soyfoods in general) is an excellent source of low-cost iron.

Soybeans are a good source of other minerals, including (per 100 g dry weight) magnesium (236 mg), potassium (1504 mg), zinc (3.8 mg), and manganese (1.2 mg). Although corresponding data are not yet available for tempeh, they are expected to be roughly equivalent on a dry-weight basis.

The *Rhizopus* molds in tempeh also produce an active enzyme called phytase, which breaks down phytates, chelating substances in soybeans that can tie up or bind a number of minerals or mineral salts (such as zinc, iron, calcium, etc.).

### Medicinal Antibiotic and Growth-Stimulating Effects

Scientific experiments (Wang et al., 1969, 1972) have shown that tempeh has a medicinal property; its Rhizopus molds produce natural, heat-stable antibacterial agents that act as antibiotics against some disease-causing organisms in much the same way that other types of mold produce antibiotics such as penicillin. Tempeh's antibacterial agents inhibit the growth of some bacteria, especially gram-positive types (those which hold a purple dye when stained by Gram's method),

such as *Staphylococcus aureus,* a food-poisoning bacterium; hence, it may be possible to increase the body's resistance to infection by eating tempeh. These agents, which safeguard the digestive tract, may be responsible for the fact that during World War II and the Japanese occupation of Southeast Asia, prisoners of war in Indonesia, Hong Kong, and Singapore suffering from dysentery and nutritional edema (an abnormal accumulation of fluid in body tissues) found that, although their systems were unable to tolerate whole cooked soybeans, they could assimilate tempeh and survive. Some reported that tempeh relieved or cured their intestinal disorders, allowing them to stay surprisingly healthy while others nearby were plagued with dysentery. Furthermore, Indonesians who eat tempeh as a regular part of their diet recognize it as a medicine for dysentery and rarely fall victim to the intestinal diseases to which they are constantly exposed. These findings may be extremely significant, particularly in poor countries where the level of sanitation is low, sources of infection abound, and medical care is often inadequate.

It has also been found that these antibiotics, in addition to minimizing infections, elicit a natural growth-stimulating effect. Their presence actually makes it possible for a lower-quality protein to aid growth in exactly the same way as if it were of a higher quality. Thus, adding a natural antibiotic to a protein source has virtually the same effect as adding a limiting amino acid (such as methionine in the case of soybeans). This is an important phenomenon in diets deficient in certain proteins and vitamins.

And there has never been recorded a case of food poisoning from any soy, grain, or soy-&-grain tempeh. In Indonesia, however, mycotoxins have been found in some tempeh made from the presscake of coconut or peanuts, as explained in detail in Appendix C of the professional edition of this book.

### Backbone of the Meatless Diet

For the rapidly increasing number of Westerners (over 10 million in America alone) who find that a meatless or vegetarian diet makes good sense, tempeh can serve as an excellent source of high-quality, low-cost protein, just as it has for centuries among millions of people throughout Indonesia practicing similar ways of eating.

### Free of Chemical Toxins

Tempeh—like other soybean products—is unique among high-protein, high-calcium foods in being relatively free of chemical toxins. By comparison, meat, fish, and poultry are a chemical feast, containing an average of 20 *times* more pesticides than legumes. Dairy foods, the next most contaminated group, contain 4½ times more. In both cases, the toxicity of animal products arises from the fact that pesticides, herbicides, and heavy metals tend to concentrate in the fatty tissues of animals at the top of food chains. For precisely this reason, since soybeans are an important legume feed crop at the base of the beef and dairy food chains, their spraying is carefully monitored by the Food and Drug Administration to keep the level of contamination to an absolute minimum.

### Low in Cost

In Indonesia, tempeh has long served as a key source of low-cost protein in the diets of both rich and poor. At the time of this writing, a 3½-ounce (100-gram) serving of commercial American-made tempeh costs only about 26 cents. If the tempeh is purchased in bulk directly from a shop, the cost may be as low as 17 cents, and if the tempeh is prepared at home, about 8 cents, including the cost of the soybeans, starter, and fuel. In Indonesia, the same-sized portion of *commercial* tempeh sells for about 6 cents. Furthermore, unlike most meats, a pound of tempeh is a pound of tempeh; no fat or bones. And, if the tempeh is served together with a complementary grain, the price drops proportionally to the increase in usable protein. As the following data show, the cost of one day's supply of usable protein (43.1 grams for an adult male) derived from or purchased in the form of tempeh is relatively low compared with that of other protein sources:

**Fig. 2.10   Price of One Day's Supply of Usable Protein from Various Foods**

| Food | Cost of one day's supply of usable protein |
|---|---|
| Soybeans, whole dry ($0.35/lb.) | $0.15 |
| Tempeh, homemade soy (.34/lb.) | .25 |
| Flour, whole-wheat (.23/lb.) | .25 |
| Milk, nonfat dried (1.39/lb.) | .36 |
| Eggs, whole medium (.77/doz.) | .40 |
| Cottage cheese, from skim milk (.69/lb.) | .42 |
| Textured soy protein (TVP or TSP) (1.34/lb.) | .42 |
| Tofu, homemade (.25/lb.) | .48 |
| Milk, whole nonfat (.37/qt.) | .62 |
| Hamburger, regular grind (.79/lb.) | .64 |
| Nutritional yeast (SCP) (1.75/lb.) | .66 |
| Rice, brown (.39/lb.) | .70 |
| Spaghetti (.52/lb.) | .77 |
| Tuna, canned in oil (1.60/lb.) | .80 |
| Tofu, regular Japanese-style (.45/lb.) | .85 |
| Tempeh, fresh soy, commercial (1.20/lb.) | .93 |
| Peanut butter, natural (1.05/lb.) | .94 |
| Chicken, breast with bone (1.29/lb.) | 1.15 |
| Bread, whole-wheat (.67/lb.) | 1.16 |
| Cheese, Swiss, domesic (2.39/lb.) | 1.20 |
| Cheese, Cheddar (2.22/lb.) | 1.22 |
| Pork loin chop, med. (1.49/lb.) | 1.57 |
| Porterhouse steak w. bone (2.49/lb.) | 2.63 |
| Yogurt, from skim milk (.55/lb.) | 2.84 |
| Lamb rib chop (2.89/lb.) | 3.51 |
| Chlorella (microalgae; 63.88/lb. in Japan) | 16.83 |

SOURCE: Prices sampled from California supermarkets and tempeh shops, January, 1978. Figures calculated from data in Lappé, *Diet for a Small Planet.*

As tempeh comes to be better known in America, there is every reason to expect its price to drop in response to the economies of larger-scale production and distribution.

*Selling tempeh in a Javanese market*

# 3

# Getting Started

Fresh soy, wheat & soy, and okara tempeh are now increasingly available throughout the United States at reasonable prices. For a list of domestic producers, see the Appendix. Most tempeh is presently sold fresh or frozen in 8- to 16-ounce plastic bags. Look for it in the refrigerated-foods section of your local natural-food or Indonesian food store. Prices generally range from $0.90 to $2.00 per pound. Dehydrated tempeh, made domestically or imported, is also available at some Indonesian stores.

When buying fresh tempeh, see that the beans and/or grains are bound together into a firm, compact cake by a dense, uniform mycelium of white mold. It should have a pleasant, clean aroma, somewhat like that of fresh mushrooms. The cake should be firm enough so that it can be held horizontally by one end without bending or breaking, and thinly sliced pieces should hold together well without crumbling. There may be some gray or black sporulation, especially near pinholes touching the tempeh surface.

## Storing and Preserving Tempeh

Tempeh is a perishable product that has its best flavor and texture when served fresh. Nevertheless, a number of traditional and modern ways have been developed to store it with only small losses in quality. Although heat or cold is used to retard or inactivate the enzymes and kill the mold, the flavor and texture nevertheless change slightly during storage for reasons not yet clearly understood. Improvement of these methods will greatly aid the development of large-scale production and marketing. During storage, the antioxidants in dried tempeh help prevent it from becoming rancid. There is virtually no danger of toxin formation in U.S. tempeh which has an acceptable fresh or even overripe flavor. The first three storage methods given below are well suited for use in most Western homes and commercial operations; the fourth, fifth, and sixth are widely practiced in traditional societies or areas where refrigeration is not available; the last three require special rather expensive equipment.

**Refrigeration:** Seal tempeh in a plastic bag, plastic wrap, or airtight container and refrigerate, preferably below 40°F. (4°C.). Do not stack lest the live mold generate heat. It will keep for 2 to 3 days, and in some cases as long as a week. Do not worry if, after 2 to 3 days, it grays or blackens slightly owing to sporulation or develops a subtle ammonia smell. Discard only if the smell should become unpleasantly strong.

**Freezing:** This is our favorite way of storing tempeh for long periods of time; it is especially good for tempeh homemade or sold in perforated plastic bags. Many research scientists advocate parboiling tempeh before freezing in order to kill the molds and stop their enzyme action. We, however, find that parboiling is a nuisance, causes loss of nutrients and flavor in the boiling water, and does not produce a better product even after several months' storage. We always freeze our cakes whole (which is easiest), but some prefer to slice them first, since weak tempeh may crumble when sliced after thawing.

Put whole tempeh (in its wrapper) or sliced tempeh in a plastic bag, seal with a rubber band, and place in a freezer. It will keep for months with only a slight loss of texture or flavor.

**Parboiling (or Steaming) and Freezing:** This method

may be the best for tempeh stored for longer than 3 months. Parboiling gives a slightly milder flavor and steaming a slightly drier consistency than fresh tempeh. Drop cakes into boiling water, return to the boil, and simmer for 5 minutes (or place in a preheated steamer for 5 minutes). Drain and cool to room temperature, cut into ½-inch-thick slices (some prefer to slice before blanching), seal tightly in plastic bags or other containers, and freeze.

**Storing at Air Temperature:** Wrap tempeh surface firmly and completely with any nonporous material (plastic, leaf, etc.) to minimize air contact, then store in a cool place having low humidity and good ventilation, preferably on a slatted rack or with two cakes propped against one another like the sides of a tent to ensure good air circulation. Serve as soon as possible. In tropical climates at 75° to 80° F. (24–27° C.) the tempeh will keep for 1 to 3 days before becoming overripe; in colder climates it may be stored for 2 to 5 days.

**Deep-Frying:** Heat oil to 375° F. (190° C.) in a wok, skillet, or deep-fryer. Drop in thinly-sliced tempeh and deep-fry for about 3 minutes, or until crisp and golden brown. Allow to cool, then seal and store in a cool, well-ventillated place. Expected storage life is one week or more. To increase the storage life to 2 to 4 weeks, sun-dry or oven-dry thinly sliced tempeh before deep-frying. This latter method is used in preparing the Tempeh Chips (Keripik Tempeh) widely sold in sealed pliofilm bags or tin cans at Indonesian markets.

**Sun-Drying:** Cut tempeh into thin (¼-inch or less) slices (if a longer shelf life is desired, parboil or blanch as described above), and place on a screen or woven bamboo trays in direct sunlight for one full day, or until well dried. Solar dryers with racks also work well. This method is widely used in Indonesia; direct sunlight, however, may destroy vitamin B$_{12}$.

**Hot-Air Dehydration:** Cut tempeh into 1-inch cubes (blanch if desired), and place on slatted racks in a circulating hot-air dryer or oven for 6 to 10 hours at 140° to 156° F. (60–70° C.); or 1½ to 2 hours at 180° to 220° F. (82–104° C.); or 30 minutes at 450° F. (232° C.). Seal in pliofilm or cellophane bags, ideally with a desiccant. Expected shelf life is 1 to 2 months; however, the tempeh may gradually develop a slightly bitter or rancid flavor, and the PER may decrease.

**Canning:** Parboil, dehydrate, or deep-fry tempeh, seal in cans (with the proper amount of distilled water for parboiled tempeh), and heat at 240° F. (115° C.) for 40 minutes.

**Freeze-Drying (Lyophilization):** Quickly freeze tempeh at 5° F. (-15° C.) or less, then evaporate off the moisture in the vacuum chamber of a freeze-dryer. Use or sell the tempeh as is or ground to a powder. Since the tempeh remains frozen during most of the process and is not heated above 80° F. (27° C.), the final product has a higher percentage of soluble nitrogen (protein) and soluble solids than if it were hot-air dehydrated, but only 52 percent as much soluble nitrogen as fresh tempeh.

## Preparatory Techniques for Tempeh

The following procedures are used regularly in cooking with tempeh. Try to master them from the outset.

**Cooling and Drying:** Homemade tempeh, fresh out of the incubator, is warm, moist, and fairly delicate. Like freshly-baked bread, it can be sliced much more easily after it has cooled; drying the surface helps it to absorb other flavors during cooking.

Remove wrapper from fresh warm tempeh and pat all surfaces gently with a clean paper or cloth towel. Set on a clean plate and allow to stand in a cool, dry place for 15 to 30 minutes, or until center has cooled to room temperature. Then proceed to slice and cook.

**Thawing:** Tempeh that is purchased frozen should either be refrozen as soon as possible for lengthy storage, or placed in the refrigerator for use within one week, or thawed for use the next day. It should not be completely thawed and refrozen, or the binding strength of the mycelium may weaken. Sliced frozen tempeh to be marinated in a salt solution may be marinated directly without thawing.

Unwrap tempeh and place on a plate in the refrigerator overnight; for faster results, unwrap and allow to stand in a clean place at room temperature. Or put into a microwave oven. We prefer not to try to slice tempeh *thinly* until it is well thawed, but some maintain that well-made frozen tempeh keeps its texture best if sliced without thawing. If thawed tempeh appears to have a weak mycelium, cut in pieces no thinner than ¼ inch, set on or pat with absorbent toweling to dry briefly, then, if deep-frying, use relatively hot oil; lower-temperature oil may cause tempeh to fall apart.

**Cutting:** Only firm, high-quality tempeh with a dense, compact mycelium should be cut into thin slices, otherwise the tempeh may crumble. When large, thin pieces are required, Indonesian chefs almost always cut diagonally rather than horizontally, in order to obtain slices with more cohesiveness and strength. In dishes for which the tempeh is to be simmered or marinated, the ends and/or sides of each piece are often cut diagonally to increase the surface area, thereby increasing permeation of the accompanying flavors. It is not necessary to scrape or cut the mold off; it disappears during cooking.

Using a sharp knife, cut as directed. When tempeh is to be diced and used plain or fried in soups or salads, ½-inch cubes are ideal.

**Scoring:** Scoring thick pieces of tempeh helps the flavors of cooking broths, marinades, and dipping sauces to permeate and season the tempeh more deeply, and may allow the cooking time to be reduced. Only firm, compact tempeh should be scored, otherwise it may crumble. Since the mycelium is, itself, quite permeable, flavors will still permeate unscored tempeh quite well.

Using a sharp knife, score upper and lower surfaces in a crosshatch pattern to a depth of 1/16 to ⅛ inch.

**Sun-Drying:** As a preparatory technique for deep-frying or shallow-frying, this gives the tempeh a crisper texture and longer shelf life.

Cut tempeh into paper-thin slices, place on a slatted rack, and dry in the sun (or in a slow oven) for 5 to 20 minutes.

**Reconstituting:** This process is used only with dried or dehydrated tempeh.

Partially fill a bowl with warm water (or, for seasoning, with a mixture of 2 teaspoons salt dissolved in ½ cup water), drop in tempeh, cover, and allow to stand for 15 to 60 minutes, or until well reconstituted. Pat off surface moisture with absorbent toweling before deep-frying or simmering.

**Steaming:** This process is used in many recipes that do not call for baking or frying.

To pan steam: Bring ½ cup water to a boil in a skillet. Add about 5 ounces tempeh, cut into ¼- to ½-inch-thick slices, cover, and cook for about 5 minutes on each side, or until water is gone. Or place on a rack over water in a preheated steamer and steam as directed in individual recipes.

**Making Overripe Tempeh:** Overripe tempeh, called for in a number of the recipes in this book, occurs of its own accord if fresh tempeh is kept in a warm room for 2 or 3 days. Or it may be made quickly and deliberately.

Remove wrapper, sprinkle tempeh surface lightly with salt, then return tempeh to wrapper (so it will not sporulate). Allow to stand for about 24 hours at 80° to 85° F. (27–30° C.) If desired, rinse off salt, then pat dry before use.

## Basic Ingredients

A detailed Glossary listing all Indonesian foods called for in the following recipes will be found at the back of this book. We recommend the use of whole (unrefined), natural foods. Since cans and bottles contribute to environmental clutter, our recipes call for ingredients which can be purchased as free as possible of nonbiodegradable packaging.

**Flour:** Since all-purpose white flour contains only about 75 percent of the protein, 36 percent of the minerals, and 25 percent of the vitamins found in natural whole-wheat flour, we generally prefer to use the latter. Dietary fiber, the roughage present in the outer layers of whole wheat, is lost in white flour.

**Honey and Sugar:** Like a rapidly growing number of people throughout the world, we try to avoid the use of white sugar (sucrose) in all its many forms. In addition to depleting the body of vitamins and minerals and causing cavities, it creates tremendous mental and physical suffering (especially when used with a meatless diet), a fact that can only be fully appreciated when one's dependence on it has been broken. The average American now consumes 112 pounds of sugar per year — 14 tablespoons a day. One teaspoon of honey imparts the same sweetness as two teaspoons of sugar at about the same price; honey too, we feel, should be used only in small quantities. It is very important that decreases in meat consumption be accompanied by proportional decreases in the use of sugar and that the fast-burning, concentrated monosaccharides and disaccharides (as in honey and sugar) be replaced by slower-burning polysaccharides (as in well-chewed grains) or less-concentrated natural sweeteners (as in fruits, maple syrup, pureed soaked dates, etc.).

**Miso:** A savory, high-protein seasoning, also called "fermented soybean paste." The three basic types are red miso (made from soybeans, rice, and salt), barley miso (made from soybeans, barley, and salt), and soybean or Hatcho miso (made from soybeans and salt). Available at natural-food stores and Oriental markets in the West. For details see our *Book of Miso* (Autumn Press).

**Oil:** For best health and flavor, use cold-pressed natural vegetable oils. Soy, corn, peanut, safflower, and small amounts of sesame work well for sautéing and salad dressings. For deep-frying, rapeseed (colza), soy, or coconut is tops.

**Rice:** Polished or white rice contains an average of only 84 percent of the protein, 53 percent of the minerals, 38 percent of the vitamins, and 30 percent of the dietary fiber found in natural brown rice, as shown in the figure below. It is a real tragedy that poorly nourished people in developing (and developed) countries continue to eat refined and devitalized white rice and that so few nutritionists are teaching the people the value of the whole food. White rice, widely eaten by early colonialists, was often first adopted as a status symbol. Since brown rice has 3 to 4 percent lipids (oils and fats) in the bran layers (*vs.* 1% for white rice), it keeps well for only 1 to 2 months in tropical climates; moreover, it takes longer (45 *vs.* 30 minutes) and requires more fuel to cook. Still, its nutritional advantages greatly outweigh these minor problems.

Fig. 3.1. Comparison of nutrients in brown and white rice

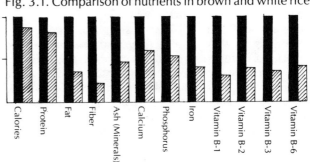

**Salt:** Natural, unrefined sea salt contains an abundance of essential minerals and trace elements that are lost in the process that brings us the 99.35 percent "pure" NaCl we now call table salt; only 77.8 percent of the solids in pure sea water are NaCl and natural salt is generally 74.0 percent NaCl. Yet salt should be consumed in moderation (½ teaspoon or 3 grams per person per day is a good average) since excess consumption is now known to be the main cause of high blood pressure (hypertension), a serious condition affecting one out of ten Americans. Daily U.S. salt consumption is 25 times more than the one-tenth of a teaspoon (10 grains) that the body requires. To reduce your salt consumption, substitute miso (12% salt) or shoyu (18% salt); these add complementary protein, energy, rich flavor, and, in the case of natural unpasteurized miso, digestion-aiding *Lactobacillus*.

*Fig. 3.2: Shoyu (natural soy sauce)*

**Shoyu:** This all-purpose Japanese fermented soy sauce, made from soybeans, cracked roasted wheat, *Aspergillus sojae* mold spores, salt, and water, is not to be confused with the low-quality domestic, synthetic soy sauce made without fermentation from hydrolyzed vegetable protein (defatted soybean meal), caramel coloring, corn syrup, salt, and water. This latter type, generally sold under Chinese brand names, constitutes two-thirds of all soy sauce now sold in America. The wonderful flavor and aroma from alcohols, organic acids, and esters produced by the fermentation process are lacking in the instant synthetic product. Once one becomes accustomed to the flavor of real shoyu, it is hard to be satisfied with imitations.

Natural shoyu, imported to America and occasionally sold as "tamari" (although real tamari is actually a completely different product), is made from whole soybeans aged in cedar vats at the natural temperature of the environment for at least one year; no preservatives are added but it has always been pasteurized. "Regular" shoyu (produced mostly by Kikkoman in America) costs about two-thirds as much as natural shoyu; defatted soybean meal is substituted for whole soybeans and the product is aged in epoxy-lined steel tanks in a slightly heated room (75° F.) for 6 months. The domestic commercial

variety unfortunately contains preservatives, but the imported low-salt variety and most shoyu sold in Japan contains 2 percent ethyl alcohol added in place of preservatives. A number of excellent traditional, natural Chinese and Southeast Asian soy sauces are now also available in the West. See also Sweet Indonesian Soy Sauce (kechap) in Appendix D. of the professional edition.

**Starches:** For best flavor, texture, naturalness, and healing properties, we prefer kudzu powder (kuzu) to other cooking starches such as arrowroot or cornstarch. For details, see our *Book of Kudzu* (Autumn Press).

**Tofu (Soybean Curd):** This delicious protein backbone of the traditional Oriental diet is now available throughout America in more than ten different forms, all of which are described in detail in *The Book of Tofu*. Popular varieties include regular tofu (7.8% protein), Chinese-style firm tofu, deep-fried tofu cutlets, burgers, and pouches, silken tofu, ever-fresh lactone silken tofu (which can be stored at room temperature for 6 months), wine-fermented tofu, and dried-frozen tofu.

## Indonesian Kitchen Utensils

The following are virtually indispensable in Indonesia, but Western substitutes are readily available:

**Coconut Grater:** A thin metal plate about 3 to 4 inches wide and 10 to 12 inches long with many tiny sharp teeth protruding from its upper surface. Unlike typical Western graters, the Indonesian *parutan* has no holes in it, so the grated coconut (or other food) stays on the surface or falls down into the bowl in which the grater is placed. A fine-toothed Western grater makes a good substitute.

*Fig. 3.3: Coconut grater*

**Leaf Wrappers:** Certain leaves, especially banana leaves, whether light green young ones or dark green old ones, can rightly be regarded as cooking utensils. Prized for the flavor they impart, they are sold in markets for this purpose after being wilted by exposure to sunlight. Ti, taro, and breadfruit leaves are also good for use in leaf cooking; all are used as we use aluminum foil, as a wrapper during steaming or grilling.

*Fig. 3.4: Mortar & pestle*

**Mortar and Pestle:** The most important piece of equipment in the traditional Indonesian kitchen is the *chowet,* a solid round grinding stone 6 to 12 inches in diameter and 1 to 4 inches deep. A beautiful marriage of the primitive and the modern, the utilitarian and the aesthetic, it is superior to a blender for grinding small quantities of chilies, seeds, fresh and dry spices, nuts, and shallots to a smooth paste. Quick and easy to use, free of breakable parts, and requiring no electricity, it is said to give a better flavor and consistency, and is much easier to clean than a blender so that not a particle is wasted. The *mutu,* a nonsymmetrical stone pestle, fits snugly into one hand during grinding. Hardwood mortars are seen occasionally; metal of any kind will not work. The Japanese *suribachi* is also generally unsatisfactory since the small amount of ground ingredients tends to get stuck and lost in the grooves. A blender may be used when the amount of ingredients to be pureed is large enough so that they do not fall below the blade and remain untouched; add oil or water together with the ingredients, then omit them later during frying or simmering. Removing small quantities of pureed spices from a blender tends to be difficult and usually results in a proportionally large loss. A white porcelain pharmacist's mortar works quite well, except that it is slightly too smooth and nonporous. Chopping, mincing, or grating never achieves quite the same effect as thorough grinding. For instructions on using a mortar, see below.

**Wok Set:** Popular now in the West, the wok *(kuali* or *wajan)* is Indonesia's standard utensil for deep-frying, stir-frying, sautéing, steaming, and, in some cases, dry-frying and grilling. Based on a design evolved by countless generations of East Asian chefs and craftsmen, it is a joy to use, as versatile and efficient as it is simple. A typical wok is about 13 inches in diameter and 3½ inches deep. When placed atop a special circular support, it may be used with electric ranges. Electric woks are also now available but electricity gives sluggish heat control. Accompanying the wok are generally a stir-frying ladle and spatula, long cooking chopsticks, a semicircular draining rack that fits around the wok's rim and allows draining foods to drip back into the wok, a mesh skimmer, and a wooden lid. Western cooks may also wish to use a deep-fat thermometer that measures temperatures up to 375° F. (190° C.).

# Basic Preparatory Techniques

## Grinding in a Mortar

Place all ingredients in a mound at center of the mortar and crush, gently at first and then more firmly, using a rolling motion of the pestle. Mix ingredients continually with the end of the pestle while mashing; 2 to 4 minutes of vigorous work usually brings the ingredients to a smooth, homogeneous puree. Scrape from the mortar with a wooden spoon or rubber spatula, then rinse with broth if desired, to retrieve all remains. Wash and dry mortar well before storing.

*Fig. 3.5: Cross section of a coconut*

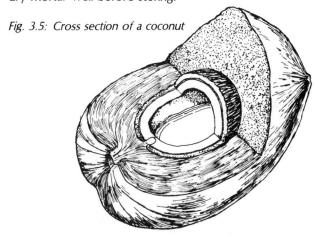

## Selecting and Opening a Coconut

Before buying a coconut, shake it to make sure it contains liquid; the more there is, the fresher the coconut. Coconuts without liquid, or those with moldy or wet "eyes" may be spoiled. Coconuts may weigh from 1 to as much as 3 pounds; an average one weighs about 1½ pounds and will yield from 2 to 3 cups of grated or chopped meat, and ½ to 1 cup of coconut water.

To open the coconut, puncture 2 of the 3 eyes (one will usually be very soft) by pushing or hammering the tip of an icepick, nail, or screwdriver through them. Drain the coconut water into a cup if you wish to save it. Coconut water is a delicious beverage but is rarely used in cooking. If you look carefully, you will notice that the coconut's surface is divided into three sections by three "ribs" that run from top to bottom, and that one of these sections is wider than the other two. Hold the coconut in one hand with the wide section down. Using the back of a cleaver, hatchet or hammer, give the rib opposite that wide section a sharp blow, striking it at somewhat of an angle near its center. Repeat several times, then rotate the coconut an inch or two and tap again. Keep on tapping and turning around the coconut on a "fault line" the same distance from one end until you hear the follow sound of the shell cracking and see a hairline split. Pry open the crack and separate the two halves.

To remove the meat, rap the shell vigorously with the cleaver, then, using a sharp, sturdy knife, cut down through the meat to the inside shell, making numerous cuts from the center to the circumference. Now pry out the meat with a very strong knife or screwdriver. (The meat can also be loosened from the shell by placing the unopened coconut in a 250° F. [130° C.] oven for 1 hour before cracking.)

## Making Coconut Milk

Using coconut milk (*santan*) is one of the keys to obtaining the fine flavors of true Indonesian cooking. It is added to sauces, soups, curries, and all types of desserts. Cooking instructions are given below. Fresh coconut purchased in the shell gives the best milk, although commercial dried, unsweetened coconut shreds or flakes also work well. Canned coconut milk is a fairly good substitute. These methods are listed with our favorites first. Tightly covered and refrigerated, coconut milk will keep for up to 5 days.

**1. Fresh coconut & blender method:** If you have a blender, this method is the tastiest and takes only about 5 minutes.

Cut coconut meat into pieces 1 inch square or smaller (you need not peel off the thin brown skin). Measure the amount required by a particular recipe (but no more than 2 cups at a time) into the blender and add an equal volume of hottest tap water (or coconut water). Puree at high speed for 1 to 2 minutes, or until fairly smooth. Transfer contents of blender to a medium-mesh strainer or sieve set over a deep bowl. Using your fingertips, press down hard on the coconut pulp in the strainer to expel as much milk as possible. Now carefully lift one handful of pulp from the strainer and, holding it over the bowl, squeeze as firmly as possible to extract any remaining milk. Reserve dry pulp in a separate bowl to use for making thin milk as described below, or discard. Repeat squeezing handfuls of pulp until the strainer is emp-

*Fig. 3.6: Making coconut milk*

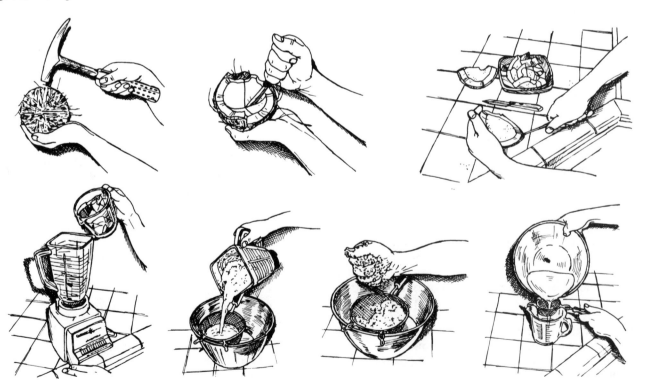

ty. Now pour milk through a small fine-mesh strainer into a measuring cup. One cup chopped (or grated) meat combined with 1 cup water should produce slightly more than 1 cup coconut milk. For a slightly richer consistency, use 1½ cups coconut to 1 cup water.

**2. Dried shredded coconut & blender method:** This method, which avoids the chore of cracking and shelling the coconut, is the easiest and quickest; the milk has a good flavor but ends up being slightly more expensive than that made from fresh coconut.

Combine 1½ cups dried shredded coconut (unsweetened) in a blender with 2¼ cups hottest tap water. Allow to stand for 5 minutes, then puree for 1 minute and proceed as above. Yields about 1¾ cups coconut milk.

**3. Fresh coconut & grater method:** The key to success in this method lies in using a grater designed for coconuts (see previous section); otherwise beware of grated fingers!

Using an Indonesian-style coconut grater or its very fine-toothed Western kitchen counterpart, grate approximately the required amount of unpeeled coconut meat into a bowl. Measure grated coconut exactly, stir in an equal volume of hottest tap water (most Indonesian cooks actually use unheated water), squeeze mixture in the bowl with one hand about 15 times (or mix vigorously with a spoon), then proceed to strain as described above. (Note: Some cooks let the mixture stand for 30 minutes before straining; others line the strainer with a double thickness of dampened cheesecloth.)

## Making Coconut Cream and Related Products

The recipes in this book all call for the use of coconut milk; however, to obtain a richer, creamier flavor, any of the following first three may be substituted in equal amounts, while the last two by-products can add subtle flavoring and nutrients to your favorite soups, stews, casseroles, or sautéed dishes.

**1. Coconut cream:** Let regular coconut milk stand refrigerated or at room temperature for 1 to 3 hours, or until the liquid separates into the rich and creamy one-third that rises to the surface above the rather thin remainder. Carefully skim off this top portion with a large spoon. God's gift to humankind, this ambrosial elixer is excellent for use in ice creams and other desserts, baked dishes, and rich sauces; or it can be whipped as a topping or pureed in a blender to make "coconut butter." Use the remaining pale, water lower portion as Skim Coconut Milk (below).

**2. Rich (or thick) coconut milk:** Combine in a blender 1 cup finely diced coconut meat with ½ cup hot water. Puree at high speed for 1 to 2 minutes, using a chopstick or fork to poke or scrape any coconut shreds from the blender's sides into its vortex. Proceed to strain, press, and re-strain as for regular coconut milk (above).

Makes about ⅔ cup of rich half-and-half. Exquisite. Indonesians make it by simply mixing freshly grated coconut with a small quantity of water and squeezing well.

**3. Simmered coconut cream:** Bring regular coconut milk to a boil in a small-diameter saucepan and simmer for 1 to 2 minutes, after which a rich cream will rise to the top. Skim off this slightly lumpy cream, using a ladle or large spoon, and use like coconut cream. We find that fresh coconut cream (above) is better tasting, smoother, and easier to make than this simmered type.

**4. Thin coconut milk:** After preparing regular or rich coconut milk, combine 1 cup of the pressed pulp with ½ to 1 cup hot water; mix well, then, if possible, puree for 1 minute in a blender. Strain and re-press. Good for use in cooked vegetable dishes, light sauces and soups, or puddings. The remaining coconut pulp may be used in granola, cookies, or breads, or as a garnish for breakfast foods or desserts.

**5. Skim coconut milk:** This is a rather thin lower portion that remains after coconut cream or simmered coconut cream has been skimmed off. Use as for thin coconut milk or serve as a chilled drink.

## Cooking with Coconut Milk

This richly flavored, slightly sweet, velvety smooth liquid, which has the look and feel of milk, sets the tone of all cuisines in Southeast Asia and the Pacific, being used in sauces, soups, curries, desserts, and a host of other delicious preparations. Indonesians do remarkable things with coconut milk (and cream) to create a depth and richness of flavor (called *gurih*) that many Westerners—including ourselves—find positively addicting. Used together with palm sugar, grated coconut, glutinous rice, and occasionally agar, it makes desserts that are out of this world. It can also be whipped for icing on Western-style cakes or simmered down to make "coconut butter," added to doughs or cream soups, and used instead of water for steaming rice. It holds and mellows the chilies and spices of the chili-hot gulai of Indonesia and the closely related curries of Hawaii, Thailand, and Vietnam. Some cooks even like to add it at the last minute to sauces and the like.

As coconut milk is cooked, its nature changes. Like dairy milk, it must be stirred constantly as it comes to a boil and kept *uncovered* during and after cooking (until well cooled) to prevent it from curdling or separating. Since it is very sensitive to high heat, be sure the flame is set at medium or lower, or cook in a double boiler. During the first stage of cooking, coconut milk becomes thick and smooth, developing a flavor that is sweet, fresh, and delicate. At the second stage, it becomes thicker and creamier, then oil begins to appear, accompanied by a smoother, richer flavor. Finally, if cooked long enough, it becomes oil, leaving a deposit on the bottom of the pan. If allowed to cool, uncovered, then sealed in a clean jar and refrigerated, it will keep for 3 to 5 days.

## Handling, Seeding, and Reconstituting Chilies

Although hot chilies are closely related to the familiar green bell pepper, they contain volatile oils that may make your skin tingle and your eyes burn. When handling fresh varieties, be careful not to touch your face or eyes and wear rubber gloves if your skin experiences irritation.

To prepare fresh chilies, rinse them under cold running water and cut or break the stems off if you wish to leave the seeds (which are by far the hottest parts) in the pods. If the chili is to be seeded, pull out the stem and the seeds attached to it, then break or cut the pod in half and scrape out the remaining seeds. (To seed small chilies, you may slice the pod first into thin diagonals and then scrape away the seeds with the point of a knife.)

To reconstitute dried chilies, soak them in plenty of hot water for about 30 minutes, then drain and allow to cool. Now split or slice open and, if desired, remove and discard the seeds.

After handling hot chilies, wash your hands thoroughly with soap and warm water.

## Making Deep-Fried Onion Flakes
### (Bawang Goreng)

Sprinkled over rice or sautéed vegetable preparations, fried onion (or shallot) flakes add crispness, a rich flavor, and a handsome brown color. Most cooks prepare ½ to 1 cup at a time and store them to be used when needed as an all-purpose topping and seasoning. Although shallots are generally used in Indonesia, large onions also work well.

Cut a peeled onion crosswise into halves, then cut each half vertically into very thin slices. Spread slices in a single layer on a screen, bamboo colander, or other porous surface and allow to air dry for 1 to 2 hours. Heat oil to 375° F. (190° C.) in a wok, skillet, or deep-fryer. Sprinkle slices over oil surface, reduce heat to low, and deep-fry, turning carefully after several minutes, so that both sides are crisp and nicely browned. Be very careful to prevent burning. Lift out flakes with a mesh skimmer and drain in a single layer on absorbent paper until cool and dry, then store (preferably refrigerate) in a well-sealed screw-top jar. Salt flakes lightly just before use. (Do not salt before storage lest they go soft.)

## Making Tamarind Water

This acidic liquid is used like lemon juice or vinegar in many Indonesian recipes; it is especially popular as a seasoning for deep-fried tempeh, used in combination with salt and spices.

Combine 2 ounces dried tamarind pulp and 1 cup boiling water in a small bowl. Mashing the pulp occasionally with a spoon, allow it to soak for 1 hour, or until pulp separates from fiber and dissolves. Rub tamarind through a fine sieve set over a second bowl, pressing down hard with the back of a spoon before discarding seeds and fiber. Sealed and refrigerated, tamarind water will keep for 1 to 2 weeks.

Another way of making tamarind water is to place a walnut-sized ball of fresh tamarind paste in a bowl, then dissolve in ½ cup hot water.

*Fig. 3.7: Javanese market ladies with chilies*

# *Rice Dishes*

Whatever else is served, rice is always the foundation of an Indonesian meal, as it is for the meals of most people throughout Asia. Thus, it is the key accompaniment for tempeh. For more details on rice, see the Glossary in the professional edition. Of all the many national rice cuisines we have tasted, that from Indonesia gets our first prize. In many cases, cooking the rice in coconut milk is the key to flavors so rich and mellow, textures so creamy smooth, and aromas so fragrant and heavenly that they defy description and comparison. Because of the pronounced nutritional superiority of natural brown rice over polished white rice, and because of the rapidly increasing popularity of brown rice in the West, each of the Indonesian recipes in this section, which ordinarily use white rice, have been adapted to brown rice. Following the basic preparation for brown rice, the recipes are listed with our favorites first.

## Brown Rice

MAKES 4 CUPS

Rice is cooked to its peak of goodness when the portion at the very bottom of the pot is golden brown and slightly crisp; the finest flavor is said to be attained when the rice is cooked in a heavy iron pot over a wood fire. The popular Indonesian equivalent is called Nasi Putih (white rice).

**2 cups brown rice, rinsed and soaked overnight in 2⅔ cups water**

In a heavy covered pot, bring water and rice to a boil over high heat. Reduce heat to low and simmer for 45 to 50 minutes, or until all water is absorbed or evaporated. Uncover pot, remove from heat, and stir rice thoroughly with a wooden spoon. (If a slightly drier consistency is desired, transfer rice to a wooden bowl before stirring). Allow to cool for several minutes, then cover pot (or bowl) with a double layer of cloth until you are ready to serve.

**To Pressure Cook:** Rinse and drain 1 cup rice. Without soaking, combine in a pressure cooker with 1¼ cups water. Bring to pressure (15 pounds), reduce heat to maintain a slow steady rocking motion of pressure regulator, and simmer for 25 minutes. Allow pressure to come down naturally for 10 to 15 minutes. Open pot and mix rice well. Allow to stand uncovered for 3 to 5 minutes, then cover with a cloth as above.

## Fragrant Spiced Coconut Rice
### *(Nasi Uduk)*

SERVES 3 TO 4

**1 cup long-grain brown rice, washed and soaked overnight in water**
**2¼ cups coconut milk**
**½ clove garlic**
**1 shallot (or ½ onion), minced**
**½ teaspoon salt**
**¼ teaspoon ground turmeric**
**¼ teaspoon ground cumin**
**½ teaspoon ground coriander**
**¼ teaspoon minced lemon rind or ½ stem lemongrass**
**¼ teaspoon terasi (optional; see Glossary)**
**¼ teaspoon kenchur powder (optional)**

Drain rice and set aside. Combine the remaining ingredients in a saucepan and bring to a boil. Add rice, return to the boil, and reduce heat to low. Cover and cook for 45 minutes. Remove from heat and allow to stand for 10 minutes, uncovered. Serve hot or cold with fried tempeh and sambals. Nice also topped with slivered omelets, Deep-Fried Onion Flakes (page 44), or slivered cucumbers.

## Rich Coconut Rice
### *(Nasi Gurih)*

SERVES 3 TO 4

**1 cup short-grain brown rice, washed and soaked overnight in water**
**2¼ cups coconut milk (page 42)**
**½ teaspoon salt**
**1¼ inches lemongrass stalk and bulb, or ½ teaspoon grated lemon rind**
**1 salam leaf or ½ bay leaf**

Drain rice, then combine with the remaining ingredients in a saucepan. Bring to a boil, stir gently, and cover. Reduce heat to low, simmer for 45 minutes, then remove from heat and allow to stand, covered, for 10 minutes. Popularly served with Sambal Goreng Tempeh Kering (page 79), plus slivered omelets and Deep-Fried Onion Flakes. Also nice with slivered cucumbers.

Other ingredients that may be simmered with the coconut milk include ¼ teaspoon ground nutmeg or mace and a dash of ground cloves. The rice may also be added to the boiling seasoned milk as in Nasi Uduk, above.

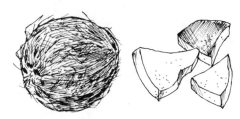

## Javanese Yellow Rice
### (Nasi Kuning)

SERVES 3 TO 4

1 cup short-grain brown rice, washed and soaked overnight in water
2¼ cups (thin) coconut milk (page 42–43)
½ teaspoon turmeric
¼ teaspoon cumin (optional)
1 tablespoon margarine or butter
½ teaspoon salt

Drain rice, then combine in a saucepan with the remaining ingredients. Bring to a boil, stir gently, and cover. Reduce heat to low, simmer for 45 minutes, then remove from heat and allow to stand, covered, for 10 miunutes. Serve hot on a large platter; garnish decoratively with slivered omelets, Deep-Fried Onion Flakes (page 44), and slivered cucumbers.

*Fig. 3.8: Nasi goreng (Indonesian stir-fried rice)*

### Indonesian Fried Rice *(Nasi Goreng)*

In Indonesia and Malaysia, Nasi Goreng is the most popular variation on plain boiled rice; in our opinion it is one of the most delicious. No two Nasi Gorengs are ever quite the same. In homes, it has long been served as a typical breakfast dish, easy for the housewife to prepare since the rice is often left over from the previous evening's meal. Recently it has become a dish featured in its own right, a favorite at restaurants for lunch or dinner, often served with elaborate garnishes. We present a related version, both containing tempeh, in Chapter 4 (see page 62).

## Festive Glutinous Yellow Rice
### (Nasi Kunyit)

SERVES 3 TO 4

Glutinous brown rice, unlike its white counterpart, must be pressure cooked to develop its prized glutinous qualities, which bind together the individual grains, imparting a unique chewy consistency.

1 cup glutinous brown rice (available at natural-food stores), washed and drained
1¼ cups coconut milk (page 42)
½ teaspoon salt
¼ clove garlic, crushed
¼ teaspoon ground turmeric
Dash of ground black pepper
1 drop vanilla or ½ pandanus leaf
Deep-Fried Onion Flakes (page 44)

Combine all ingredients (except the onion flakes) in a pressure cooker, bring to full pressure (15 pounds), reduce heat to maintain a slow steady rocking motion of pressure regulator, and simmer for 25 minutes. Remove from heat and allow pressure to come down naturally. Serve hot, garnished with the fried onion flakes.

## Coconut Glutinous Rice
### (Nasi Lemak)

SERVES 3 TO 4

1 cup glutinous brown rice, rinsed and drained
1¼ cups coconut milk (page 42)
Pinch of salt

Combine all ingredients in a pressure cooker, bring to full pressure (15 pounds), reduce heat to maintain a slow steady rocking motion of pressure regulator, and simmer for 25 minutes. Remove from heat and allow pressure to come down naturally. Open lid and stir briefly; adjust seasoning. Serve hot. Popular on festive occasions.

## Boiled Glutinous Rice
### (Nasi Ketan)

SERVES 3 TO 5

1 cup glutinous brown rice, rinsed and drained
1 cup water
Pinch of salt
Garnish: ½ cup freshly grated coconut, seasoned to taste with salt

Combine rice, water, and salt in a pressure cooker, bring to full pressure (15 pounds), reduce heat until as low as possible to just maintain a slow steady rocking motion of pressure regulator, and simmer for 25 minutes. Remove from heat and allow pressure to come down naturally. Open cooker and shape rice into 5 or 6 round patties. Serve hot, topped with a sprinkling of the lightly salted grated coconut. Simple but nice. For variety, substitute coconut milk for the water.

## Compressed Rice in Banana Leaves (Lontong)

SERVES 6

Popular in Indonesian and Malaysian restaurants or at parties, served as cold snack after a movie or as a weekend treat, Lontong is a favorite accompaniment for Saté with Peanut Sauce, Opor Tempeh, or Sambal Goreng Tempeh (pages 98, 95, 77). It is usually prepared the day before it is served since cooling imparts its prized firm texture; refrigerated, it will keep for 3 days.

Fold banana leaves into sausage shapes (or substitute cheesecloth) about 6 inches long and 1½ inches in diameter. Fill bags about two-thirds full with 1 pound washed short-grain (oval) rice and cook for 2 to 3 hours in boiling water, or until the rice swells to capacity, becoming a firm lump. Cool well, then, just before serving, unwrap leaves and cut rice into ½-inch-thick rounds. Use to accompany the dishes listed above.

**For Western Kitchens:** Combine 1 cup short-grain brown rice (rinsed) and 2 cups water in a pressure cooker, bring to full pressure (15 pounds), reduce heat to just maintain a slow steady rocking motion of pressure regulator, and cook for 40 minutes. Remove from heat and allow pressure to come down naturally. Open cooker and stir rice vigorously with a wooden spoon, then press into a pie plate or cake tin to a depth of about 1 inch. Cover rice surface with a piece of paper, aluminum foil, or a lightly oiled banana leaf. Top with a pressing lid and 2-pound weight; refrigerate for several hours, or until very firm. Using a moistened knife, cut rice into 2-inch squares and serve as described above. (Another method of firming rice is to pack hot rice firmly into deep tapered or cylindrical glasses, chill, remove from glasses, and cut into rounds. These are nice dusted with corn flour and fried.)

## Compressed Rice in Woven Coconut-Leaf Packets (Ketupat)

Weave strips of young coconut leaves into 4-inch square bags, fill each bag three-fourths full of short-grain or glutinous rice, and boil for 2 to 3 hours in water or coconut milk. Serve as described above. The glutinous-rice-and-coconut-milk version is popular as a sweet dish; it can also be prepared in a hollow bamboo lined with young banana leaves and placed near an open fire to cook.

# Other Basic Recipes

The following sauces, dressings, omelets, and other basic preparations are often served with tempeh, so we have grouped them all together here. They will be called for frequently in later recipes.

## Creamy Tofu Dressings, Dips, and Mayonnaise

MAKES ABOUT 1 CUP

Quick and easy to prepare, this is a delicious way of using tofu, with almost unlimited possible variations. Each is thick, rich, and full of flavor, yet remarkably low in fats and calories. The perfect answer for those who love dressings but dislike their oily qualities.

**6 ounces (170 grams) tofu**
**1½ to 2 tablespoons lemon juice or vinegar**
**2 to 4 tablespoons oil**
**½ teaspoon salt**
**Choice of seasonings (see below)**

Combine all ingredients in a blender and puree for 20 seconds, or until smooth (or mash all ingredients and allow to stand for 15 to 30 minutes before serving). If desired, serve topped with a sprinkling of parsley or a dash of pepper. Refrigerated in a covered container, these preparations will stay fresh for 2 to 3 days; the consistency will thicken (delectably).

SEASONINGS. CHOOSE ONE:

* **Curry:** ½ teaspoon curry powder and 2 tablespoons minced onions. Top with a sprinkling of 1 tablespoon minced parsley.
* **Garlic & Dill:** ½ to 1 clove minced garlic and ½ teaspoon dill seeds. Top with a sprinkling of parsley.
* **Onion:** ¼ cup minced onion. Excellent on most deep-fried tempeh preparations.
* **Roquefort:** 1 tablespoon Roquefort cheese, 1 extra tablespoon oil, ½ clove garlic, ¼ teaspoon each fresh mustard and pepper.
* **Tofu Tartare Sauce:** 1 more tablespoon oil, ½ teaspoon honey, ¼ teaspoon each prepared mustard and turmeric, dash of paprika. After pureeing, mix in 2 tablespoons minced parsley and 3 to 4 tablespoons chopped dill pickles.
* **Tofu Mayonnaise:** Use only the tofu, lemon juice, oil, and salt; omit seasonings.

## Tartare Sauce

MAKES ½ CUP

5 tablespoons mayonnaise (egg or tofu; see above)
2 tablespoons minced green pepper
1 tablespoon minced parsley
1 teaspoon minced onion
1 teaspoon lemon juice
¼ teaspoon curry powder

Combine all ingredients, mixing well. Delicious on Breaded Tempeh Cutlets (page 59).

## Tangy Cocktail Sauce

MAKES ½ CUP

⅓ cup ketchup
5 teaspoons lemon juice
1 teaspoon shoyu (natural soy sauce)
1 teaspoon finely minced onion
1 tablespoon minced parsley

Combine all ingredients, mixing well. Delicious on Deep-Fried Tempeh (page 53) or Tempeh Shish Kebab (page 70). For variety add ¼ to ½ teaspoon hot mustard, 1 teaspoon horseradish, or ½ teaspoon crushed anise or ground roasted sesame seeds.

## Ketchup & Worcestershire Sauce

MAKES 6 TABLESPOONS

4 tablespoons ketchup
2 teaspoons Worcestershire sauce

Combine ingredients, mixing well. Serve with Deep-Fried Tempeh (page 53).
For a richly flavored sautéed version, sauté 1 small minced onion in 2 tablespoons margarine or oil for 4 to 5 minutes. Add 5 tablespoons ketchup and 1 tablespoon Worcestershire sauce and, stirring constantly, bring just to a boil. Serve over (10 ounces) Deep-Fried Tempeh.

## Sweet Ketchup & Curry Sauce

MAKES ¾ CUP

½ cup tomato puree
2½ tablespoons honey
1½ teaspoons curry powder
¼ to ½ teaspoon salt

Combine all ingredients in a small saucepan and, mixing well, bring just to a boil. Remove from heat and cool to room temperature. Serve on fried tempeh.

## Sweet Indonesian Soy Sauce
### *(Kechap Manis)*

MAKES 2 CUPS

This is a simple method for obtaining the flavor of kechap manis using Japanese shoyu as a basis. The method for making real Indonesian fermented kechap is described in Appendix D of the professional edition.

1 cup dark-brown or palm sugar
1 cup water
¾ cup shoyu (natural soy sauce)
6 tablespoons dark molasses
¼ teaspoon ground laos
¼ teaspoon ground coriander
¼ teaspoon freshly ground black pepper

Combine the sugar and water in an enameled or stainless-steel saucepan and bring to a boil, stirring until sugar dissolves. Increase heat to high and cook, uncovered, for 5 minutes, or until syrup reaches 200° F. (93° C.) on a candy thermometer. Reduce heat to low, stir in the remaining ingredients, and simmer for 3 minutes, then strain through a fine-mesh sieve set over a bowl. Tightly covered, kechap manis may be kept at room temperature for 2 to 3 months.

VARIATION

*Simplified Recipe: Combine in a small saucepan: 1 cup shoyu, ½ cup molasses, and 3 tablespoons dark-brown sugar. Bring to a boil over medium heat and stir until sugar dissolves. Keep in a covered jar.

## Wasabi-Shoyu or Lemon-Shoyu

MAKES ABOUT ⅛ cup

2 tablespoons shoyu
1 teaspoon wasabi paste (Japanese horseradish) or 2 tablespoons lemon juice

Combine ingredients, mixing well. Serve as a seasoning or dipping sauce for Deep-Fried Tempeh (page 53).

## Peanut Sauce (Saus Kachang)

MAKES ½ CUP

This rich and slightly spicy delicacy, served with Gado-Gado, Saté, Pechel, fried tempeh, and hot or cold cooked vegetables is, in our opinion, one of Indonesia's great contributions to world cuisine, and is especially suited to American tastes. Indonesians swear that using whole peanuts, freshly fried and ground in a mortar, gives the best flavor, but we find that peanut butter also gives excellent results. Of the more than twenty recipes we have tested, the following slightly Westernized version is our favorite.

1 tablespoon oil
¼ onion, minced
½ clove garlic, minced
¼ cup peanut butter, preferably crunchy, or fried whole peanuts
¼ cup milk (coconut, soy, or dairy)
1 tablespoon brown or palm sugar, or 1½ teaspoons honey
1 teaspoon lemon juice
1 teaspoon shoyu (natural soy sauce) or Worcestershire sauce
Dash of salt (optional)
Dash of chili powder or Tabasco sauce

Heat a wok or skillet and coat with the oil. Add onion and garlic, and sauté for 3 to 4 minutes, or until lightly browned, then combine in a blender with the remaining ingredients and puree for 1 minute.

For variety add 5 macadamia nuts or small quantities of any of the following: coriander seeds, grated ginger-root, paprika, terasi, grated lemon rind, lemongrass, kenchur root, laos, or lime leaf.

*Balinese mask*

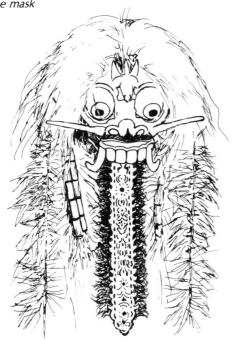

*Indonesian mortar and pestle*

## Noodles

SERVES 2 OR 3

4½ to 5 ounces (135 grams) dry buckwheat or whole-wheat noodles

Bring 2 to 3 quarts water to a rolling boil over high heat. Scatter noodles slowly over surface of water and return to the boil. Lower heat until water is boiling actively but does not overflow. Cook uncovered for about 5 minutes, or until noodles are tender but not soft. Pour into a colander placed in the sink and drain briefly, then transfer to a large container filled with circulating cold water. Stir with chopsticks for several minutes until they cool to the temperature of the water, then transfer noodles back into colander; drain well and serve, as is or doused with boiling water.

## Paper-Thin Omelets

MAKES ABOUT 8

In Indonesia, these omelets are generally slivered and served atop rice dishes, together with Deep-Fried Onion Flakes (page 44), thinly sliced chilies, sambals, and the like.

4 eggs
¼ teaspoon salt
1 teaspoon ground roasted sesame seeds (optional)
1 to 2 teaspoons oil

In a small bowl, combine eggs, salt, and sesame; mix well. Heat a small skillet and coat lightly with oil, pouring off any excess. Pour about one-eighth of the egg mixture into the skillet, swishing it around quickly so that it just covers the bottom of the pan. Cook over high heat for about 20 to 30 seconds on one side only to form a thin omelet. Transfer omelet to a plate and allow to cool. Prepare 8 omelets, oiling the pan lightly after every 3 or 4. Sliver to use as a garnish.

## Our Favorite Tempeh Recipes

Among the many dishes in this book, there are certain ones we enjoy again and again, and like to serve to guests as an introduction to tempeh cookery. The Western recipes represent adaptations of tempeh to Western tastes and ingredients. Most of our very favorites, which call for fried or sautéed tempeh, have a savory flavor and/or crisp texture. We have also included a list of our nonfried Western favorites. The Indonesian recipes, traditional favorites in their native country, are also easily made in your kitchen; instructions for basic preparations (such as coconut milk) are given in this book, and ingredients such as turmeric, lemongrass, and tamarind are now available at a growing number of specialty shops, supermarkets, and natural-food stores. All are described in the Glossary. We have starred those recipes we suggest you make your very first selections.

### Western Favorites

Seasoned Crisp Tempeh (page 54)
*Coriander & Garlic Crisp Tempeh (page 55)
Delectably Crunchy Tempeh Condiment (page 56)
*Crisp Tempeh Chips (page 57)
Breaded Tempeh Cutlets (page 59)
*Tempeh, Lettuce, & Tomato Sandwiches (page 59)
Tempeh in Pita-Bread Sandwiches (page 60)
Pizza with Tempeh (page 60)
Tortillas with Tempeh & Guacamole (page 61)
Curried Stir-Fried Noodles with Tempeh (page 62)
Waldorf Salad with Nutlike Tempeh Chunks (page 64)
Tempeh & Mushroom Sauce (page 66)
*Savory Tempeh in Tomato & Herb Sauce (page 66)

### Nonfried Favorites

Creamy Tempeh Dip with Garlic & Dill (page 63)
Tempeh Mock Tuna Spread (page 63)
Tempeh & Tomato Sauce for Sloppy Joes or Pizza (page 66)
Baked Tempeh with Nutritional Yeast Gravy (page 68)
Delectable Tempeh & Lasagna (page 69)
*Baked Tempeh & Tomato Sauce (page 69)

### Indonesian Favorites

Tempeh Bachem (Tempeh Cutlets) (page 76)
Sambal Goreng Tempeh (Spicy-Fried Tempeh in Coconut Milk) (page 77)
Kering Tempeh (Sweet & Crispy Tempeh Sauté) (page 78)
Sambal Goreng Tempeh Kering (Crunchy Chili-Fried Tempeh Topping) (page 79)
Menjeng Tempeh (Tempeh & Coconut Croquettes) (page 79)
Kotokan Tempeh (Tempeh Cubes in Coconut Milk) (page 83)
Bistik Tempeh (Beefsteak Tempeh in Brown Sauce) (page 84)
Rendang Tempeh (Tempeh Cutlets in Coconut-Milk Sauce) (page 84)
Tempeh Masak Taucho (Tempeh Slices in Rich Miso & Coconut-Milk Sauce) (page 84)
Tempeh Asam Manis (Sweet & Sour Tempeh) (page 85)
Tempeh Bumbu Rujak (Tempeh with Rujak Spice Sauce) (page 85)
Opor Tempeh (Tempeh & Tofu Chunks Stewed in Coconut Milk) (page 95)

## Serving Tempeh Throughout the Day

The following serving suggestions may help in menu planning. Many people wishing to move to meatless diets ask about what to use as a dinner entree in place of meat. We suggest a grain dish accompanied by tempeh and cooked vegetables or a salad.

**Breakfast**
Tempeh with grains
Tempeh with eggs

**Lunch**
Tempeh in sandwiches or burgers
Tempeh in salads
Tempeh dressings
Tempeh with grains

**Dinner**
Tempeh hors d'oeuvres
Tempeh in soups
Tempeh baked or broiled
Tempeh in sauces

# 4

# Western-Style and Oriental Tempeh Recipes

Tempeh's unusual versatility lends itself well to a wide range of Western-style and Oriental recipes. Those presented here are but a sampling of the many delicious possibilities, intended to introduce you to basic principles of preparation and invite you to further experimentation. In general, the *sections and the recipes within each section have been listed with our favorites first.*

Most recipes simply call for "tempeh" without specifying which type. While most people will probably be using soy tempeh, other types (okara, wheat & soy, etc.) usually also work well. Please note that all ingredients needed for a recipe must be assembled and prepared in advance. Thus, when deep- or shallow-fried tempeh is listed as an ingredient, the frying should be done before the basic recipe preparation is started. For notes on eating tempeh without further cooking, see the introductory section to Chapter 5, Indonesian Tempeh Recipes; we serve only cooked tempeh.

## *Tempeh Deep- or Shallow-Fried, Stir-Fried and in Hors d'Oeuvres*

Over the centuries, Indonesians have found that deep- or shallow-frying tempeh best evokes its prized savory flavor and crisp, crunchy texture. In preparing recipes using these and related techniques, there are cer-

tain principles that will help you achieve best results:

* When using tempeh as an ingredient in your favorite recipes (such as salads, sauces, soups, casseroles, etc.), try frying it first. Then add it to these dishes just before they are served to preserve its crisp texture.

* When served in its own right, tempeh works best as an hors d'oeuvre, a snack, or an accompaniment or seasoning for brown rice. Here it is especially important to serve it freshly deep-fried, crisp and hot at its peak of flavor.

* The thinner tempeh is sliced before deep-frying, the crisper it will be.

* A brief drying (such as sun-drying for 5 to 15 minutes) before deep-frying gives a crisper texture and longer storage life.

* Tempeh, like other deep-fried foods, should be served in relatively small quantities (1½ to 3 ounces per serving) to promote good digestibility.

There are at least nine basic techniques that can be used in the process of frying tempeh: each has numerous variations and yields distinctly different products.

1. *Deep- or shallow-frying:* Quick and easy without batter; yields crisp slices.
2. *Pan-frying:* In margarine, oil, or butter, with a little water for steaming; yields savory slices.
3. *Seasoned brine:* Dip slices in a plain or seasoned 10 percent salt-water solution, then fry; yields Seasoned Crisp Tempeh or Coriander and Garlic Crisp Tempeh.
4. *Stir-frying or sautéing:* Yields crumbly condiments or tempeh and vegetable sautés.
5. *Presimmering:* Simmer in seasoning liquid then fry; yields rich fillets.

51

6. *Seasoned brine and flour:* Dip slices in plain or seasoned salt-water solution, dust with flour, and fry; yields crisp chips.
7. *Seasoned batter:* Coat slices with seasoned batter and fry; yields batter-fried crisps.
8. *Bound breading:* Dip slices in seasoning solution, dust with flour, dip in lightly beaten egg, and roll in bread crumbs, then fry; yields breaded cutlets.
9. *Croquettes:* Mash tempeh, mix with flour, seasonings, and vegetables, shape into patties, and deep-fry.

## About Deep-Frying

Throughout East Asia, deep-frying has been one of the most popular cooking techniques since ancient times. There are good reasons for this: (1) it imparts a delicious rich and savory flavor, transforming even the simplest of ingredients or leftovers into prize creations; (2) it is faster and uses much less fuel than baking, a cooking technique rarely used in Asia: (3) it serves as an excellent source of essential (usually unsaturated) oils in diets which have long contained very little meat or dairy products; (4) in tropical climates where unwanted microorganisms propagate quickly on foods, it sterilizes them, making them much safer to eat; and (5) by greatly reducing the foods' moisture content (and sterilizing them), it increases their shelf life.

Some Westerners still have the image of deep-fried foods as being necessarily "greasy," yet this is simply a result of poor technique. Anyone who has enjoyed Indonesian tempeh chips or Japanese tempura knows that deep-frying can yield foods that are delicate, light, and crisp. Others have stated that deep-fried foods are hard to digest; yet, if the foods are prepared properly as described below and if they are served in moderate portions (typically 1½ to 3 ounces per person at a meal), they are generally found to be easily digested, much more easily than meats and dairy products. Finally, it is pointed out that overheating deep-frying oil can impair its nutritional value and saturate some of the fats, thereby possibly raising serum cholesterol levels; these problems are easily avoided, even by a beginner, by simply keeping most oils at 375° F. (190° C.) and coconut oil at 400° F. (210° C.) or below and by using oil for deep-frying no more than two or three times. Virtually all nutritionists agree that many people in North America are now consuming much too much fat, especially saturated fats from meats, dairy products, and eggs. Some *40 percent* of the calories in our diet now comes from fats! By consuming fewer animal products and using moderate amounts of savory soy protein foods such as tempeh, properly deep-fried to be light and crisp, we can substantially lower fat intake while still enjoying a satisfying, nutritious, and balanced diet.

**Utensils:** Please read about the wok on page 41. If you do not have a wok (the best pot for deep-frying), a heavy skillet 2½ to 3 inches deep and 10 to 12 inches in diameter, a heavy 3- to 4-quart kettle, or an electric deep-fryer will also work well.

**Oils:** Any typical vegetable oil gives good results. Indonesians prefer coconut oil since it is inexpensive and abundant, imparts a wonderful flavor, and has a high smoke point; it can be heated to 400° F. before it will begin to smoke or break down. However it is high in saturated fats. We and most Japanese prefer rapeseed or colza oil made from the seeds of *Brassica napus,* a plant related to cabbage. Widely grown in Canada, it was once outlawed in the United States since the seeds contained goiter-causing substances plus the undesirable erucic acid, both of which have now been bred out. Other good oils for deep-frying include soy, corn, peanut, safflower, and cottonseed, or a combination of two or more of these. The addition of 10 to 30 percent sesame oil to any of these will give the foods a delicious nutty flavor.

Generally the deep-frying container should contain oil to a depth of 1½ to 2 inches, which usually requires 1 to 2½ cups of oil for a typical wok. Having sufficient oil helps to keep the temperature stable and provides a large oil surface. However, some cooks like to use relatively little oil when deep-frying foods dipped (without batter) in seasonings, since a portion of the seasonings inevitably goes into the oil and can flavor subsequent batches. Used deep-frying oil should be kept in a sealed jar and stored in a cool dark place or refrigerated. Using a little of this oil in other cookery imparts a rich flavor and helps to use up the oil. When deep-frying, try to use about one part fresh oil to one part used. Changing the oil frequently gives best results. Dark or thick oil has a low smoke point and imparts a poor flavor; foods deep-fried in used oil are not as light and crisp as they could be. Pour oil from the storage jar carefully so that any sediment remains at the bottom of the jar. Then add fresh oil to fill the wok or skillet to the desired depth.

Maintaining the oil at the proper temperature is the most important part of deep-frying. At first it may be easiest to measure the temperature with a deep-frying thermometer. Experienced chefs judge the oil's temperature by its subtle crackling sound, aroma, and appearance; they will generally drop in a small piece of tempeh and wait until it rises to the oil surface and begins to sizzle in a lively manner, indicating that the rest of the tempeh can now be added. If the oil begins to smoke, it is too hot. Overheating shortens the life of the oil and imparts a poor flavor to foods cooked in it.

Keeping the oil clean is another secret of successful deep-frying. This is especially important when using batter or bound-breading coatings. Use a mesh skimmer or a perforated metal spatula or spoon to remove all particles of food and batter from the oil's surface. Skim after every two or three batches of ingredients have been cooked. Place the small particles of deep-fried batter skimmed from the oil into a large colander or bowl lined with absorbent paper and allow to drain thoroughly; these may be used later as tasty additions to soups, salads, sautéed vegetables, noodles, or other grain dishes. Some Indonesian cooks put a slice of raw onion in the oil during deep-frying to improve the oil's flavor, then remove the onion as soon as the frying is finished.

Deep-frying changes the nutritional composition of tempeh: 100 grams of fresh tempeh weighs only 63 to 83 grams after deep-frying, largely because of a loss of moisture. Amara, in 1976, found that, whereas the percentages of moisture, protein, and fat in fresh tempeh are respectively 60.4, 19.5, and 7.5, in deep-fried tempeh they average 50.0, 22.5, and 18.0. By comparison, in cooked hamburger they are 54.2, 24.2, and 20.3, and in deep-fried tofu burgers 64.0, 15.4, and 14.0.

**Serving:** To ensure that deep-fried foods are served at their peak of flavor and crispness, prepare them just before you are ready to serve the meal, preferably after your guests have been seated at the table. If you have a large quantity of ingredients to deep-fry and wish to serve them simultaneously, keep freshly cooked pieces warm in a 250° F. (120° C.) oven.

**Cleanup:** After all foods have been deep-fried, allow the oil to cool in the wok or skillet, then pour it through a mesh skimmer or fine-weave strainer held over a funnel into your used-oil container. Seal the jar and discard any residue in the skimmer. Wipe all utensils with absorbent paper (washing is unnecessary) and store in a sealed plastic bag. Some cooks like to wash the hot wok immediately with water and a scrub brush (never use soap), place it back on the fire until the inside is just dry, then wipe it quickly with a dry dishtowel.

## Deep-Fried Tempeh                    SERVES 3 TO 4

Please begin by reading About Deep-Frying (above). This technique, so popular in East Asia, is not difficult to learn; when properly executed it gives crisp, light, and golden-brown results every time.

*Fig. 4.1: Deep-frying tempeh*

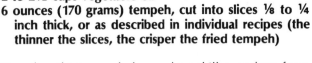

**2 to 2½ cups vegetable oil**
**6 ounces (170 grams) tempeh, cut into slices ⅛ to ¼ inch thick, or as described in individual recipes (the thinner the slices, the crisper the fried tempeh)**

Pour the oil into a wok, heavy deep skillet, or deep-fryer to a depth of 1¼ to 2 inches. Heat over high heat until temperature registers 375° F. (190° C.) on a deep-frying thermometer or until oil just begins to crackle and become fragrant, but does not smoke. Reduce heat to medium and place tempeh slices one by one into the oil or put in smaller pieces on a mesh skimmer. Tempeh should sink momentarily, then float to the surface, where it sizzles in a lively manner. Stirring tempeh gently with long chopsticks or tongs, deep-fry for 2 to 3 minutes, or until crisp and golden brown. Drain pieces briefly on a draining rack (smaller pieces on a mesh skimmer or slotted spoon), then transfer onto absorbent (paper) toweling placed on a large dish or tray and allow to drain for several minutes. Now arrange tempeh on plates, or serve like Japanese tempura on a bamboo colander or shallow bowl lined with neatly folded white paper. Use as an ingredient in other recipes as instructed or serve crisp and hot with any of the following:

**\*Seasonings:** Sprinkle on one or more immediately after frying:
    Shoyu (natural soy sauce)
    Salt
    Ketchup
    Worcestershire sauce
    Teriyaki sauce
    Lemon-shoyu (equal parts lemon juice and shoyu)

**Dips or Toppings:** Spoon on or dip just before serving:
    Creamy Tofu Dip (page 47)
    Creamy Tempeh Dip (page 63)
    Tempeh Tartare Sauce with Dill (page 64)
    Tangy Cocktail Sauce (page 48)
    Ketchup & Worcestershire Sauce (page 48)
    Sweet Ketchup & Curry Sauce (page 48)
    Sweet Indonesian Soy Sauce (kechap) (page 48)
    Peanut Sauce (page 49)
    Thousand Island or Roquefort Dressing (your favorite)

## Shallow-Fried Tempeh

SERVES 3 TO 4

Many Americans may find this technique easier than deep-frying: it can be done quickly and easily in a typical skillet; it requires the use of much less oil and the oil heats up much faster; and it is quite similar to the familiar techniques of pan-frying or sautéing. However, it is often difficult to cook the sides of each piece evenly; small cubes or croutons are harder to turn; the process may take longer, especially when frying a lot of tempeh; and the finished product is generally not quite as crisp as deep-fried tempeh.

**¼ to ⅓ cup vegetable oil or 2¼ tablespoons butter**
**6 ounces (170 grams) tempeh, cut into slices ⅛ to ¼ inch thick, or as described in individual recipes**

Heat the oil in a skillet over medium heat; it should be fragrant and on the verge of starting to crackle but should not be smoking. Put in tempeh and shallow-fry for 1½ to 2 minutes on each side, turning with long chopsticks, tongs, or a spatula. When crisp and golden brown, drain briefly on a draining rack and/or absorbent paper toweling, then use either as an ingredient in other recipes as instructed or serve crisp and hot with any of the seasonings, dips, or toppings listed under Deep-Fried Tempeh, above.

*Fig. 4.2: Shallow-frying tempeh*

## Pan-Fried Tempeh

SERVES 2 TO 4

We generally prefer deep- or shallow-frying to pan-frying. Although this technique uses the least amount of oil, is easy, and gives tempeh with the subtle flavor and aroma of fried chicken, it takes considerably longer (especially when frying a lot of tempeh), does not brown the sides of each piece, works poorly for croutons or small pieces which cannot be turned easily, and yields tempeh which is not as crisp as that from deep- or shallow-frying.

**3 tablespoons oil or butter**
**6 ounces (170 grams) tempeh, cut into pieces roughly 2½ by 2½ by ¼ inch**
**4 tablespoons (lightly salted) water for steaming**

Heat 1½ tablespoons oil or butter in a skillet, add tempeh slices, and fry on one side until nicely browned. Now sprinkle 1 tablespoon water around the cakes, cover skillet, and steam for 1 to 2 minutes, or until water is gone. Repeat sprinkling and steaming. Melt the remaining 1½ tablespoons oil or butter in the skillet, turn over the tempeh, and fry the second side. Sprinkle on 1 tablespoon remaining water, cover, and steam; repeat as before. Serve as for Deep-Fried Tempeh (above).

*Fig. 4.3: Seasoned crisp tempeh with dip*

## Seasoned Crisp Tempeh

SERVES 2 OR 3

Our very favorite tempeh recipe, this quick and easy delicacy, its close relative Coriander & Garlic Crisp Tempeh (below), and their numerous variations are the only tempeh recipes you need to know. Please begin by reading about deep-frying, shallow-frying, and pan-frying, above.

**½ cup water**
**1 teaspoon salt**
**6 ounces (170 grams) tempeh, cut into slices ⅛ to ¼ inch thick, or as described in individual recipes (the thinner the slices, the crisper the fried tempeh)**
**Oil for deep- or shallow-frying**

Combine the water and salt in a bowl, mixing well. Dip in tempeh slices, then drain briefly on absorbent (paper) toweling or on a rack, patting surface lightly to absorb excess moisture and thus prevent spattering during frying. Heat the oil to 375° F. (190° C.) in a wok, skillet, or deep-fryer. Slide in tempeh and deep- or shallow-fry for 3 to 4 minutes, or until crisp and golden brown. (Or pan-fry as described above.) Then drain briefly on fresh toweling and serve immediately. Delicious as is, as an ingredient in other tempeh recipes, as an accompaniment for brown rice, or as an hors d'oeuvre or crisp finger food with any of the toppings or dips listed under Deep-Fried Tempeh (above).

**\*Tempeh Croutons:** Cut tempeh into ⅜- to ½-inch cubes and proceed to deep- or shallow-fry as above. Try these when crunchy and hot on top of soups, stews, sauces, casseroles, curries, noodle dishes, or rice; cool for use in salads. Refrigerated in a well-sealed jar, they will keep for up to 1 week, but may develop a slightly bitter or smoked flavor while losing much of their crispness.

**\*Tempeh French Fries:** Cut tempeh to the shape of French fries (½ by ½ by 2½ inches) and deep-fry as above. Nice with Tempeh Burgers.

**\*Tempeh Bacon:** Very thin slices of tempeh fried until nicely browned have much the same crispness, savory flavor, and appearance as bacon. Use these in salads, egg dishes, soups, etc.

**\*Deep-Fried Tempeh Shish Kebab:** Cut tempeh into ½- to 1-inch cubes, deep-fry, and skewer with other ingredients for use as Tempeh Shish Kebab (page 70).

*Fig. 4.4: Tempeh shish kebab*

## Coriander & Garlic Crisp Tempeh   SERVES 2 OR 3

We and generations of cooks in Indonesia have experimented with many natural seasonings to use with deep- or shallow-fried tempeh. All agree that coriander and garlic are the most delicious. Try experimenting with other combinations or with the variations listed below.

Mix ½ teaspoon ground coriander and 1 clove of crushed garlic with the salt and water in the recipe for Seasoned Crisp Tempeh, above. Then proceed as instructed.

VARIATIONS

**\*Curried Crisp Tempeh:** Use 1¼ teaspoons curry powder in place of the coriander and garlic.

**\*Garlic Crisp Tempeh:** Omit the coriander in the preparation above. For added tang, substitute several tablespoons lemon juice for the water.

**\*Tangy Lemon Crisp Tempeh:** Substitute ¼ cup lemon juice and 1 teaspoon salt for the water and salt in Seasoned Crisp Tempeh.

**\*Tamarind Crisp Tempeh:** Substitute tamarind water (see Basic Preparatory Techniques) for the lemon juice in the variation above.

*Fig. 4.5: Coriander & garlic crisp tempeh*

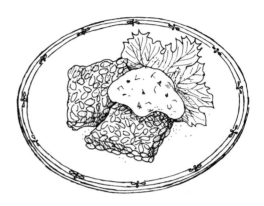

## Crispy Tempeh Bits   MAKES ½ CUP

Remarkably similar to the commercial products Imitation Bacon Bits, Bac\*os, or Bac-o-bits (which are also made from soy protein), these little nuggets of goodness have a rich flavor and crunchy texture like those of crisp bacon or walnuts. They make a nice topping or garnish for (or ingredient in) egg dishes, salads (especially Waldorf), rice or noodles, and soups; they can also be used like Tempeh Croutons (above).

**3 ounces (85 grams) tempeh, cut into pieces ¾ by 2 by ⅛ inch**

Prepare as for Seasoned Crisp Tempeh or Coriander & Garlic Crisp Tempeh (above), but after frying and draining, cool to room temperature and crumble each piece. Use crisp and fresh, or refrigerate in a sealed jar.

## Tempeh Pancakes or Waffles   SERVES 1 OR 2

**4 ounces (115 grams) tempeh, cut into slices 2½ by 3 by ¼ inch, shallow- or deep-fried (see above) and lightly salted**
**Butter**
**Maple syrup or honey**

Arrange freshly fried tempeh on platters, spread with margarine or butter, and top with maple syrup. Serve immediately while crisp and piping hot.

## Tempeh Fondue

SERVES 3 OR 4

Cut 6 ounces (170 grams) tempeh into ½-inch cubes and deep-fry for 3 minutes, or until crisp and golden brown. Then impale on slender skewers and use as the main ingredient or with other ingredients in your favorite fondue recipe.

*Fig. 4.6: Tempeh fondue*

## Tempeh Chops with Applesauce

SERVES 4

¼ cup water or apple juice
4 small apples, cored and chopped
¼ cup raisins
1 teaspoon lemon juice
Dash of cinnamon
6 ounces (170 grams) tempeh, cut into 4 slices ⅜ inch thick, shallow- or deep-fried, or pan-fried (see above), and lightly salted

Combine the first three ingredients in a saucepan, bring to a boil, and simmer, covered, over low heat for 20 minutes. Mix in lemon juice and cinnamon, remove from heat, and allow to stand for at least several hours while the flavors marry. Chill if desired, then spoon in dollops onto freshly fried tempeh. Serve immediately while tempeh is at its peak of crispness.

Or simply top crisp tempeh with dollops of applesauce.

## Delectably Crunchy Tempeh Condiment

SERVES 2

3 tablespoons oil
¼ onion, minced
1 clove of garlic, crushed or minced
1 bay leaf
2 tablespoons brown sugar
¼ teaspoon salt
¼ teaspoon chili powder or Tabasco sauce

½ teaspoon paprika
1½ teaspoons shoyu (natural soy sauce)
6 ounces (170 grams) tempeh, cut into pieces 1 by ¼ by ¼ inch (or crumbled), and deep- or shallow-fried (pages 52–54)

Heat a large wok or skillet and coat with the oil. Add onion and stir-fry or sauté until lightly browned. Mix in the next seven ingredients and stir-fry for 30 to 60 seconds. Add tempeh pieces, stir-fry for 1 minute more, and remove from heat. Remove bay leaf and serve hot or cold. Delicious as a topping for brown rice, baked potatoes, curries, salads, etc. The Indonesian relative of this recipe is Sambal Goreng Tempeh Kering (page 79).

## Curried Tempeh Condiment

MAKES 1¼ CUPS

2 tablespoons of butter
1 onion, minced
8 ounces (227 grams) tempeh, cut into half-inch cubes, and deep- or shallow-fried (pages 52–54)
1½ teaspoons curry powder
1 cup water
1 teaspoon shoyu (natural soy sauce)
⅛ teaspoon salt

Melt margarine in a skillet. Add onion and sauté for 3 to 4 minutes, or until slightly browned. Stir in fried tempeh plus the remaining ingredients and simmer uncovered for 15 minutes, or until fairly dry. Nice on brown rice or in Indian curry recipes.

## Tempeh & Vegetable Sauté

SERVES 4 TO 5

This is a nice way to use leftover tempeh or vegetables. Experiment freely with other ingredient combinations.

2 tablespoons oil (oil from frying tempeh works well)
½ onion, thinly sliced
½ cup sliced carrots
½ cup sliced mushrooms
2 cups slivered cabbage
1 green pepper, thinly sliced
6 ounces (170 grams) tempeh, cut into pieces 1 by 1½ by ¼ inch, dipped in seasoning solution, and deep- or shallow-fried to make Coriander & Garlic Crisp Tempeh (page 55)
4 teaspoons shoyu (natural soy sauce)
Dash of white pepper (optional)

Heat the oil in a wok or skillet. Add onion and carrots, and sauté for 2 minutes. Add mushrooms and sauté for 2 minutes. Add cabbage and green pepper, and sauté for 2 to 3 minutes more, or until cabbage is tender. Add tempeh, shoyu, and pepper, mixing well. Serve hot or cold.

## Rich Tempeh Fillets
SERVES 4

**6 ounces (170 grams) tempeh, cut into pieces 1 by 2 by
¼ inch**
**¼ onion, grated or very finely minced**
**3 tablespoons dark-brown sugar or 1½ tablespoons
honey**
**½ teaspoon salt**
**1¼ cups water, milk, or coconut milk (page 42)**
**Oil for deep- or shallow-frying**

Combine the first five ingredients in a saucepan and
bring to a boil. Simmer uncovered on medium heat for
10 to 15 minutes, or until all liquid has evaporated. Remove from heat and allow to cool to room temperature.
Heat the oil to 375° F. (190° C.) in a wok, skillet, or
deep-fryer. Slide in tempeh and deep- or shallow-fry for
2 to 3 minutes, or until nicely browned. Drain briefly
and serve while crisp and hot. Also good cooled and
diced in Carrot & Cabbage Salad (page 65). The Indonesian counterpart of this recipe is Tempeh Bachem.

VARIATION

**\*Sweet & Tangy Fillets:** Simmer the tempeh for 10 to 15
minutes in a mixture of 2 teaspoons lemon juice, 2 tablespoons brown sugar, ¼ teaspoon salt, and ½ cup water. Then drain and fry as above.

## Crisp Tempeh Chips
SERVES 4 TO 6

The basic recipe using wheat flour gives chips that
are full-bodied and crunchy; the variation using rice flour
or corn starch yields the crispiest ones of all.

**½ cup water, milk, or coconut milk (page 42)**
**¼ teaspoon coriander powder**
**1 clove garlic, crushed**
**1 teaspoon salt**
**Oil for deep- or shallow-frying**
**6 ounces (170 grams) tempeh, cut into pieces ¾ by 2
by ⅛ inch and lightly scored**
**4 to 6 tablespoons whole-wheat flour**

Combine the first four ingredients in a shallow bowl, mixing well. Heat the oil to 375° F. (190° C.) in a wok, skillet,
or deep-fryer. Dip tempeh in seasoning mixture for a
few seconds, dust in flour, and slide into oil. Deep- or
shallow-fry for 3 to 4 minutes, or until crisp and golden
brown. Drain on absorbent toweling. Serve immediately.
Nice as an hors d'oeuvre served, if desired, with Creamy
Tofu (or Tempeh) Dip (page 47 or 63), or as a side dish
or seasoning/topping for brown rice.

VARIATIONS

**\*Crunchiest Tempeh Chips:** Cut tempeh into very thin
strips and, for best texture, sun-dry for 15 minutes. Use
½ teaspoon coriander, 2 cloves garlic, and 6 tablespoons rice flour, cornstarch, arrowroot, or kudzu powder. Proceed as above.
**\*Curry Chips:** Substitute 1¼ teaspoons curry powder
for the coriander and garlic in the basic recipe. Especially
good as a seasoning for brown rice.
**\*Ginger Chips:** Substitute 1½ teaspoons powdered ginger for the coriander and garlic in the basic recipe. If
dusted with cornstarch, this preparation works especially
well as a meaty ingredient in a Sweet & Sour Sauce (page
67 or 85).

## Batter-Fried Tempeh Crisps
SERVES 3 TO 4

**¼ cup (whole-wheat) flour or 7 tablespoons
(brown) rice flour**
**1 clove garlic, crushed**
**¼ teaspoon coriander powder**
**¼ teaspoon salt**
**¼ teaspoon turmeric (optional)**
**½ cup water, milk, or coconut milk (page 42)**
**Oil for deep- or shallow-frying**
**6 ounces (170 grams) tempeh, cut into pieces ¾ by
2½ by ⅛ inch, preferably thinner**

Combine the first six ingredients in a bowl, mixing to
form a batter. Heat oil to 375° F. (190° C.) in a wok,
skillet, or deep-fryer. Dip tempeh slices into batter, slide
into oil, and deep- or shallow-fry for 3 to 4 minutes, or
until crisp and golden brown. Serve immediately as an
hors d'ouevre or side dish. Nice accompanied by
Creamy Tofu (or Tempeh) Dip (page 47 or 63). The
Indonesian equivalent of this recipe is Keripik Tempeh.

VARIATIONS

**\*To give the tempeh a wispy, delicate coating, add ¼ to
½ teaspoon baking soda to the batter. During storage,
baking soda is said also to help the crust keep its crispness and freshness longer. However, it can cause some
spattering during deep-frying.
**\*Curried Batter:** ½ cup unbleached white flour, 11 tablespoons water, ½ teaspoon salt, and ½ teaspoon
curry powder. Mix quickly and lightly.
**\*Tempura Batter:** 1 cup ice-cold water, 1¼ to 1½ cups
unbleached white flour, ½ teaspoon salt and, if desired,
1 egg yolk or whole egg. Mix quickly and lightly. Serve
tempeh topped with a sprinkling of shoyu.
**\*Indonesian Batters:** Coconut milk, rice flour, salt, coriander, and garlic. See Keripik Tempeh (page 78).
**\*Fritter Batter:** Your favorite.

## Tempeh-Filled Pot Stickers or Gyoza

MAKES 33

Now quite popular at Chinese restaurants in the West, these half-moon-shaped delicacies are called Steamed Dumplings (chiao-tzu) in their simplest steamed or boiled form, and Pot Stickers (kuo-t'ieh or kuo-teh) in their more popular crispy fried form. Both represent nice ways to use leftover tempeh; we feel they would also make excellent commercial products, especially for the vegetarian market.

*Fig. 4.7: Making tempeh-filled pot stickers or gyoza*

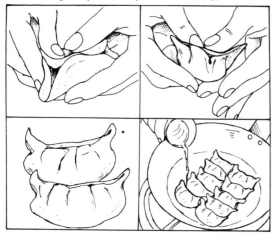

6 ounces (170 grams) tempeh, cut into ½-inch cubes, and deep- or shallow-fried (page 52–54)
3½ ounces (100 grams) cabbage or Chinese cabbage (2½ to 3 leaves), parboiled and minced
2 tablespoons chopped chives
¼ cup minced leek (or scallion) whites
1 clove garlic, crushed or grated
½ teaspoon grated gingerroot
¼ to ½ teaspoon salt
Dash of pepper (optional)
1 tablespoon cornstarch, arrowroot, or kudzu powder
33 wonton or gyoza skins or wrappers, each 3¼ inches in diameter (storebought or homemade)
4 to 6 tablespoons oil
6 to 8 tablespoons water
Dipping Sauce. For each person:
    1½ tablespoons shoyu (natural soy sauce)
    1 teaspoon (rice) vinegar
    ½ teaspoon sesame oil or rayu (red-chili oil)
    ¼ teaspoon hot mustard (optional)

Mash tempeh, then combine with the next eight ingredients, mixing well. Place 1½ teaspoons of the mixture at the center of each of 11 wonton skins. (Wrap the remaining skins temporarily with plastic wrap to prevent their drying.) Moisten the perimeter of each skin with water, fold over like a turnover, and seal the edges as shown above, then arrange on a plate and cover with plastic wrap. Proceed to fill the remaining 22 skins in the same way.

Heat a large skillet, pour in 2 to 4 tablespoons oil (typically used deep-frying oil), heat briefly, then pour off this oil and reserve. Immediately coat skillet with 1½ tablespoons new oil (which prevents pot stickers from sticking), arrange 11 dumplings in the skillet as shown, and fry for about 2 minutes, or until crisp and golden brown. Add 2 to 3 tablespoons water, cover skillet, and cook for 2 to 3 minutes, or until water is gone. Proceed to fry and steam all dumplings. Serve immediately, while crisp, hot and tender, accompanied by the dipping sauce.

## Deep-Fried Tempeh, Vegetable, & Curry Patties

SERVES 4

Know in Japan as *kaki-agé*, this technique offers a nice way to use tempeh which has a crumbly texture due to improper development of the mycelium during fermentation.

½ cup unbleached white flour
½ cup milk (soy or dairy)
¼ teaspoon curry powder
¼ to ½ teaspoon salt
4¾ ounces (135 grams) tempeh, crumbled (about 1 cup)
½ cup slivered carrots
¾ cup minced onions
Oil for deep-frying

Combine the first four ingredients in a bowl, mixing quickly and lightly, then stir in the next three ingredients. Heat the oil to 375° F. (190° C.) in a wok, skillet, or deep-fryer. Place 2 to 3 tablespoons of the batter-and-tempeh mixture on a (wooden) spatula. Using chopsticks or a butter knife, flatten the mixture to form a 2½- to 3-inch round, then carefully slide round into hot oil. Repeat, making 8 or 9 rounds, until all mixture is used. Deep-fry each round for 2 to 3 minutes on one side, then turn and fry for 1 to 2 minutes more on the second side until crisp and golden brown. Drain briefly on absorbent toweling; serve hot, topped, if desired, with ketchup and/or Worcestershire sauce, or on a bun or toast like a burger.

For variety, presteam the tempeh for 25 minutes, then mix with sautéed garlic and onion seasoned with shoyu, plus beaten egg, cornstarch, salt, pepper, and minced parsley. Fry in margarine or oil like pancakes and serve as above. Or coat tempeh with a bound breading, then deep-fry.

## Tempeh Crisps or Krupuk

Containing cassava flour (yuca), mashed or pureed tempeh, salt, and seasonings, these crunchy, big, light-as-air deep-fried delicacies resembling Indonesian shrimp crisps (*krupuk*) are now sold commercially as "Tempeh Chips" by Bali Foods in Los Angeles (see Appendix).

## Breaded Tempeh Cutlets

SERVES 4

One of our favorite tempeh recipes, this would make an excellent commercial product. We generally make a large batch at home, freeze the leftovers, and then reheat them in a frying pan or oven just before serving. Bread-crumb flakes, called *panko* in Japanese, are similar to bread crumbs except that each particle has been rolled under pressure to form a tiny, thin flake. We and most Japanese chefs prefer them to bread crumbs for deep-frying. Available at Japanese food markets, a fairly good substitute is made by putting croutons of dried bread into a blender until they form a medium-coarse meal.

5½ ounces (160 grams) tempeh, cut into 8 cutlets, each 1½ by 2 by ⅜ inch, scored in a crosshatch pattern on upper and lower surfaces
1 teaspoon of salt dissolved in ½ cup water
¼ cup white flour
1 egg, lightly beaten
7 tablespoons bread crumbs, bread-crumb flakes, or cornmeal
Oil for deep-frying
Topping. Choose one:
    Tartare Sauce (page 48 or 64)
    Ketchup & Worcestershire Sauce (page 48) or plain Worcestershire sauce
    Creamy Tofu Dressings (page 47)
    Sweet Indonesian Soy Sauce (Kechap Manis, page 48)
    A squeeze of lemon juice and a sprinkling of salt

Dip tempeh slices in salt-water solution, then drain for several minutes to remove excess moisture. Dust tempeh with flour, dip in egg, and roll in bread crumbs, then set on a rack to dry.

Heat the oil to 375° F. (190° C.) in a wok, skillet, or deep-fryer. Slide in tempeh and deep-fry for about 3 minutes or until crisp and golden brown. Serve immediately, accompanied by your choice of toppings.

VARIATION

**\*Tempeh Cutlets Simmered with Ginger & Shoyu:** Cut 6¼ ounces (180 grams) tempeh into 6 pieces, 1½ by 2½ by ⅜ inch. Combine in a saucepan with 1½ tablespoons shoyu, ½ teaspoon freshly grated or 1 teaspoon powdered ginger, and ¼ cup water. Bring to a boil and simmer, uncovered, for about 4 minutes, or until all liquid has been absorbed or evaporated. Drain and cool, then proceed to coat with flour, egg, and bread crumbs, and deep-fry as in the basic recipe. Serve each portion accompanied by a parsley sprig and lemon wedge.

*Gingerroot*

# Tempeh with Sandwiches, Burgers, Pizza, Tortillas, and Tacos

These preparations are already favorites in some parts of America. Sun-dried and ground, spray-dried, or freeze-dried tempeh powder can also be used in amounts of 10 to 15 percent to add complementary protein to breads, crackers, biscuits, and other baked goods.

## Tempeh, Lettuce, & Tomato Sandwiches

MAKES 2

In this delectable dish, which is already a favorite at The Farm Food Company Delicatessen in California, the tempeh adds the savoriness and crispness of bacon, plus the richness of cheese.

1½ tablespoons of butter
2 slices whole-grain bread
Spread, premixed
    2 tablespoons mayonnaise (tofu or egg, page 47)
    2 teaspoons ketchup
    ¼ teaspoon mustard
2 small lettuce leaves or ¼ cup alfalfa sprouts
4¼ ounces (120 grams) tempeh, cut into 2 slices, each 3½ by 2½ by ⅜ inch, dipped into salt water and shallow- or deep-fried to make Seasoned Crisp Tempeh (page 54); or use Breaded Tempeh Cutlets (page 59) topped with Worcestershire sauce and/or ketchup
2 large tomato slices
2 to 4 slices dill pickle or some sauerkraut (optional)

Melt the butter in a large skillet. Drop in bread slices, then turn over immediately; fry on both sides for 1 to 2 minutes, or until nicely browned. Coat one side of each slice of bread with one-half the spread, then top with lettuce and a slice of tempeh. Coat top of each tempeh slice with the remaining spread, then crown with a tomato slice and, if desired, dill pickles. Serve immediately as an open-faced sandwich.

For variety, toast the bread and then spread with butter. Use Tempeh Tartare Sauce (page 64) as a spread. Also makes a good triple-decker club sandwich or a closed sandwich.

## Tempeh Reuben Sandwich

SERVES 2

A hefty, hearty, wonderful whole meal developed by the cooks at the Rochester Zen Center.

**3½ ounces (100 grams) tempeh, cut into two slices, each 3½ by 3½ by ¼ inch thick**
**2 to 3 tablespoons oil**
**4 tablespoons water**
**2 teaspoons shoyu (natural soy sauce)**
**2 slices of Swiss cheese**
**4 thin slices of pumpernickel bread, or two Kaiser or onion rolls, cut into halves (warmed, toasted, or cold, and margarined or buttered, as you like)**
**¼ to ½ cup sauerkraut**
**½ to 1 teaspoon (hot) mustard**
**4 dill pickle slices**

Use oil and 3 tablespoons (unsalted) water to make pan-fried tempeh (page 54); after steaming second side, sprinkle shoyu over tempeh, add 1 tablespoon water, and steam for 30 seconds more. Cover each piece of tempeh with a slice of cheese, re-cover skillet, and cook for 3 to 4 minutes more, or until cheese melts. Place tempeh on 2 slices of the bread, mound with sauerkraut, and complete each sandwich with the other piece of bread or roll that has been spread with mustard. Serve with dill pickles on the side.

For variety, omit cheese; put ¼ cup alfalfa sprouts or a slice of tomato under the tempeh on each sandwich.

## Pizza with Tempeh

MAKES 2 NINE-INCH PIZZAS

Crisp slices of fried tempeh together with tempeh pizza sauce, served on pizza in place of the traditional anchovies, create a meatless delicacy with savory flavor and fine texture. For wheat-free tempeh pizza, see page 69.

**4 tablespoons margarine or butter**
**½ onion, thinly sliced**
**1¼ cups sliced mushrooms**
**1 green pepper, sliced**
**1 cup Tempeh & Tomato Sauce for Pizza (page 66) or your favorite**
**2 nine-inch pizza crusts (use your favorite recipe) or wheat-flour tortillas**
**3½ ounces (100 grams) tempeh, cut into pieces each ½ by ½ by ¼ inch, dipped in salt water, and shallow- or deep-fried to make Seasoned Crisp Tempeh (page 54)**
**1 cup grated (Mozzarella) cheese**
**Parmesan cheese and/or Tabasco sauce (optional)**

Preheat oven to 350°F. (175°C.). Melt 2 tablespoons margarine or butter in a skillet. Add onion and mushrooms and sauté for 3 to 4 minutes. Add green pepper and sauté for 2 minutes more, then remove from heat. Spread the remaining 2 tablespoons margarine, then the pizza sauce, and finally the sautéed vegetables on the two pizza crusts. Top with a sprinkling of the fried tempeh and Mozzarella cheese. Bake for 10 to 15 minutes, or until crust is nicely browned and cheese is melted. Serve hot, topped if desired with a sprinkling of Parmesan and/or Tabasco.

## Tempeh in Pita-Bread Sandwiches

SERVES 2 TO 4

Also known as "pocket bread," pita is a round, hollow Mid-Eastern flat bread that can be cut crosswise into halves, each of which is opened to form a sort of pouch. Stuffed with a variety of foods, it makes a novel and delicious dripless sandwich.

**1½ tablespoons margarine or butter**
**2 pita breads, each cut crosswise into halves and opened to form pouches**
**½ cup Tartare Sauce (tofu or tempeh: page 48 or 64)**
**¼ cup ketchup**
**3 to 4 lettuce leaves, torn into large pieces**
**½ cup alfalfa sprouts**
**4 large tomato slices**
**4 ounces (113 grams) tempeh, cut into pieces 2 by 3 by ⅜ inch, dipped into seasoning solution, and shallow- or deep-fried to make Coriander & Garlic Crisp Tempeh (page 55)**

Melt the margarine or butter in a large skillet. Add pita-bread halves and fry for 1 to 2 minutes on each side, or until crisp and golden brown, then remove from heat. Open each pouch and spread inside with Tartare Sauce and ketchup, then stuff with lettuce, sprouts, tomato, and tempeh slices. Serve immediately, while pita and tempeh are crisp.

*Fig. 4.8: Tempeh pita bread sandwich*

*Fig. 4.9: Tempeh burger*

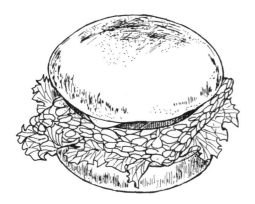

## Tempeh Burger
MAKES 1

We like Tempeh Burgers best when served open-face style using only one-half a bun. We will write the recipe in the traditional way here and you can adapt it if you wish. The addition of a slice of cheese also makes a delicious tempeh cheeseburger. Watch out, McDonald's!

1 whole-wheat bun, sliced horizontally into halves and toasted
1 tablespoon margarine or butter
1½ tablespoons mayonnaise or Tartare Sauce (page 48 or 64)
1½ tablespoons ketchup
¼ to 1 teaspoon mustard
1 small lettuce leaf
1 large, thin slice of onion (optional)
2 thin slices dill pickle
1 slice tempeh, 3 by 3 by ¼ inch (1-1¼ ounces; 28-35 grams), made into Coriander & Garlic Crisp Tempeh or Rich Tempeh Fillet (pages 55 and 57)
1 large thin slice tomato

Coat cut surfaces of bun with margarine. On one side spread mayonnaise, ketchup, and mustard; on this stack consecutively lettuce leaf, onion, dill pickles, tempeh, and tomato. Top with the remaining half of the bun. Serve hot; nice with Tempeh French Fries (page 55).

## Tortillas with Tempeh & Guacamole
SERVES 4 TO 8

5 tablespoons margarine or butter
8 corn tortillas (6-inch diameter)
4 to 6 lettuce leaves, chopped, shredded, or torn
1 cup alfalfa sprouts
1 to 1½ tomatoes, diced
1¼ cups Tempeh Guacamole (page 64) or your favorite guacamole
9 ounces (225 grams) tempeh, cut into ⅜-inch cubes, dipped into seasoning solution, and shallow- or deep-fried to make Coriander & Garlic Crisp Tempeh (page 55)
Taco sauce (homemade or commercial; optional)

Melt 1 to 2 teaspoons margarine in a skillet. Add 1 tortilla and fry for 30 seconds on each side, or until warm and lightly browned, then transfer from skillet to serving plate. Repeat until all tortillas have been fried. Place an equal portion of the lettuce, sprouts, and tomatoes atop each tortilla, then spoon on a generous dollop of the guacamole. Top with a sprinkling of crisp tempeh cubes and, if desired, some taco sauce. Serve immediately while tempeh retains its crispness.

Tempeh is also nice in burritos.

*Fig. 4.10: Tortilla with tempeh & guacamole*

## Tacos with Tempeh

Into crisply deep-fried folded corn tortillas pack a mixture of shredded lettuce, deep-fried tempeh slices, chopped tomato, and minced onion. Top with hot sauce (salsa) and grated cheese or nutritional yeast. Other nice fillings include black olives, avocadoes, sliced hard-boiled eggs, and mushrooms.

# Tempeh with Grains and Beans

Grains have been the staple or primary food in virtually all traditional societies, a key fact that many moderns have forgotten. Beans have been a key source of complementary protein. Tempeh adds rich flavor, crunchy texture, and complementary protein to a great variety of tasty recipes, especially those featuring rice or noodles. For other baked recipes with grains, see Tempeh in Baked or Broiled Dishes. Note that many of Indonesia's favorite rice recipes are given under Rice Dishes in Chapter 3. In Indonesia, most tempeh dishes are served over a bowl of rice.

### Curried Stir-Fried Noodles with Tempeh
SERVES 2 OR 3

2 tablespoons oil
1 egg, lightly beaten
¼ onion, thinly sliced
½ cup slivered carrots
1 cup chopped cabbage
¼ cup thinly sliced green onions or scallions
3 ounces (85 grams) (buckwheat) noodles, cooked and well drained (page 49)
¼ to ⅜ teaspoon curry powder
2 teaspoons shoyu (natural soy sauce)
¼ teaspoon salt
6 ounces (170 grams) tempeh, cut into ⅜-inch cubes, dipped into seasoning solution, and deep-fried to make Coriander & Garlic Crisp Tempeh (page 55)

Heat a wok or large skillet and coat with the oil. Add egg and scramble for 30 seconds. Add onion and carrots, and stir-fry for 3 minutes. Add cabbage and green onion, and sauté for 2 minutes. Add all remaining ingredients except the tempeh and stir-fry vigorously for 1 to 2 minutes more. Mix in crisp tempeh and remove from heat. Delicious hot or cold.

VARIATION

* **Singapore Stir-Fried Noodles with Tempeh:** In place of the curry powder, shoyu, and salt use 2 tablespoons ketchup, 1 teaspoon Worcestershire sauce, 2 teaspoons soy sauce, a dash of Tabasco sauce, and, if desired, 1 cup spinach or green beans, chopped and steamed for 3 minutes (makes ¼ cup). Prepare as in the basic recipe above. Wonderful!

### Japanese-Style Fried Noodles with Tempeh
SERVES 4

2 tablespoons oil
⅓ cup slivered carrots
1¼ cups chopped cabbage
⅓ cup chopped green pepper
½ cup chopped green onions or scallions
4½ ounces (125 grams) (buckwheat) noodles, cooked for 6 to 8 minutes (page 49)
6 ounces (170 grams) tempeh, cut into ⅜-inch cubes, dipped into salt water, and deep- or shallow-fried to make Seasoned Crisp Tempeh (page 54)
1 to 1½ tablespoons shoyu (natural soy sauce)
Dash of salt

Heat a wok or skillet and coat with the oil. Add carrots and sauté for 1 minute. Add cabbage and green pepper, and sauté for 1 minute. Add green onions and sauté for 30 seconds. Add cooked noodles and sauté for 1 minute. Add deep-fried tempeh and shoyu, mix well, and sauté for 30 seconds more. Season to taste with salt. Serve immediately, while tempeh is at its peak of crispness.

### Stir-Fried Rice with Tempeh
SERVES 3 OR 4

2 tablespoons oil
1 clove garlic, minced
1 egg, lightly beaten
1½ cups cooked (brown) rice, well cooled or chilled
⅓ cup chopped green onion
5½ ounces (156 grams) tempeh, cut into pieces each ⅜ by ⅜ by 1 inch, dipped into salt water, and deep- or shallow-fried to make Seasoned Crisp Tempeh (page 54)
2 teaspoons shoyu (natural soy sauce)

Heat a wok or skillet and coat with the oil. Add garlic and sauté for 15 seconds. Add egg and scramble well. Add rice and stir-fry for 2 minutes. Add green onion, fried tempeh, and shoyu, mixing well. Serve immediately.

## Tempeh Jambalaya

SERVES 4 TO 5

5 tablespoons margarine, butter, or oil
2½ to 3 cups sliced mushrooms
5 ounces (142 grams) tempeh, cut into ⅜-inch cubes
   and deep- or shallow-fried (page 52–54)
¾ cup cooked brown rice
1 green pepper, chopped
¼ onion, chopped
½ stalk celery, chopped
½ cup chopped tomatoes
Dash of chili powder or Tabasco sauce
¼ teaspoon paprika
½ teaspoon salt
2 tablespoons minced parsley

Preheat oven to 350° F. (175° C.). Melt 3 tablespoons margarine in a large skillet. Add mushrooms and sauté for 3 to 4 minutes, then combine in a bowl with all remaining ingredients except the parsley. Add 2 tablespoons melted margarine, mixing well. Transfer to a small, lightly oiled baking dish and bake for 30 to 35 minutes. Serve garnished with the parsley.

## Fried Tempeh a la Boston-Style Baked Beans

SERVES 3

6 ounces (170 grams) tempeh, cut into ⅜-inch cubes
   and deep- or shallow-fried (page 52–54)
2 tablespoons minced onions
2 tablespoons ketchup
1 teaspoon dark molasses
1 teaspoon Worcestershire sauce
½ teaspoon vinegar
¼ teaspoon dry mustard
¼ teaspoon curry powder
¼ cup water
Dash of salt

Preheat oven to 350° F. (175° C.). Combine all ingredients, mixing well, then spoon into a small, lightly oiled gratin dish or baking pan. Bake for 15 minutes. Serve hot or cold.

VARIATIONS

Use small cubes of Seasoned Crisp Tempeh (page 54) in place of bacon or meats in your favorite bean recipes, with refried pinto beans, cooked soybeans, and the like. Add them near or at the end of cooking.

# Tempeh in Dips, Spreads, and Dressings

Here are some nice ways to get thick, creamy textures without the use of lots of high-calorie oil and expensive, cholesterol-rich dairy products. Also good ways to make use of tempeh having a weak mycelium.

## Creamy Tempeh Dip with Garlic & Dill

MAKES 1½ CUPS

6 ounces (170 grams) tempeh, cut into pieces each 1 by
   1 by ½ inch, steamed for 20 minutes (page 39) and
   cooled for 30 minutes
¼ cup oil
2 tablespoons lemon juice or vinegar
½ teaspoon salt
½ cup water
1 clove garlic, crushed or minced
¼ teaspoon dill seeds
2 tablespoons minced parsley

Combine the first seven ingredients in a blender and puree for 30 to 60 seconds, or until smooth. Serve chilled, topped with a sprinkling of the parsley. Delicious with chips or with fresh carrot, celery, or jícama slices. Refrigerated in a covered container, it will stay fresh for up to 3 days.

VARIATION

* Curry: Substitute ¼ teaspoon curry powder for the garlic and dill. Tastes remarkably like an egg-based dip.

## Tempeh Mock-Tuna Spread

MAKES 1¼ CUPS

6 ounces (170 grams) tempeh, diced
1 tablespoon water
¼ cup mayonnaise
1 tablespoon minced onion
½ teaspoon salt
2 tablespoons minced parsley

Steam tempeh for 20 to 25 minutes (page 39). Transfer to a bowl and immediately, while tempeh is still hot, add water and mash thoroughly, then allow to cool to room temperature. Add remaining ingredients and mix well. Delicious on crackers or in sandwiches.

## Tempeh Guacamole

MAKES 2¼ CUPS

**6 ounces (170 grams) tempeh, diced, steamed for 20 minutes (page 39), and allowed to cool to room temperature**
**1 avocado**
**¼ cup oil**
**3 tablespoons lemon juice**
**½ teaspoon salt**
**¾ cup water**
**1 clove garlic, crushed or minced**
**2 tablespoons minced onion (optional)**
**Dash of Tabasco sauce (optional)**

Combine all ingredients in a blender and puree for 1 minute (turning off blender and mixing occasionally if necessary), or until smooth and very thick. Serve as a dip (with corn chips) or in Tortillas with Tempeh & Guacamole (page 39).

*Fig. 4.11: Tempeh guacamole*

## Creamy Tempeh Dressing or Dip with Curry

MAKES 2 CUPS

**6 ounces (170 grams) tempeh, cut into 1-by-1-by-½-inch pieces, steamed for 20 minutes (page 39) and cooled for 30 minutes**
**4 to 6 tablespoons oil**
**2 to 3 tablespoons lemon juice**
**½ teaspoon salt**
**¾ cup water**
**¼ to ½ teaspoon curry powder**
**2 to 3 tablespoons minced parsley**

Combine the first six ingredients in a blender and puree for 30 to 60 seconds, or until smooth. Mix in parsley, chill, and serve in dollops over your favorite tossed green salads. Refrigerated in a covered container, it will stay fresh (and thicken slightly) for up to 3 days. Tastes remarkably like an egg-based dressing.

VARIATIONS

**\* Garlic & Dill:** Substitute 1 clove crushed or minced garlic and ¼ to ½ teaspoon dill seeds for the curry powder.
**\* Russian:** Substitute ¼ cup ketchup, 1 tablespoon grated horseradish, and 1 teaspoon grated onion for the curry powder.

**\* Tempeh Tartare Sauce with Dill Pickles:** Use 4 tablespoons oil and 2 tablespoons lemon juice. Omit curry powder. Combine the first five ingredients plus ¼ teaspoon turmeric in a blender and puree until smooth. Then mix in ⅓ cup minced dill pickles and 3 tablespoons minced parsley. Serve as a dip or dressing.
**\* Mock Deviled Egg Salad:** Substitute ¼ teaspoon turmeric and 2 tablespoons minced onion for the curry powder. Omit parsley or sprinkle on top.

## Tempeh in Salads

The possibilities for experimentation are as endless as they are good-tasting and good-textured.

### Crisp Tempeh Croutons in Salads

This is our favorite way of using tempeh in tossed green salads. Simply prepare Seasoned Crisp Tempeh (page 54), preferably cut into ½-inch cubes. Cool to body temperature and just before serving, while still crisp, sprinkle over salads.

### Waldorf Salad with Nutlike Tempeh Chunks

SERVES 3

Tasting and looking remarkably like walnuts, the tempeh adds a crunchy texture and fine flavor.

**5 ounces (140 grams) tempeh, cut into ⅜-inch cubes, dipped in salt water, and deep- or shallow-fried to make Tempeh Croutons (page 55), then cooled to room temperature**
**1 apple, peeled and diced into ½-inch cubes**
**3 lettuce leaves or 1 stalk of celery, chopped**
**¼ cup raisins**
**¼ cup mayonnaise (tofu or egg; page 47)**
**1 tablespoon lemon juice**

Combine all ingredients, mixing well. Serve immediately, while tempeh is at its peak of crispness.

## Tempeh & Curried Rice Salad

SERVES 3

Here is a delicious way to make full use of the protein complementarity potential of tempeh and brown rice.

1 cup cooked brown rice, cooled (page 45)
1 tomato, diced
7 green beans, cut diagonally into ³⁄₄-inch-long slices, parboiled for 4 minutes, drained and cooled (about ¹⁄₂ cup)
½ cup sliced cucumber rounds
1 leaf of lettuce, torn or shredded
¼ cup mayonnaise (tofu or egg; page 47)
1 tablespoon minced onion
¼ to ½ teaspoon curry powder
1½ tablespoons minced parsley
1½ teaspoons lemon juice
⅛ teaspoon salt
6 ounces (170 grams) tempeh, cut into ½-inch cubes, dipped into salt water, and deep- or shallow-fried to make Tempeh Croutons (page 55), then cooled to room temperature

Combine all ingredients but the tempeh, mixing well, then gently stir in the tempeh cubes. Chill for 20 to 30 minutes; serve in a large salad bowl lined, if desired, with lettuce leaves. For a crunchier texture, serve immediately without chilling.

## Tempeh in Chinese-Style Carrot & Sprout Salad

SERVES 5 TO 6

2 cups slivered carrots
3 cups mung-bean sprouts
1 cucumber, slivered
6 ounces (170 grams) tempeh, cut into pieces each ¼ by ¼ by 2½ inches, dipped into salt water, and deep- or shallow-fried to make Seasoned Crisp Tempeh (page 54)

Sesame-Vinegar Dressing
    2 tablespoons mild rice or distilled white vinegar
    2 tablespoons sesame oil
    1½ tablespoons shoyu (natural soy sauce)
    ½ teaspoon grated gingerroot
    Dash of cayenne pepper

Bring 8 cups water to a boil in a pot. Drop in carrots and parboil for 2 to 3 minutes. Drop in bean sprouts and cucumber, and parboil for 30 seconds more. Drain (reserving water for use as a soup stock) and rinse vegetables under cold running water for 5 seconds. Redrain thoroughly in a colander or strainer, then combine vegetables in a bowl with tempeh and dressing, mixing lightly. Chill for 1 to 2 hours before serving. A refreshing summertime salad.

## Tempeh Mock Chicken Salad

SERVES 2

7 ounces (200 grams) tempeh, diced into ³⁄₈-inch cubes and deep-fried until crisp and golden brown or steamed for 20 minutes, then cooled to room temperature
½ cup mayonnaise (tofu, soymilk, or egg)
2 tablespoons minced onion
2 tablespoons chopped dill pickles
¼ cup minced parsley
1 teaspoon dried sweet basil
Dash of garlic powder or crushed garlic
Dash of salt
2 to 3 cups torn or shredded lettuce

Combine the first eight ingredients, mixing well. Serve mounded on the lettuce.

## Tempeh & Mushroom Salad

SERVES 3 OR 4

6 ounces (170 grams) tempeh, cut into pieces, each ³⁄₈ by ³⁄₈ by ½ inch, dipped into salt water, and deep- or shallow-fried to make Seasoned Crisp Tempeh (page 54), then allowed to cool for 5 minutes
2 cups chopped, torn, or shredded lettuce
1 small tomato, diced
5 mushrooms, thinly sliced
½ cucumber, cut lengthwise into halves, then crosswise into thin slices
3 tablespoons chopped parsley
French Dressing
    ½ cup oil
    2 tablespoons lemon juice
    2 tablespoons vinegar
    ½ teaspoon salt
    ½ teaspoon prepared mustard

Combine the first six ingredients in a salad bowl. Add dressing and toss lightly. Serve immediately so that tempeh remains crisp.

## Carrot & Cabbage Salad with Diced Tempeh Fillets

SERVES 4 TO 5

6 ounces tempeh, made into Rich Tempeh Fillets (page 57)
½ cup grated carrot
3 cups slivered cabbage
¼ cup mayonnaise (egg or tofu; page 47)
2 tablespoons minced parsley
Dash of salt and pepper
4 lettuce leaves (optional)

Cool tempeh fillets to room temperature and cut into ¼-inch cubes, then combine with the next five ingredients, mixing well. Serve in individual bowls, mounded, if desired, on lettuce leaves.

## Tempeh in Sauces

Fried tempeh slices or cubes may be used as a basic ingredient in, or as a crunchy garnish for, most Western-style sauces. In addition to the following, our favorites, you might wish to try Tomato, Chili-Pepper Hot, Peanut, or Miso-Peanut Butter, to name but a few.

### Tempeh & Mushroom Sauce                    SERVES 3 TO 4

This delicious preparation can be served as is, as an entrée, or used as a topping for pasta or grain dishes.

**5 tablespoons margarine or butter**
**2 cups thinly sliced mushrooms**
**3 tablespoons whole-wheat flour**
**1½ cups milk (soy or dairy)**
**½ teaspoon salt**
**6 ounces (170 grams) tempeh, cut into pieces ¾ by ¾ by ⅜ inch, and deep- or shallow-fried (page 52–54)**

Melt 2 tablespoons margarine in a skillet. Add mushrooms and sauté for 4 to 5 minutes, or until lightly browned, then transfer to a bowl. Melt the remaining 3 tablespoons margarine in the skillet. Add flour and cook for 30 seconds. Add milk a little at a time, stirring constantly, then cook for 3 to 4 minutes, or until sauce develops a smooth, thickened consistency. Mix in salt, remove from heat, and stir in sautéed mushrooms and fried tempeh. Serve as is, as an entrée, or use over cooked pasta, rice, or other grains. Delicious hot or cold.

### Savory Tempeh in Tomato & Herb Sauce              SERVES 4 TO 5

**Oil for deep-frying**
**10 ounces (283 grams) tempeh, cut into slices each 1 by 2 by ¼ inch thick**
**3 to 4 tablespoons whole-wheat flour**
**2 tablespoons margarine**
**1 onion, minced**
**1 clove of garlic, minced**
**2 cups tomato sauce (commercial or homemade)**
**2 tablespoons honey**
**1 teaspoon shoyu (natural soy sauce)**
**½ teaspoon basil**
**¼ teaspoon oregano**
**Dash of thyme**
**Dash of marjoram**
**Dash of cayenne, chili powder, or Tabasco sauce**

Heat the oil to 350° F. (175°C.) in a wok, skillet, or deep-fryer. Dust tempeh slices in flour and deep-fry for 4 to 5 minutes, or until crisp and nicely browned. Drain well.

Melt the margarine in a skillet. Add onion and garlic and sauté for 4 to 5 minutes. Mix in remaining ingredients and deep-fried tempeh, bring to a boil and, stirring occasionally, simmer covered for 30 minutes. Good hot or cold.

### Tempeh & Tomato Sauce for Sloppy Joes or Pizza         MAKES ABOUT 2 CUPS

**3½ ounces (100 grams) tempeh, diced**
**¾ cup (200 grams) chili sauce, your favorite or a commercial variety**
**¾ cup (200 grams) tomato puree (with salt)**
**1 clove of garlic, minced**
**¼ onion, minced**
**Dash of pepper (optional)**

Combine all ingredients in a blender and puree for only 15 seconds. Transfer to a heavy pot, bring to a boil, and simmer, covered, over very low heat, stirring occasionally, for 15 minutes. Be careful not to allow bottom to burn. Serve hot or cold on bread or toast for sloppy joes, or on pizza as a topping. For variety, add minced green pepper after pureeing.

### Spaghetti Sauce with Tempeh                    SERVES 4

Prepared with fresh tomatoes rather than the usual tomato sauce ketchup, this sauce has a deep rich flavor and can be made in countries where tomato sauce or ketchup is not readily available.

**2 tablespoons oil**
**1 clove of garlic, minced or crushed**
**1½ cups minced onion**
**⅔ cup chopped green pepper**
**½ cup sliced mushrooms**
**4 cups chopped well-ripened tomatoes**
**1 bay leaf**
**⅛ to ¼ teaspoon oregano**
**1 teaspoon salt**
**8 ounces (225 grams) tempeh, cut into ⅜-inch cubes and made into Tempeh Croutons (page 55)**
**2 tablespoons margarine or butter**

Heat a wok or skillet and coat with the oil. Add garlic and onion, and sauté for 5 minutes. Add green pepper and mushrooms and sauté for 2 minutes more. Add the next four ingredients and simmer, uncovered, stirring every 3 minutes, for 20 minutes. Meanwhile deep-fry the tempeh croutons. When the sauce has become very thick, add tempeh and margarine, mixing well. Remove bay leaf. Serve over spaghetti or buckwheat noodles (soba). Delicious hot or cold. Top, if desired, with a sprinkling of grated Parmesan cheese. Or, substitute fried tempeh for shrimp in your favorite recipe for Creole.

## Curry Sauce with Tempeh

SERVES 4

**1 cup thinly sliced onions**
**1 cup diced potatoes**
**1 cup thinly sliced carrots (half moons)**
**1 cup chopped cabbage**
**1½ cups water or soup stock**
**2 tablespoons margarine or butter**
**2 tablespoons white flour**
**1½ to 2 teaspoons curry powder**
**1 teaspoon salt**
**8 ounces (225 grams) tempeh, cut into ⅜-inch cubes, and made into Tempeh Croutons (page 55)**

Combine the first five ingredients in a saucepan, bring to a boil, and simmer, covered, for 15 minutes. Meanwhile, melt the margarine in a skillet. Add flour and, stirring constantly, cook over low heat for about 1 minute, or until well blended. Stir in curry powder and salt, then remove from heat. When vegetables are done, add contents of skillet and tempeh, mixing well, and simmer for another 2 minutes. Serve over rice, topped if desired with any or all of the following: sliced bananas (sprinkled with lemon juice), grated coconut, raisins, diced apples or pineapple, peanuts, almonds or cashews, chopped hard-boiled eggs, and chutney.

## Tempeh in Onion White Sauce

SERVES 3 OR 4

**4½ tablespoons margarine or butter**
**¾ cup sliced onions**
**1 cup sliced mushrooms**
**3½ tablespoons whole-wheat flour**
**1½ cups milk (soy or dairy)**
**¼ to ½ teaspoon salt**
**8 ounces (227 grams) tempeh, cut into pieces each ½ by ½ by ¾ inch, dipped into salt water, and deep- or shallow-fried to make Seasoned Crisp Tempeh (page 54)**

Melt 1½ tablespoons margarine or butter in a skillet. Add sliced onions and mushrooms, and sauté for 4 to 5 minutes, then remove from skillet. Melt 3 tablespoons margarine in the skillet. Add flour and stir constantly over low heat for 1 to 2 minutes, or until nicely blended and cooked. Stir in milk a little at a time, then increase heat to medium and cook, whisking or stirring, for 3 to 4 minutes more, or until sauce develops a smooth, fairly thick consistency. Season with salt, remove from heat, and mix in deep-fried tempeh and sautéed vegetables. Nice either as a side dish or over brown rice or pasta. Or use as the basis for Tempeh Gratin (page 69).

## Sweet & Sour Tempeh

SERVES 3 OR 4

In this rendition of the traditional Chinese favorite, the trick lies in cooking the vegetables quickly so that they retain their fresh crispness, which complements the tender crunchiness of the deep-fried tempeh.

**1½ tablespoons oil**
**½ cup sliced onions**
**½ cup sliced mushrooms**
**½ cup diagonally-cut leeks or green onions, white parts only, cut about ⅜ inch thick**
**½ cup chopped green peppers (1-inch triangles or squares)**
**½ cup sliced green beans (1¼-inch lengths; parboiled for 2 minutes)**
**½ cup sliced carrots (thin half moons; parboiled for 2 minutes)**
**Sauce (premixed)**
    **2 tablespoons honey**
    **2 teaspoons vinegar**
    **1½ tablespoons shoyu (natural soy sauce)**
    **1 tablespoon cornstarch, arrowroot, or kudzu powder**
    **½ cup water or soup stock**
**½ teaspoon grated gingerroot**
**6 ounces (170 grams) tempeh, cut into pieces each ½ by ½ by ¾ inch, and deep- or shallow-fried (page 52–54)**

Heat a wok and coat with the oil. Add onions and sauté for 2 minutes. Add the next five ingredients and sauté for 1 minute, or until lightly cooked but still crisp and tender. Stir in premixed sauce and gingerroot and cook, stirring constantly, for 1 minute, or until nicely thickened. Gently stir in tempeh cubes and remove from heat. Serve as is, or over rice or noodles. Delicious hot or chilled.

### Baked Tempeh with Nutritional Yeast Gravy

SERVES 4

Nutritional yeast, a natural food loaded with protein, vitamins, and minerals, is sold at most natural and health food stores.

3 tablespoons margarine or butter
1 onion, chopped
2 tablespoons whole-wheat flour
3 tablespoons nutritional yeast
1 cup water
¼ teaspoon salt
¼ teaspoon garlic powder or 1 clove of garlic, crushed
1 tablespoon shoyu (natural soy sauce)
5 ounces (142 grams) tempeh, cut into pieces 2 to 1 by ⅛ to ¼ inch thick

Preheat oven to 350° F. (175° C.). Melt the margarine in a skillet. Add onion and sauté for 3 to 4 minutes. Add whole-wheat flour and nutritional yeast, and cook for 3 to 4 minutes. Add water a little at a time, stirring to make a thick, smooth sauce. Mix in salt, garlic powder, and shoyu, bring to a boil, and remove from heat, then stir in the tempeh and transfer to an oiled 8-inch-square baking pan. Bake for 20 minutes. Serve hot or cold, as a side dish or over brown rice or pasta.

### Tempeh & Onion Sauce

SERVES 4

1½ tablespoons oil
5 onions, thinly sliced
1 tablespoon margarine or butter
1½ tablespoons shoyu (natural soy sauce)
6 ounces (170 grams) tempeh, cut into ½-inch cubes, dipped into salt water, and deep- or shallow-fried to make Tempeh Croutons (page 55)
1 egg, lightly beaten
3 tablespoons grated Parmesan cheese

Heat a large casserole or heavy pot and coat with the oil. Add onions, cover, and simmer over very low heat for about 3½ hours, stirring once every 30 minutes for the first 2½ hours, then once every 10 to 15 minutes during the last hour. When onions are a dark golden brown and very soft, stir in margarine and shoyu, and simmer for 2 minutes, then, for best flavor, cover and allow to stand (or refrigerate) overnight while the flavors marry. Next morning, add freshly fried tempeh, bring to a boil, and simmer for 5 minutes. Stir in egg, sprinkle cheese on top, and cook for 1 minute, or until egg becomes firm. Serve hot or cold over buttered toast or cooked brown rice.

### Tempeh with Sesame-Gingerroot Sauce

SERVES 2 OR 3

1 tablespoon sesame butter, tahini, or ground roasted sesame seeds
1½ tablespoons red, barley, or Hatcho miso
1 tablespoon honey
½ teaspoon fresh grated or 1 teaspoon powdered gingerroot
¼ cup water
5¼ ounces (150 grams) tempeh, cut into slices each ¾ by 2½ by ⅜ inch, deep- or shallow-fried (page 52–54), and lightly scored on top and bottom

Combine the first five ingredients in a wok or skillet and bring to a boil. Add deep-fried tempeh and simmer for 1 to 2 minutes, or until sauce thickens slightly. Delicious hot or cold.

For variety, add sautéed eggplant, zucchini, or onion slices, to make a close relative of the Japanese dish Shigiyaki.

### Deep-Fried Tempeh with Ankaké Sauce

SERVES 2

2 tablespoons oil
1 small onion, thinly sliced
1 green pepper, sliced or chopped
2 tablespoons shoyu (natural soy sauce)
1 tablespoon honey
1½ teaspoons grated gingerroot
½ cup water or soup stock
2 teaspoons cornstarch, arrowroot, or kudzu powder, dissolved in 2 tablespoons water
4½ ounces (130 grams) tempeh, cut into ¼-inch-thick slices, and deep- or shallow-fried (page 52–54)
2 cups hot cooked brown rice (page 45), or 3 ounces (buckwheat) noodles, cooked (page 49)
2 tablespoons roasted sesame seeds, whole or ground

Heat a wok or skillet and coat with the oil. Add onion and green pepper, and sauté for 3 to 4 minutes. Add the next four ingredients and bring to a boil. Stir in dissolved cornstarch and cook for 1 minute more, or until nicely thickened. Gently mix in deep-fried tempeh and remove from heat. Divide cooked rice or noodles between two bowls, pour on the sauce (or serve in an accompanying bowl), and serve topped with a sprinkling of sesame seeds.

## Tempeh in Baked or Broiled Dishes

Use fried tempeh in or atop almost any casserole; it is especially good layered with mushrooms, fresh tomatoes, and/or tomato sauce. Try seasoning with miso or shoyu (natural soy sauce). Broiled tempeh tastes best if it is first deep-fried or at least brushed with oil.

### Delectable Tempeh & Lasagne     SERVES 4 TO 6

Lasagne are the handsome 2-inch-wide, rather thick noodles often having a wavy ridge along one or both edges. The tasty and nutritious whole-wheat type is now available at most natural-food stores.

3½ tablespoons olive oil
2 onions, chopped
2 cups thinly sliced mushrooms
3 (well-ripened) tomatoes, chopped
1 clove of garlic, crushed
1 green pepper, seeded and chopped
1¼ teaspoons salt
1 bay leaf
¼ teaspoon oregano
3 ounces (85 grams) (whole-wheat) lasagne noodles
½ cup ricotta cheese or mashed firm tofu
⅓ cup crumbled mozzarella cheese
8 ounces (226 grams) tempeh, cut into large slices each ⅜ inch thick, dipped into salt water, and shallow- or deep-fried to make Seasoned Crisp Tempeh (page 54), then sliced into thin strips
3 tablespoons grated Parmesan cheese

Heat 2 tablespoons oil in a heavy pot. Add onions and sauté for 3 to 4 minutes. Add mushrooms and sauté for another 3 to 4 minutes. Add tomatoes, garlic, and green pepper, and sauté for 30 seconds. Stir in 1 teaspoon salt, the bay leaf and oregano, reduce heat to low, and simmer, covered, stirring every 5 to 10 minutes, for 1 hour. Remove from heat and, for best flavor, allow to stand overnight while flavors marry. Remove bay leaf.

Combine in a large pot: 4 to 8 quarts water, 1½ teaspoons olive oil, and a dash more of salt. Bring to a boil, drop in lasagne, return to the boil, and, stirring occasionally, simmer for 15 to 20 minutes. Drain and separate lasagne strips.

Preheat oven to 350° F. (175° C.). In an 8-by-8-by-2¼-inch baking dish, make a layer of one-third of the sauce (about 1 cup) and one-half each of the lasagne, ricotta cheese, Mozzarella cheese, and tempeh. Repeat, make identical layers, then top with the remaining one-third of the sauce. Top with a sprinkling of Parmesan cheese and bake for about 40 minutes. Let stand briefly before cutting and serving.

### Baked Tempeh & Tomato Sauce     SERVES 4 TO 5

Tomato Sauce
  3 tablespoons margarine or butter
  1 onion, minced
  2 cloves of garlic, minced
  1 cup tomato sauce (commercial or homemade)
  2 tablespoons honey
  1 tablespoon shoyu
  ⅛ teaspoon oregano
  1 tablespoon mustard
  2 teaspoons lemon juice
  Dash of cayenne pepper or Tabasco sauce
9 ounces (255 grams) tempeh, cut into slices each 1 by 2 by ⅛ inch

Preheat oven to 350° F. (175° C.). To prepare the sauce, melt the margarine in a skillet. Add onion and sauté for 5 minutes, or until nicely browned. Add garlic and sauté for 1 minute. Mix in remaining sauce ingredients, bring to a simmer, and cook for 30 seconds more, then remove from heat.

In an oiled (or margarined) 8-inch square baking pan, make a layer using one-third of the tomato sauce. On this neatly arrange half the tempeh slices. Repeat with a second layer each of sauce and tempeh, then top with a layer of sauce. Bake for 30 minutes. Nice as an accompaniment for cooked brown rice or a topping for spaghetti.

VARIATIONS

*Wheat-free Tempeh Pizza: Arrange ⅛-inch-thick slices of tempeh in a single layer to cover the bottom of a margarined or oiled baking pan. Cover it with a thick layer of your favorite tomato or pizza sauce. Top with slices of mushrooms, green peppers, onions, etc., then sprinkle grated Mozzarella cheese on top. Bake in a preheated 350° F. oven for 20 minutes, or until cheese bubbles. Serve hot or, for best flavor, refrigerate overnight, allowing the flavors to "marry," then serve reheated or cold.

*Tempeh Baked in Barbecue Sauce: Cover the bottom of a baking pan with a thin layer of your favorite barbecue sauce, top with a layer of thin slices of deep-fried (or uncooked) tempeh, then cover with a thick layer of barbecue sauce. Bake in a preheated 350° F. oven for 30 minutes.

### Tempeh Gratin     SERVES 3 OR 4

Prepare Tempeh in Onion White Sauce (page 67). Place in an even layer in well-oiled gratin dishes or in a casserole, sprinkle with 2 tablespoons grated Parmesan cheese, and bake in a preheated 350° F. oven for 10 minutes, or until surface is nicely browned. Serve piping hot.

## Tempeh Shish Kebab with Teriyaki Sauce

SERVES 4

**Teriyaki Sauce**
- 2 tablespoons shoyu (natural soy sauce)
- 1½ tablespoons sake or white wine
- 1 tablespoon honey
- ½ teaspoon grated gingerroot
- 1 clove of garlic, crushed
- 1½ teaspoons sesame oil
- ⅛ teaspoon prepared mustard
- 8 stainless-steel or bamboo skewers (soak bamboo in salt water before use)
- 8 pineapple chunks (1 inch on a side) or ½ onion cut into 8 wedges
- ½ tomato, cut into 8 wedges
- 4 small mushrooms, each cut vertically into halves
- ½ green pepper, seeded and cut into 8 chunks
- 6 ounces (170 grams) tempeh, cut into pieces each 1 by 1 by ⅜ inch, and steamed for 20 minutes (page 39) or deep-fried

Combine Teriyaki Sauce ingredients, mixing well, then pour into a shallow baking pan or tray. Add remaining ingredients and marinate for at least 30 minutes on each side (or, for a deeper flavor, overnight). Skewer vegetables and tempeh and, basting occasionally, grill over a barbecue or oven burner or under an oven broiler, until nicely browned and fragrant. Serve piping hot.

For variety, brush with currant jelly before broiling.

## Tempeh Cheese Soufflé

SERVES 6

- 3 tablespoons margarine
- 1 onion, minced
- Dash of garlic powder or 1 clove of garlic, crushed
- 5 ounces (140 grams) tempeh, cut into strips each 2 by ⅛ by ⅛ inch
- 3 tablespoons whole-wheat flour
- 1 cup milk (soy or dairy)
- ½ teaspoon salt
- Dash of pepper
- 4 egg yolks, beaten
- 1 cup grated mild Cheddar cheese
- 4 egg whites

Preheat oven to 350° F. (175° C.). Melt 1 tablespoon margarine in a skillet. Add onion and sauté 4 minutes. Mix in garlic and tempeh, then remove from heat.

Melt 2 tablespoons margarine in another skillet. Add 2 tablespoons flour and sauté for 1 minute. Add milk gradually, stirring to make a white sauce. Pour sauce into onion-tempeh mixture, then stir in salt, pepper, egg yolks, and 1 cup grated cheese. Coat a soufflé pan with a little margarine, then dust with 1 tablespoon whole-wheat flour. Beat egg whites until stiff, then fold in the tempeh mixture, pour into the pan, and bake for 25 minutes, or until set.

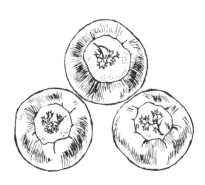

*Fig. 4.12: Tomatoes stuffed with tempeh*

## Tempeh-Stuffed Tomatoes

SERVES 5

- 5 firm (medium to large) tomatoes, unpeeled
- ¾ teaspoon salt (approximately)
- 2 tablespoons oil
- 1 onion, chopped
- 2 cups thinly sliced mushrooms
- 2 tablespoons margarine or butter
- 2 tablespoons whole-wheat flour
- 1 cup milk (soy or dairy)
- 5 ounces (142 grams) tempeh, cut into ½-inch cubes, dipped into salt water, and shallow- or deep-fried to make Seasoned Crisp Tempeh (page 54)
- 2 tablespoons minced parsley
- 2½ tablespoons grated Parmesan cheese (optional)

Preheat oven to 350° F. (175° C.). Cut a large hollow in the stem end of each tomato, reserving the removed portions for use in other recipes. Using a shaker, sprinkle the hollow of each tomato lightly with salt (about ⅛ teaspoon total), then invert tomatoes on a rack or shallow pan and allow to drain for 15 minutes. Meanwhile, heat a skillet and coat with the oil. Add onion and mushrooms, and sauté for 4 to 5 minutes; remove from heat. Melt the margarine in a second skillet. Add flour and cook for 30 seconds. Add milk a little at a time, stirring constantly, then cook for 3 to 4 minutes, or until sauce develops a smooth, thickened consistency. Mix in first ½ teaspoon salt, then the fried tempeh and the sautéed onion and mushrooms. Use mixture to stuff tomatoes. Place tomatoes on a rack in a pan with enough water to keep them from scorching (or, if they are a little ripe, put them in a greased muffin tin to prevent falling apart); bake for 10 to 15 minutes. Serve topped with a sprinkling of parsley and, if desired, the Parmesan cheese. Delicious hot or cold.

VARIATION

**\*Stuffed Green Peppers:** Remove seeds and membranes from 3 green peppers, then parboil them for about 10 minutes, or until nearly tender. Fill cases with tempeh & vegetable white sauce described in above recipe and bake in a lightly oiled pan for 10 to 15 minutes.

## Broiled Tempeh

As basic techniques for browning and crisping tempeh, we definitely prefer shallow-frying and deep-frying to broiling, since the latter requires more time and much more energy, is more difficult to control since the food cannot be seen without opening the broiler, and unless carefully done the surface can be singed while the center is left undone. In Indonesia, onchom is broiled more often than tempeh; a brazier is used in preference to gas since it imparts a pleasant barbecued flavor and aroma. Serve with any of the toppings or dips for Deep-Fried Tempeh (page 53).

**Oven broiler:** Cut 8 ounces tempeh into large slices, each about ⅜ inch thick. Brush both sides of each slice with a total of 3 tablespoons oil, or fry in oil until barely browned. Broil cakes under medium heat for about 3 minutes on each side, checking occasionally, until nicely browned and fragrant.

**Charcoal brazier:** Light a charcoal fire in an oblong brazier and allow to stand until fire has turned to coals. Skewer tempeh and grill, turning occasionally, for 3 to 4 minutes on each side, or until surface is lightly singed.

## *Tempeh in Soups and Stews*

Use Crisp Tempeh Croutons or diced Seasoned Crisp Tempeh as a garnish for, or ingredient in, your favorite soups, adding it just before serving. Unfried tempeh, which has a milder, less meatlike flavor, should be simmered in the soup for 15 to 30 minutes. Miso or shoyu adds rich deep flavors, and seasonings such as cumin and coriander work especially well in accompanying broths. Nice also in tomato stews and Jambalaya. Dehydrated tempeh and spray-dried or freeze-dried tempeh are all at their best in soups and stews.

### Creamy Corn Chowder with Tempeh
SERVES 4 TO 5

1 cup fresh corn kernels (from 1 ear of corn)
6 ounces (170 grams) tempeh, cut into pieces each ¾ by ¾ by ½ inch
2 cups milk, preferably coconut milk (page 42)
1 onion, sliced
1 cup green beans, slivered diagonally
1½ cups diced potatoes
¼ teaspoon coriander
¼ teaspoon turmeric
½ clove of garlic, crushed
¾ to 1 teaspoon salt
Dash of chili powder or Tabasco

Combine all ingredients in a soup pot, cover, and bring to a boil, then simmer for 15 minutes. Delicious hot or cold.

### Tempeh & Onion Miso Soup
SERVES 3

1 onion, thinly sliced
2 cups water or soup stock
2½ tablespoons red, barley, or Hatcho miso
4½ ounces (130 grams) tempeh, cut into pieces ¾ by 1¼ by ¼ inch, and deep- or shallow-fried (page 52–54).

Combine onion and water in a saucepan, bring to a boil, and simmer, covered, for 10 minutes. Add miso creamed in a little of the hot broth, then add tempeh and return just to the boil. Serve immediately.

### Tempeh Cream-of-Tomato Soup
SERVES 4

1 tablespoon oil
1 onion, diced
1 tomato, diced
2 cups milk (soy or dairy)
1 teaspoon salt
⅛ teaspoon oregano
⅛ to ¼ teaspoon marjoram
Dash of cayenne pepper
6 ounces (170 grams) tempeh, diced
2 tablespoons minced parsley

Heat a pot and coat with the oil. Add onion and sauté for 3 to 4 minutes, or until transparent. Add tomato and sauté for 2 to 3 more minutes. Add the next six ingredients and bring to a boil. Cover and cook for 20 minutes, then remove from heat and allow to cool for 1 hour. Transfer to a blender and puree for 1 minute, or until smooth. Serve as is, chilled, or reheated, topped with a sprinkling of parsley.

## *Tempeh with Eggs*

Crisply fried tempeh works nicely like bacon or ham in various egg recipes. As we learn to use soyfoods as a basic protein source, we find our interest in eggs (and dairy products) decreases; we include this section for those in transition and for confirmed inveterate egg lovers.

### Scrambled Eggs with Tempeh Bits
SERVES 1 OR 2

1½ tablespoons margarine or butter
2 eggs, lightly beaten
¼ cup Crispy Tempeh Bits (page 55)
Dash of salt

Melt the margarine in a skillet, add the eggs and then the tempeh bits, and scramble for 2 to 3 minutes, or until just firm. Season lightly with salt; serve hot.

## Fried Egg over Crisp Tempeh Slices

SERVES 1

**2 tablespoons oil**
**2¼ ounces (65 grams) tempeh, cut into slices 1¼ by 2 by ¼ inch**
**1 egg**
**Dash of salt**
**Dash of pepper**

Heat a skillet and coat with the oil. Add tempeh slices and fry on one side for 1 to 2 minutes, or until crisp and golden brown. Turn slices, break egg over the top, cover, and fry for 3 to 4 minutes, or until egg is done. Sprinkle with salt and pepper. Serve hot.

## Tempeh, Egg & Onion Sauté

SERVES 3 TO 4

This richly flavored preparation is a close relative of two popular Japanese dishes, Katsu-domburi (Katsudon) and Tamago-ni.

**1 tablespoon oil**
**1 onion or large leek, sliced**
**1½ tablespoon shoyu (natural soy sauce)**
**1 tablespoon honey**
**¾ cup water or stock**
**6 ounces (170 grams) tempeh, cut into pieces each ¾ by ⅜ by ½ inch, dipped into salt water, and deep- or shallow-fried to make Seasoned Crisp Tempeh (page 54)**
**2 eggs, lightly beaten**
**1 to 2 cups cooked brown rice (page 45)**

Heat a skillet and coat with the oil. Add onion and sauté for 4 minutes. Stir in the next four ingredients, bring to a boil, and simmer uncovered for 3 to 4 minutes. Add eggs and cook, stirring constantly, for 30 seconds, or until eggs are just firm. Serve hot or cold over cooked rice.

## Tempeh Omelet

SERVES 1 OR 2

Tempeh adds the savoriness and crispness of bacon plus the richness of cheese.

**2 ounces (57 grams) tempeh, cut into ⅜-inch cubes, dipped into salt water, and deep- or shallow-fried to make Seasoned Crisp Tempeh (page 54)**
**2 eggs, lightly beaten**
**Dash of salt**
**1½ tablespoons margarine or butter**

Combine the first three ingredients, mixing lightly. Melt the margarine in a large skillet, pour in tempeh-egg mixture, and cook over low heat for 1 to 3 minutes. When omelet has an even consistency, fold and serve.

## Egg Foo Yung with Tempeh

SERVES 3 OR 4

**5½ ounces (160 grams) tempeh, cut into ½-inch cubes, dipped into salt water, and deep- or shallow-fried to make Seasoned Crisp Tempeh (page 54), then cooled to body temperature**
**¾ cup thinly sliced leeks**
**2 cups mung-bean sprouts**
**3 eggs, lightly beaten**
**1 teaspoon shoyu (natural soy sauce)**
**¼ teaspoon salt**
**2 tablespoons oil**
**Sauce. Choose one (optional):**
　**Worcestershire or Ketchup & Worcestershire (page 48)**
　**Ankaké Sauce (page 68)**

Combine the first six ingredients in a bowl, mixing gently. Heat a skillet and coat with the oil. Pour in one-third to one-fourth of the mixture and fry on one side like an omelet for 2 to 3 minutes, or until firm and nicely browned. Turn with a spatula and fry on the second side for 1 to 2 minutes, then remove from skillet. Repeat until all batter is used. Serve hot or chilled, topped, if desired, with a sauce.

# Tempeh Beverages

In recent years there have been a number of studies on the production of soymilk-like beverages from tempeh. Murata, in 1971, freeze-dried tempeh, ground it very fine, then mixed it in a blender with water, sweetening, and chocolate. Sastroamidjojo, working in 1966 at the UNESCO-sponsored soymilk factory in Yogyakarta, Indonesia, used tempeh-like soaked soybeans to extract the milk, which was reported to have a slightly bitter flavor. Arbianto, in 1977, harvested tempeh when its nutritional value was at a maximum, extracted a milk from it, and then inoculated this milk with *Lactobacillus* to make various fermented foods including cheese, yogurt, and various dips. In each case, the process extended the shelf life of the tempeh nutrients. We have made tempeh soymilk using the recipe for Homemade Soymilk given in our *Book of Tofu* and simply substituting tempeh for soaked soybeans. However, the resulting milk has a poor flavor and aroma with a bitter aftertaste and thin consistency. The okara is pasty, very fine grained, and starchy. Thus we strongly prefer to make regular soymilk from whole soybeans; it is easier, less expensive, and much better flavored.

# 5

# Indonesian Tempeh Recipes

Learning to cook Indonesian style takes a little practice but the results are unquestionably well worth the time and effort. After mastering a few basic principles and locating a good nearby source of ingredients, you will be glad you started this adventure in wonderful flavor. We have made every effort to make the learning experience easy and enjoyable while still allowing you to get to know the real thing. We have devoted a separate chapter to the use of tempeh in Indonesian cookery for several reasons:

1. Most of these recipes use a number of distinctive Indonesian ingredients (such as fresh chilies and coconuts, shallots, and numerous spices), which are not yet widely used in Western cookery. Yet all are available at Indonesian specialty shops in the West (although often in their dried state only), and in a growing number of supermarkets and gourmet stores. Conimex is the largest distributor in America.

2. Several of the basic kitchen utensils (especially the widely used stone mortar and pestle, and, to a lesser extent, the wok) are not yet found in most Western kitchens, although their popularity is increasing. Yet alternatives are easily improvised, for example, using a blender in place of a mortar (see Indonesian Kitchen Utensils).

3. Some of the basic cooking techniques (such as leaf wrapping) require special materials and explanation. Again, Western improvisations (such as using aluminum foil) give fairly good results.

4. Many Indonesian tempeh dishes are meant to be served atop rice and are seasoned and spiced (usually with chilies) accordingly. In all of the recipes calling for chilies, we have reduced the amount substantially to suit Western tastes and to adapt the dishes for use either as as entrée or as an accompaniment for rice.

5. A number of the recipes call for the use of overripe tempeh, which is not yet sold in the West. Yet it can be made at home as described on page 39 and, in many cases, regular soy tempeh can be substituted.

Tempeh is widely used in the full panorama of Indonesian cuisine, and in a great many of the best-known dishes it is used interchangeably with meat or poultry.

*Fig. 5.1: Indonesian village kitchen*

Thus there are Rendang Beef and Rendang Tempeh and Saté Chicken and Saté Tempeh. The number of such "pairs" is surprisingly large. Tempeh (or its close counterpart, tofu) is generally used for day-to-day home cooking by all classes of people, and most extensively by the lower income groups; meat is reserved more for special occasions, or for restaurant menus. Nevertheless, tempeh is also used on festive occasions and is served in at least several dishes in most restaurants. Thus, because of the principle of interchangeability, this chapter will introduce you not merely to Indonesia's most popular tempeh recipes, but to a rich array of the country's favorite dishes and cooking styles.

It will also introduce you, to a lesser extent, to Malaysian cookery, since many of the dishes from the two countries are the same or strikingly similar. In both cuisines you will find peanut sauces, fried rice (Nasi Goreng), saté, and coconut-milk curries, to mention but a few.

Many of the recipes are closely related and interrelated. Depending on the amount of coconut milk or soup stock added, for example, a chili-fried sambal might become either a sauce, a stew, or a soup. To help give more of a sense of these relationships, so well known to Indonesians, we have pointed out a number of them in the introductions to individual recipes and tried to group these recipes together.

During our travels in Indonesia we asked many people of different backgrounds to list for us their seven favorite tempeh recipes in order of preference. The results of this survey are shown below. Recipes with the highest

frequency ranking were mentioned by the most people, whereas those with the highest quality ranking were listed by people as their very favorites. Most of these are served at either Javanese, Sundanese, or Sumatran restaurants, or at the popular streetside kitchenettes or stands called *warung*.

Fig. 5.3 Indonesia's favorite tempeh recipes

| Recipe Name | Frequency Ranking | Quality Ranking |
| --- | --- | --- |
| Tempeh Goreng | 1 | 2 |
| Tempeh Bachem | 2 | 4 |
| Keripik Tempeh | 3 | 6 |
| Sayur Lodeh | 4 | 7 |
| Sambal Goreng Tempeh | 5 | 3 |
| Terik Tempeh | 6 | 5 |
| Sambal Goreng Kering Tempeh | 7 | 1 |

In recipe titles, if the word tempeh comes first, it often signifies that tempeh is the main and featured ingredient; if second or last, that the tempeh plays a supporting role.

In virtually all recipes, tempeh is regarded as a raw ingredient which must be cooked, either fried, deep-fried, sautéed, simmered, grilled, or the like. However, a few Indonesians have told us that, although it was uncommon, when they were young, they had occasionally eaten tempeh uncooked, lightly sprinkled with salt, as soon as it had been made while it was still warm from the incubation. Since the soybeans are cooked and then fermented, this way of eating is thought to be nutritious. On the other hand, other Indonesians have told us that they have never heard of eating tempeh in this way.

All the recipes in this chapter are well known by name throughout Indonesia; most appear in at least one cookbook, and we learned them directly from Indonesian cooks in Java. About 35 of the tempeh recipes and 7 onchom recipes are given in the country's 1,123-page national cookbook, *Mustika Rasa* (Gems of Taste). Here they and all other Indonesian recipes are grouped into eight basic categories: staple foods (especially rice, corn, sagu, and cassava); souplike dishes, containing plenty of liquid; dishes without liquid (including all sautéed or steamed preparations), deep-fried dishes; grilled or roasted dishes; sambals and sauces; snacks; and beverages. The technique of oven baking, for example, is rarely used since it is thought to require too much fuel and time; in its place roasting in leaf wrappers, steaming, or deep-frying are used with excellent results. Likewise, (except in West Java), vegetables are rarely served raw, as in salads; throughout Southeast Asia, as in most traditional societies, cooking is considered vitally important in ridding vegetables or other raw foods of parasites or pathogenic microorganisms that might cling to them.

While many of these recipes are prepared in all three provinces of Java, the seasoning tends to be different in each area. The Sundanese cooking of West Java

*Fig. 5.2: Grinding spices with a mortar* (chowet)

combines Sumatran spiciness and solidity with Javanese sophistication and lightness; the cooks tend to add their spices after cooking, like their vegetables raw, and generally do not like sweet dishes accompaning their rice. In Central Java, almost every dish that calls for chili and spices (which means most of them) also demands a large amount of palm sugar; the cuisine is a mix of sweet, sour, and hot flavors. The sugar gets the fire of the chilies and spices under control without actually quenching it and enables one to recognize and appreciate the spectrum of flavors that might otherwise be missed. The East Javanese are fond of a number of set combinations of spices and aromatic nuts and roots, while the Balinese, further east, like their spice flavors vivid, the tastes of their ingredients clear and recognizable. In the following tempeh recipes, we have included seasoning combinations from throughout Java, using those found in the areas where the dish itself is most popular.

In order to understand the role of tempeh in Indonesian cookery, it is important to understand what constitutes a typical meal. Rice is virtually always the main course and central element; the meal is planned around it. Steamed white rice is used to fill the main plate — which looks like a large white Western soup plate — and then helpings of accompanying side dishes are arranged around and on top of the mound of rice. At most meals, a souplike vegetable dish (sayur) and a chili-and-spice sauce (sambal), either of which may contain tempeh, will appear, both served *over* the rice. Likewise, most tempeh dishes are meant to be served as an accompaniment for rice, although a few deep-fried delicacies are served in a separate dish as snacks or finger foods. The rice is sometimes mixed with the other foods just before each bite. A failure to understand this practice of eating almost everything with rice has encouraged the Western notion that Indonesian foods are too chili-hot and spicy to cope with. The foods *are* hot, to be sure, but one must remember that every dish and every sambal has been designed to be spread over a quantity of bland rice.

Indonesians traditionally eat with the fingers of the right hand — it is considered impolite to use the "unclean" left hand. Although spoons and forks are now also widely used, it is usual to revert to tradition on ceremonial occasions. At such meals, the diner is provided with a finger bowl to moisten the fingers before eating so that the rice will not stick to them. The food is formed into small balls and placed in the mouth. No trace of food should touch the palm of the hand, and the fingers should not enter the mouth.

Although Indonesians love spicy-hot food, they don't usually serve food piping hot from the stove, perhaps in part because of their hot, humid climate; saté, soto, and sometimes rice are exceptions. A spoon is the main eating utensil and a knife is used only in the kitchen to precut the foods. Fruit is the typical dessert. Before the meal, it is customary and polite to say "Selamat makan," "Blessings upon this meal." The following sections and the recipes within each section are generally listed with our *favorites first*.

Selamat makan!

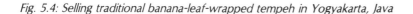

Fig. 5.4: Selling traditional banana-leaf-wrapped tempeh in Yogyakarta, Java

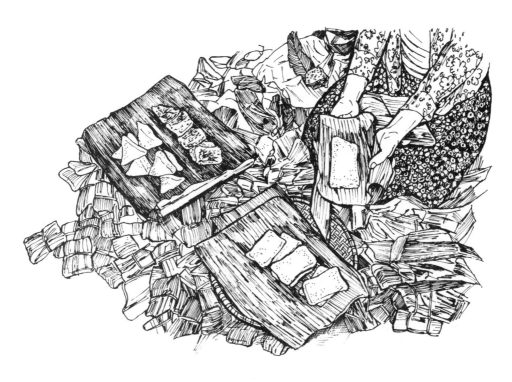

## Tempeh Fried, Deep-Fried, and in Hors d'Oeuvres

To this category belong Indonesia's most popular and, in our opinion, most delicious tempeh recipes. Most are quick and easy to prepare, require ingredients generally available in the West, and make delicious hors d'oeuvres, side dishes, or accompaniments for brown rice. Please begin by reading the section entitled About Deep-Frying in Chapter 4.

*Fig. 5.5: Deep-frying tempeh keripik in batter*

### Deep-Fried Tempeh with Seasonings *(Tempeh Goreng)*
SERVES 1 OR 2

From Northern Sumatra all the way down to Bali and points south, this is Indonesia's most popular way of serving both regular and okara tempeh. Thin slices are marinated in a mixture of salt, water, and seasonings (especially garlic and coriander or, occasionally, turmeric) and then deep-fried or pan-fried in coconut oil until they have a savory flavor and crisp texture remarkably like those of southern-fried chicken. In poorer areas, where people often cannot afford spices, the tempeh is simply dipped in salt water or rubbed lightly with salt before being fried. Generally served hot as an accompaniment or side dish for rice (usually topped with a fiery hot chili sambal), Tempeh Goreng may also be served as a snack or hors d'oeuvre (usually topped with a sprinkling of salt), or used as an ingredient in soups or sauces. The snack form is widely available in restaurants, at roadside stands, and from vendors on trains. In Indonesian hospitals, it is used as a key protein source in diabetic diets. The word *goreng* means "to fry or deep-fry," and the title Tempeh Goreng can be used in either of three

senses: to refer to all tempeh preparations that are either deep-fried or pan-fried; to refer (as is most common) to the specific recipe described here; or to refer to the slices fried in a thick batter, generally known as Tempeh Goreng Tepung, close relatives of which are Tempeh Keripik and Tempeh Kemul. Since this recipe is as quick and easy to prepare as it is delicious, we think it will become as popular in the West as it is in Indonesia.

2¼ ounces (65 grams) soy tempeh or okara, cut into 4 slices each 1½ by 2½ by ⅜ inch, and scored if desired
½ teaspoon coriander seeds
½ clove of garlic
½ candlenut (optional)
¼ to ½ teaspoon salt
2 tablespoons water
½ to 1 cup (coconut) oil for deep-frying

Arrange tempeh slices on a plate. Combine the next five ingredients in a mortar or blender and crush or puree until smooth, then pour over the tempeh and allow to soak for 1 minute.

Heat oil to 400° F. (210° C.) in a wok, skillet, or deep-fryer. Slide in tempeh and deep-fry for about 5 minutes, or until crisp and golden brown. Serve immediately as a snack or side dish (topped, if desired, with a fresh chili, chili sauce, or sweet Indonesian soy sauce), or over rice.

VARIATIONS

*\*Yogyakarta Tempeh Goreng:* Use whole, unsliced 25-gram cakes of tempeh. Score upper and lower surfaces in a crosshatch pattern before deep-frying.
*\*Balinese Tempeh Goreng:* Cut tempeh into pieces 1 by 3½ by ¾ inch. Combine in a mortar: coriander seeds, fresh turmeric, kenchur root, garlic, and salt (no water), then grind until smooth. Dip tempeh in mixture, then deep-fry.
*\*Tempeh Goreng Tepung:* Dip tempeh slices into a batter made of 3 to 4 tablespoons wheat flour, 1 lightly beaten egg, ½ teaspoon salt, and a pinch of coriander. Deep-fry until crisp and golden brown.

### Tempeh Cutlets *(Tempeh Bachem)*
SERVES 1 OR 2

Simmered in a delectable mixture of coconut milk, palm sugar, and spices, then deep-fried until crisp and golden brown, these juicy cutlets have a wonderful flavor and texture that make them ideal for use in Western-style burgers. One of Indonesia's two most popular tempeh recipes (the other is Tempeh Goreng), it is especially well known in Central Java, where rectangular and triangular cakes are sold at many restaurants and roadside stands. Okara tempeh is often used in place of its regular counterpart. The word *bachem* means "to simmer in a small quantity of liquid."

2 shallots or ¼ minced onion
½ to 1 teaspoon salt
½ cup coconut milk, made from ½ cup each freshly grated coconut and water or substitute tamarind water (page 42 or 44)
1½ tablespoons dark-brown or palm sugar
2¾ ounces (75 grams) regular or okara tempeh, cut into 3 slices 2½ by 1½ by ½ inch
½ cup (coconut) oil for deep-frying

Combine the shallots or onion and salt in a mortar and grind until smooth, then combine in a saucepan together with the coconut milk and sugar, mixing well. Bring to a boil, add tempeh and simmer uncovered, stirring occasionally, for about 15 minutes, or until almost all the liquid has evaporated. Heat the oil to 350° F. (175° C.) in a wok, skillet, or deep-fryer. Slide in tempeh and deep-fry for about 4 minutes, or until nicely browned. Serve as is or over rice.

VARIATIONS

* 1 teaspoon bruised laos root, 1 salam leaf, and ½ teaspoon tamarind paste may be added with the coconut milk, then the laos and salam removed before serving.
*Tempeh Ungkep: Prepare as in the basic recipe, but serve without frying. Popular in Bali. "Ungkep" means to simmer covered with seasonings in a small quantity of water.

*Tamarind paste and pods*

## Crisp Tempeh Cutlets
### (Tempeh Bachem Kering)
SERVES 2

When Tempeh Bachem is prepared so as to have a particularly crisp consistency, the word *kering* ("dry") may be attached to the end of the name; when served without deep-frying, in its juicier form, the word *basah* ("moist") may be used. Close relatives of the latter preparation are Kering Tempeh and Tempeh Ungkep. The crisp preparation is especially popular in Central Java.

4 ounces (115 grams) tempeh, cut into pieces 1¼ by 2 by ¾ inch, then scored on upper and lower surfaces
1 teaspoon tamarind paste
1 salam leaf
1 teaspoon laos root, bruised until partially mashed
¾ to 1¼ teaspoons salt
2 or 3 shallots, thinly sliced
3½ to 4 tablespoons (50 grams) dark-brown or palm sugar
3 cups water or coconut milk (page 42)
2 tablespoons (coconut) oil

Combine the first eight ingredients in a wok or skillet and bring to a boil. Simmer, covered, for about 30 minutes, or until almost all of the liquid has evaporated. Stir in oil and remove from heat. Remove salam leaf and laos root. Serve as an hors d'ouevre or side dish, or over rice.

## Spicy-Fried Tempeh in Coconut Milk
### (Sambal Goreng Tempeh)
SERVES 3 TO 4

Sambal Goreng Tempeh can be prepared in either a moist or liquid (*basah*) form or in a rather crisp, dry (*kering*) form. This recipe, in which tempeh is simmered in spicy coconut milk, is the former type, and thus technically should be grouped under Sauces. The next, in which the tempeh is prefried and then sautéed with spices and palm sugar, is the latter. Our survey of favorite Indonesian recipes showed that both types are highly regarded, being ranked respectively third and first in terms of fine flavor. We heartily agree. Indonesia's popular Sambal Goreng, a close relative made without tempeh, uses only the first five ingredients listed below, plus terasi.

2 shallots
½ clove of garlic
1 red chili (3 inches long), seeded
1 teaspoon salt
3 tablespoons oil
4½ ounces (130 grams) tempeh (regular, okara, or partially overripe), cut into pieces ¾ by ⅜ by ⅜ inch
1 teaspoon laos root, slightly mashed
1 salam leaf
1 cup coconut milk, made from 1 cup each freshly grated coconut and water (page 42)

Combine the first four ingredients in a mortar and grind until smooth. Heat a wok or skillet and coat with the oil. Add contents of mortar and sauté for 30 seconds. Add the remaining ingredients, bring to a boil, and simmer, uncovered, for 4 minutes. (The consistency should be fairly liquid.) Serve over rice.

VARIATIONS

*Sambal Godok Tempeh: Prepare as for Sambal Goreng Tempeh, but simmer the condiments in the coconut milk instead of sautéing, and use overripe tempeh.
* Add or substitute deep-fried tofu, diced tomatoes or green peppers, or palm sugar.

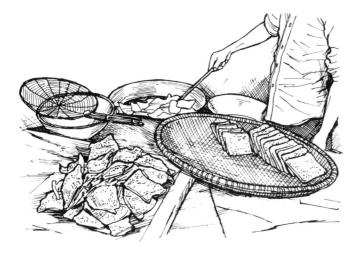

*Fig. 5.6: Deep-fried tempeh keripik in a Javanese market*

Combine the first three (or five) ingredients in a mortar and grind until smooth. Add coconut milk and both flours, mixing until smooth. Heat oil to 400° F. (210° C.) in a wok, skillet, or deep-fryer. Dip tempeh slices into batter then slide them into oil and deep-fry for 3 to 4 minutes, or until crisp and golden brown.

VARIATIONS

**\*Okara Tempeh Keripik** (Tempeh Gembus Keripik): Prepare the batter using ¾ teaspoon coriander seeds, 1 candlenut, ¼ to ½ teaspoon salt, ⅓ to ½ cup coconut milk, and ½ cup rice flour. Cut okara tempeh into ¼-inch-thick slices, then dip into batter and deep-fry.
\* Sun-dry tempeh slices for 10 to 15 minutes, then dust with 2 tablespoons tapioca (or rice) flour and dip in the basic batter. Deep-fry or bake.
\* Use any of the following additional ingredients in the batter: eggs, baking soda, tamarind paste, salam leaf, or palm sugar; the latter three are relatively uncommon.
\* Make a simple batter using only the rice flour, coconut milk, garlic, ground coriander seeds, and salt.

## Sweet & Crispy Tempeh Sauté (Kering Tempeh)

SERVES 3

This delicious topping is a close relative of Sambal Goreng Tempeh, except that it contains no coconut milk and has a somewhat drier consistency (the word *kering* means crisp or dry). Prefrying the tiny pieces of tempeh imparts a prized crispiness, and the use of palm sugar mellows the chilies. Some recipes call for the use of peanuts.

⅔ cup (coconut) oil for deep-frying
4½ ounces (125 grams) tempeh, cut into slices ⅜ by 1¾ by ⅛ inch, and dried in the sun (or a very slow oven) for 30 minutes
4 shallots, thinly sliced
1 clove of garlic, thinly sliced
1 red chili (3¼ inches long), thinly sliced and seeded
½ teaspoon tamarind paste, dissolved in ½ cup water
1 teaspoon laos root, bruised
1 salam leaf
5 inches of lemongrass stalk and bulb
3½ tablespoons dark-brown or palm sugar
½ teaspoon salt

Heat the oil to 375° F. (190° C.) in a wok, skillet, or deep-fryer. Slide in tempeh and deep-fry for 3 to 4 minutes, or until crisp and golden brown, then remove from wok, drain, and set aside. Remove all but 2 tablespoons oil from wok. Add shallots and garlic, sauté for 2 minutes, then remove from wok and set aside. Add chili, sauté for 1 minute, then remove and set aside. Add tamarind water and the remaining ingredients, bring to a boil, and simmer for 2 to 3 minutes. Add fried tempeh, shallots, garlic, and chili, mixing well. Remove laos root, salam leaf, and lemongrass. Serve over rice or with Lontong (page 47).

## Tempeh Chips (Keripik Tempeh)

SERVES 1 OR 2

Indonesia's third-most-popular way of serving tempeh, these delightfully crisp chips are prepared by dipping thinly sliced tempeh squares, rectangles, or triangles into a thin batter of coconut milk, rice flour, and spices, then deep-frying them until golden brown. The freshly prepared chips, rich with the fragrance of coconut oil, are sold piping hot at many marketplace stands; an order of seven 4-inch squares goes for about 24 cents (U.S.). Large commercial makers from Malang and Purwokerto also sell the same-sized chips packaged thirty in a sealed cellophane bag alongside crackers and staple foods at nicer food stores throughout the country. Retailing for only 34 cents, the chips will last for up to one month, after which time they develop a slightly bitter flavor but are still quite acceptable. In commercial products, baking soda is often used in the batter to help retain crispness for a long time. In Central Java, okara tempeh is widely used instead of its regular soy counterpart. In some areas the word *keripik* is spelled *kripik*.

½ clove of garlic
¼ teaspoon coriander seeds
¼ to ½ teaspoon salt
½ candlenut (optional)
⅛ teaspoon fresh or powdered turmeric (optional)
½ cup coconut milk (page 42) or water
¼ cup rice flour
2 tablespoons wheat flour, or substitute rice flour
1 cup (coconut) oil for deep-frying
3½ ounces (100 grams) tempeh, cut into ⅛-inch-thick slices, then into any of the following shapes:
4-inch squares, 2 ½-inch triangles, or 3-by-6-inch rectangles

## Crunchy Chili-Fried Tempeh Topping *(Sambal Goreng Tempeh Kering)*

SERVES 1 OR 2

Although this recipe is considered to have the finest flavor of any Indonesian tempeh recipe, it is often served only on special occasions. In some market stands, it is prepared fresh and sold wrapped in banana leaves; a 1-cup portion costs about 16 cents (U.S.). Consisting primarily (about 95%) of tiny crisp slices of deep-fried tempeh coated with a sweet-and-hot spice mixture, it adds life and depth to any dish with which it is used. In Sumatra, little or no sugar is used to mellow the incendiary chilies.

½ cup (coconut) oil for deep-frying and sautéing
2½ ounces (70 grams) tempeh cut into slices 1½ by ½ by ⅛ inch
1 to 2 red chilies, thinly sliced (seeded if desired)
3 shallots, thinly sliced
1 clove garlic, thinly sliced
¼ teaspoon terasi
1 teaspoon tamarind paste
1½ tablespoons dark-brown or palm sugar
¼ teaspoon laos root
3 tablespoons water
¼ to ½ teaspoon salt
1 salam leaf

Heat the oil to 375° F. (190° C.) in a wok, skillet, or deep-fryer. Slide in tempeh and deep-fry for about 4 minutes, or until crisp and golden brown, then remove and drain. Now deep-fry consecutively for 3 minutes each the chili, shallots, and garlic; remove each ingredient when crisp and golden brown, and set aside to drain before adding the next ingredient.

Combine the terasi, tamarind paste, sugar, and laos root in a mortar; bruise laos and grind other ingredients until smooth. Heat a wok and coat with 1 tablespoon of the deep-frying oil. Add ground spices and water, mixing well. Add salt and salam leaf and bring to a boil. Add deep-fried tempeh, chili(es), shallots and garlic, and cook, stirring constantly, for 1 minute. Remove salam leaf and laos root. Serve over rice. Especially popular with Coconut Rice (Nasi Uduk or Nasi Gurih; page 45).

A delicious and popular variation contains crisp fried peanuts and/or tiny fried dried fish (*teri*). It has a very dry consistency, being cooked the second time until the tempeh is crisp; the flavor is sweet and not especially hot.

## Tempeh & Coconut Patties *(Menjeng Tempeh)*

SERVES 2

Popular in Central and East Java, this recipe generally features overripe tempeh mixed with spices, shaped into patties or balls, and deep-fried. In some areas peanut & coconut-presscake tempeh (tempe menjes) is used in place of regular soy tempeh.

2¾ ounces (80 grams) tempeh, preferably overripe
1 red chili, thinly sliced
2 fiery dwarf red chilies
⅓ lime leaf
⅛ teaspoon kenchur root
2 teaspoons dark-brown or palm sugar
½ clove of garlic
½ to ¾ teaspoon salt
1½ ounces (40 grams) freshly grated coconut
1 to 1½ cups (coconut) oil for deep-frying

Place tempeh in a preheated steamer and steam for 15 minutes. Remove and set aside. Combine the next seven ingredients in a mortar and grind until smooth. Add steamed tempeh and grind for about 4 minutes, or until thoroughly mixed, then stir in coconut. Using two tablespoons and your fingertips, shape mixture into 6 balls or patties. Heat the oil to 375° F. (190° C.) in a wok, skillet, or deep-fryer. Slide in balls and deep-fry for 4 to 5 minutes, or until crisp and golden brown. Serve hot as an hors d'oeuvre or as a side dish with rice.

For variety, substitute white pepper for the kenchur root and lime leaf. Dip patties into lightly beaten egg before deep-frying.

## Tempeh Patties *(Perkedel Tempeh)*

MAKES 3, SERVES 1

There are many Indonesian recipes for Perkedel and all are prepared in the form of balls or patties; other main ingredients include sweet corn, crab, pork, and beef. Stuffed chilies are also classed as a variety of Perkedel. These moist croquettes are similar to Menjeng Tempeh (above), except that the tempeh is not steamed.

½ clove of garlic
1 red chili, minced (optional)
¼ teaspoon coriander seeds
⅛ teaspoon kenchur root
½ lime leaf
¼ to ½ teaspoon salt
2½ ounces (70 grams) tempeh, thinly sliced
½ to 1 cup (coconut) oil for deep-frying

Combine the first six ingredients in a mortar and grind until smooth. Add tempeh slices and grind for about 4 minutes more, or until very well mixed. Using 2 tablespoons and your fingertips, shape the mixture into 3 patties, each about 2 inches in diameter and ½ inch thick. Heat the oil to 375° F. (190° C.) in a wok, skillet, or deep-fryer. Slide in patties and deep-fry for 4 to 5 minutes, or until crisp and golden brown. Serve as is or as a side dish with rice.

*Kenchur root*

## Tempeh Fried in Thick Batter *(Tempeh Kemul)*

SERVES 1 OR 2

The word *kemul* means "blanket," referring to the thick rice-flour batter in which this tempeh is fried. The use of rice flour rather than wheat flour imparts a prized crispness. Sometimes known as Tempeh Goreng Tepung ("tempeh fried in batter"), it becomes Mendoan Tempeh when immature tempeh is deep-fried more quickly, so that the inside retains some of its moisture. Some Indonesians like to eat this as a snack with a fresh raw chili perched on top.

¼ teaspoon cumin seeds
½ teaspoon coriander seeds
½ clove of garlic
⅛ teaspoon fresh or powdered turmeric
¼ to ½ teaspoon salt
2 tablespoons water
2 tablespoons wheat flour
3 tablespoons rice flour
½ to 1 cup (coconut) oil for deep-frying
2 to 2¼ ounces (65 grams) tempeh, cut into 4 slices
    each 2½ by 1¼ by ⅜ inch

Combine the first six ingredients in a mortar and grind until smooth. Add wheat and rice flour, mixing until smooth. Heat the oil to 350° F. (175° C.) in a wok, skillet, or deep-fryer. Dip tempeh slices in batter then slide into oil and deep-fry for about 4 minutes, or until crisp and golden brown. Serve immediately as a snack or side dish, or over rice.

## Crisp & Tender Tempeh Snacks *(Mendoan Tempeh)*

SERVES 3 TO 4

A specialty of the Banyumas and Purwokerto areas of Central Java, Mendoan (also called Mendo) is made by dipping thin tempeh slices into a rice-flour-and-spice batter, then frying them quickly to yield a crisp surface enclosing a tender, slightly moist interior. Immature soy tempeh is widely used in place of its mature counterpart. Close relatives are Keripik Tempeh and Tempeh Kemul. Many Indonesians like to eat Mendoan as a snack with a raw chili perched atop each piece, or dipped in a mixture of sweet Indonesian soy sauce (kechap) and ground chilies.

2 cloves of garlic
¼ teaspoon coriander seeds
½ teaspoon salt
5 to 6 tablespoons rice flour, or substitute wheat flour
¾ to 1 cup coconut milk (page 42), or substitute water
1 cup (coconut) oil for deep-frying
7 ounces (200 grams) immature or mature tempeh, cut
    into slices 1½ by 2 by ⅜ inch

Combine the first three ingredients in a mortar and grind until smooth. Combine the flour and coconut milk to make a medium-thick, smooth batter, then mix in the contents of the mortar. Heat the oil to 375° F. (190° C.) in a wok, skillet, or deep-fryer. Dip tempeh slices into batter, then slide into oil and deep-fry for about 3 minutes, or until they become lightly browned but are still tender and not too crisp. Serve hot as a snack or hors d'oeuvre, or as an accompaniment for rice.

For variety, serve topped with any of the sauces used for Saté; add ¼ teaspoon kenchur root to the batter ingredients ground in the mortar.

## Fried Tempeh & Tofu Salad with Peanut Sauce Dressing *(Lengko-Lengko)*

SERVES 3

1 to 1½ cups (coconut) oil for deep-frying
4¼ ounces (120 grams) tempeh, cut into pieces 1½ by
    1½ by ¾ inch
8 ounces (230 grams) tofu, cut into 1½-inch cubes
2½ tablespoons (30 grams) (raw) peanuts
1 red chili (3 inches long), cut lengthwise into halves
2 shallots, cut into halves
½ to 1 teaspoon salt
1 tablespoon dark-brown or palm sugar
½ cup hot water
2 tablespoons sweet Indonesian soy sauce (kechap)
½ cucumber (75 grams), sliced diagonally into thin
    ovals
1¾ ounces (50 grams) mung-bean sprouts, parboiled
    for 2 to 3 minutes and drained
2 tablespoons minced chives
5 to 6 tablespoons Deep-fried Onion Flakes (page 44)

Heat the oil to 375° F. (190° C.) in a wok, skillet, or deep-fryer. Slide in the tempeh and deep-fry for 3 minutes, then remove from wok, drain, and set aside. Slide in tofu and deep-fry for 3 minutes, then drain and set aside. Slide in peanuts and deep-fry for 1½ minutes, then remove with a mesh skimmer, drain, and set aside. Slide in red chili and shallots, deep-fry for 30 seconds, remove with a mesh skimmer and drain, then combine in a mortar with the peanuts and grind until smooth. Add salt, sugar, water, and kechap, mixing well, to form a peanut sauce.

Cut tempeh and tofu into ¾-inch cubes and combine in a salad bowl with the cucumber, sprouts, and chives; mix lightly. Pour peanut sauce over the top, then serve topped with a sprinkling of Deep-Fried Onion Flakes.

## Mellow Tempeh Cutlets
### (Tempeh Goreng Dengan Bumbu) SERVES 1 OR 2

2½ ounces (70 grams) tempeh cut into slices 1½ by 3 by ⅜ inch, then scored diagonally on the upper and lower surfaces to aid absorption
3 shallots
1½ teaspoons white sugar or ¾ teaspoon honey
½ teaspoon coriander seeds
½ teaspoon tamarind paste
¼ teaspoon salt
2 tablespoons water
1 cup (coconut) oil for deep-frying

Arrange tempeh slices on a plate. Combine the next five ingredients in a mortar and grind until smooth, then mix in the water and pour over the tempeh slices; allow to marinate for 5 minutes. Heat the oil to 375° F. (190° C.) in a wok, skillet, or deep-fryer. Slide in tempeh and deep-fry for 7 to 8 minutes, or until crisp and golden brown. Serve as is or over rice.

## Tempeh, Tofu, & Chili Sauce
### (Sambal Goreng Krechek Tempeh) SERVES 6 TO 10

A close relative of Oseng-Oseng Tempeh (but without the soy sauce), this recipe takes its name from krechek, dried cow's skin, which, when fried, becomes crisp like cracklings. A small amount (50 grams) of fried krechek is traditionally added after the tofu and simmered for 5 minutes. We omit it in this recipe (with little or loss of flavor) since this book uses no meat.

¼ cup oil
5 shallots
2 laos roots, each 3 inches long
2 cloves of garlic
5 red chilies
4 whole salam leaves
⅓ cup fresh peté beans, slivered lengthwise (or substitute fresh green soybeans or baby limas)
3 cups coconut milk, made from 3 cups each freshly grated coconut and water (page 42)
7½ ounces (210 grams) tempeh (regular or overripe), cut into ½-inch cubes
1½ teaspoons dark-brown or palm sugar
1 to 2 teaspoons salt
1½ cups tofu (½-inch cubes)

Heat a wok or skillet and coat with the oil. Add the next three ingredients and sauté for 1 minute. Add chilies and salam leaves and sauté for 3 minutes. Add the next five ingredients and one-half the tofu, and simmer for 2 minutes. Add rest of tofu and simmer for 2 minutes more. (Plenty of liquid should be left.) Remove salam leaves. Serve over rice.

## Fried Tempeh with Turmeric
### (Tempeh Goreng Biasa) SERVES 2

1½ teaspoons fresh or powdered turmeric
3 candlenuts
Dash of pepper
½ teaspoon salt
3 tablespoons oil
2½ ounces (70 grams) tempeh, cut into 6 slices, each 1½ by 3½ by ⅛ inch

Combine the first four ingredients in mortar and grind until smooth. Heat a wok or skillet and coat with the oil. Add contents or mortar and sauté for 30 seconds. Add tempeh and sauté for 3 minutes, then scoop out of wok, leaving most of the sauce in the wok. Serve tempeh as an hors d'oeuvre or over rice. Discard sauce.

## Deep-Fried Okara Tempeh
### (Onchom Goreng) SERVES 2

1 clove of garlic
1 teaspoon coriander seeds
¼ to ½ teaspoon salt
¼ cup water
1 cup (coconut) oil for deep-frying
8 to 10 slices of okara tempeh or onchom, each 1½ by 3 by ¼ inch

Combine the first three ingredients in a mortar and grind until smooth. Add water, mixing well. Heat the oil to 375° F. (190° C.) in a wok, skillet, or deep-fryer. Dip tempeh or onchom slices into mixture then slide into oil and deep-fry for about 5 minutes, or until crisp and golden brown. Serve immediately as an hors d'oeuvre, side dish, or over rice.

VARIATION

**\*Bandung Batter-Fried Onchom (Onchom Goreng):** Grind 2 candlenuts, ¼ teaspoon turmeric, and ¼ teaspoon cumin together with the ingredients above. Add ⅜ cup rice flour or wheat flour plus enough water to make a medium-thick batter and ½ teaspoon salt. Dip slices in batter and deep-fry.

## Indonesian Salad with Peanut & Coconut-Milk Sauce (Gado-Gado)

SERVES 4 TO 6

One of Indonesia's most famous recipes, Gado-Gado, consists of a basis of cooked (or in some cases both cooked and raw) vegetables and sliced fried tempeh or tofu with a spicy peanut sauce poured over the top, then garnished with a sprinkling of Deep-Fried Onion Flakes, shrimp crisps, slivered omelets, or hard-boiled eggs. The term Gado-Gado (the Indonesians write it gado²) means "a mixture," and ingredients in this mixture and its sauce can vary widely depending on the season and the chef. Tempeh is widely used in some areas (especially Bali and Central Java) and not used in others (West Java). Unlike so many Indonesian dishes, Gado-Gado is *not* eaten as a topping or side dish for rice but as a main course in its own right, in much the same fashion as a large luncheon salad in the West.

A close relative of Gado-Gado is Pechel, steamed vegetables and tempeh with spicy peanut sauce, which uses simpler, lower-priced ingredients in the base and sauce.

**Basic Ingredients. Choose six of the following, listed in order of popularity:**
**4 hard-boiled eggs, peeled and halved or sliced**
**½ to 1 cup green beans, cut into 1-inch lengths and lightly boiled**
**12 to 16 ounces (340 to 450 grams) tofu, cubed or thinly sliced, preferably deep fried**
**1 to 2 cups shredded regular or Chinese cabbage, blanched**
**1 cucumber, unpeeled, cut crosswise into ¼-inch-thick slices**
**2 to 4 small new (waxy) potatoes, boiled and sliced**
**8 ounces (230 grams) fresh spinach or kangkung, coarsely chopped and steamed for 12 to 15 minutes**
**4 to 8 ounces (115 to 230 grams) fresh bean sprouts, steamed for 12 to 15 minutes**
**4 to 6 ounces (110 to 170 grams) tempeh, cut into thin slices or cubes and deep-fried (page 52)**
**1 cup Peanut Sauce (page 49)**

**Toppings. The first two are always used; the rest are optional:**
**4 large shrimp crisps (krupuk)**
**½ cup Deep-Fried Onion Flakes (page 44)**
**2 chopped hard-boiled eggs**
**½ cup slivered omelets**

Arrange your choice of basic ingredients on a large serving platter and allow to cool to room temperature, then arrange the freshly deep-fried tempeh on top. Serve accompanied by the Peanut Sauce (in a bowl or sauceboat) and the toppings in separate bowls. Invite your guests to serve themselves the basic ingredients, topping them with the sauce and a liberal sprinkling of the toppings.

VARIATIONS

*Pechel: Use mung-bean sprouts, cabbage, kangkung, and tempeh as the basis. Make a simplified version of the sauce, using fried peanuts, chilies, terasi, brown sugar, salt, and water.
* Alternate basic ingredients include lettuce, carrots, shallots, green peas, and Ketupat or Lontong (compressed rice boiled in banana leaves) (page 47).

*Gado-Gado*

## Tempeh Croquettes with Sauce (Kroket Tempeh)

MAKES 9

*Kroket* is the Indonesian spelling of "croquette," and this modern preparation was modeled after its Western counterpart. We prefer to serve the croquettes topped with a sauce, although they are usually served plain in Indonesia.

**4 ounces (115 grams) tempeh**
**2 shallots**
**½ to ¾ teaspoon salt**
**½ clove of garlic**
**⅛ teaspoon white peppercorns (or ground pepper)**
**1 green onion or scallion, minced**
**⅓ cup minced celery leaves**
**2 teaspoons margarine**
**½ cup bread crumbs**
**1 egg, lightly beaten**
**1 cup (coconut) oil for deep-frying**
**½ to 1 cup Peanut Sauce (page 49), Ketchup & Worcestershire Sauce (page 48), or Worcestershire sauce**

Place tempeh in a preheated steamer and steam for 15 minutes, then cool briefly, mash thoroughly in a mortar, and set aside. Combine the next four ingredients in a mortar and grind together until smooth. Add green onion, celery leaves, margarine, and tempeh, mixing well. Shape mixture into 9 cylindrical croquettes, each 1 inch in diameter and 1½ to 2 inches long. Roll in bread crumbs, dip into beaten egg, and roll in bread crumbs again. Heat the oil to 375° F. (190° C.) in a wok, skillet, or deep-fryer. Place croquettes on a mesh skimmer, set gently into oil, and deep-fry for about 3 minutes, or until crisp and golden brown. Serve hot, accompanied by the Peanut Sauce.

### Deep-Fried Tempeh Patties (Mendol)

SERVES 2 OR 3

This specialty of East Java often features partially overripe tempeh, used to impart a cheesy texture and rich flavor.

1 shallot
½ clove of garlic
½ teaspoon kenchur root
½ teaspoon salt
¼ lime leaf
1 red chili (2½ inches long), seeded
5 ounces (145 grams) fresh or partially overripe tempeh
1 cup (coconut) oil for deep-frying

Combine the first six ingredients in a mortar and grind until smooth. Add tempeh and mash until well mixed. Shape mixture into patties 2 inches in diameter and ¾ inch thick. Heat the oil to 375° F. (190° C.) in a wok, skillet, or deep-fryer. Slide in patties and deep-fry for 3 to 4 minutes, or until crisp and golden brown. Drain and serve.

### Deep-Fried Tempeh Balls (Ento-Ento)

MAKES 12

4 shallots
1 red chili
½ clove of garlic
½ teaspoon salt
5½ ounces (160 grams) tempeh
½ cup (coconut) oil for deep-frying
1 egg, lightly beaten
1 cup Peanut Sauce (page 49) or sweet Indonesian soy sauce

Combine the first four ingredients in a mortar and grind until smooth. Add tempeh and mash together with other ingredients. Shape mixture into 1-inch balls. Heat oil to 375° F. (190° C.) in a wok, skillet, or deep-fryer. Dip balls in beaten egg then deep-fry for about 3 minutes, or until lightly browned. Serve plain or topped with the sauce.

### Overripe Tempeh & Kangkung in Coconut Milk (Sambal Goreng Kangkung Tempeh Busuk)

SERVES 3 OR 4

2 tablespoons oil
½ to 1 red chili, thinly sliced and seeded
½ to 1 green chili, thinly sliced and seeded
2 shallots, thinly sliced
1 teaspoon laos root
½ clove of garlic, thinly sliced
3¼ ounces (90 grams) overripe tempeh, cut into pieces ½ by ½ by ⅜ inch

1 cup coconut milk, made from 1 cup each freshly grated coconut and water (page 42)
½ teaspoon tamarind paste
1 salam leaf
¼ teaspoon dark-brown or palm sugar
½ teaspoon salt
2 ounces (55 grams) kangkung leaves, or substitute spinach

Heat a wok or skillet and coat with the oil. Add the next five ingredients and sauté for 1 minute. Add tempeh and sauté for 1 minute. Add the next five ingredients, mixing well, then bring to a boil and simmer, stirring constantly, for 2 minutes. Add kangkung and simmer, stirring, for 4 minutes more. Remove laos root and salam leaf. Serve over rice.

## Tempeh in Sauces and Sautéed Preparations

The Indonesian term for "sauce" is *kuah* and the more widely used Dutch term is *saos*. Yet neither appears in most recipe titles since Indonesian sauces are an integral part of their respective dishes and are almost never served separately. But what indescribable flavors! The smooth, almost velvety richness of Indonesian coconut-milk "sauces" is a treat many Westerners have yet to discover. With coconuts available in many parts of North America at reasonable prices, these recipes are easy to prepare.

### Tempeh Cubes in Coconut Milk (Kotokan Tempeh)

SERVES 3

½ teaspoon coriander seeds
2 shallots
½ clove garlic
½ to ¾ teaspoon salt
1 candlenut
4 ounces (112 grams) tempeh, cut into ½-inch cubes
1⅓ cups rich coconut milk (page 42)
1 salam leaf

Combine the first five ingredients in a mortar and crush until smooth, then combine with the remaining ingredients in a saucepan. Bring to a boil and simmer, covered, stirring occasionally, for 8 to 10 minutes. (A fair amount of liquid should remain.) Remove salam leaf. Serve over rice.

## Beefsteak Tempeh in
##     Brown Sauce *(Bistik Tempeh)*   SERVES 1 OR 2

The delicious Indonesian sauce that accompanies Beefsteak Tempeh has a rich and hearty flavor, and a gravylike consistency. A traditional relative of this modern recipe is Tempeh Semur.

½ cup fresh coconut milk, made from ½ cup each grated coconut and water (page 42)
1 shallot
½ clove garlic
⅛ teaspoon nutmeg
¼ teaspoon white peppercorns (about 10) or a dash of white pepper
¼ to ½ teaspoon salt
1½ tablespoons oil
2 to 2¼ ounces tempeh (60 grams), cut into 4 slices 1 by 1½ by ⅜ inch
2 teaspoons sweet Indonesian soy sauce (kechap)
1½ teaspoons dark-brown or palm sugar (optional)
1 teaspoon vinegar (optional)

Prepare coconut milk and set aside. Combine the next five ingredients in a mortar and grind until smooth. Heat a wok or skillet and coat with the oil. Add contents of the mortar together with tempeh and sauté for 1 minute. Add soy sauce, coconut milk and, if desired, the sugar and vinegar. Bring to a boil and simmer uncovered for about 6 minutes, or until all liquid has evaporated. Serve as is or over rice.

## Tempeh Cutlets in
##     Coconut-Milk Sauce
##       *(Rendang Tempeh)*   SERVES 4 TO 6

The original Rendang, native to the city of Padang in West Sumatra, consists of bite-size chunks of beef which have been cooked in heavily spiced coconut milk until all of the sauce has been absorbed or evaporated. The abundance of chilies and low moisture content help Rendang to keep for 3 to 4 days without refrigeration. We find the tempeh version of this classic to be one of Indonesia's most delicious tempeh recipes. A juicier variety is called Kalio Tempeh.

6 red chilies, seeded if desired
4 teaspoons laos root
1 teaspoon gingerroot
½ to 1 teaspoon salt
5 cloves garlic
5 shallots
2 stalks lemongrass, each tied in a loop (see above)
3 cups coconut milk, made from 3 cups each freshly grated coconut and water (page 42)
16 ounces (450 grams) tempeh, cut into pieces 1½ by 1½ inch by ½ inch

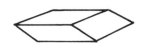

Combine the first four ingredients in a mortar and grind until smooth. Add garlic and shallots; crush and mix well. Add lemongrass and bruise firmly. Combine coconut milk and tempeh in a wok and bring to a boil. Add contents of mortar, return to the boil, and simmer, stirring every 2 minutes, for 8 to 10 minutes, or until most of the liquid has evaporated. Remove lemongrass stalks. Serve as a side dish or over rice.

Other popular seasonings include candlenuts, cumin, cinnamon, coriander, lime leaf, and salam leaf.

*Lemongrass*

## Tempeh Slices in Rich Miso
##     & Coconut-Milk Sauce
##     *(Tempeh Masak Taucho)*   SERVES 3 TO 4

In Indonesia, *masak* means "to cook" and *taucho* is Indonesian soybean miso. Thus the literal translation of this recipe's title would be "tempeh cooked with taucho."

¼ cup (coconut) oil
3 shallots, thinly sliced
2 cloves garlic, thinly sliced
7 ounces (200 grams) tempeh, cut into slices ¾ by 2½ by ⅜ inch
1 green chili, thinly sliced
1½ tablespoons taucho (or substitute Japanese soybean miso)
¼ cup water
2½ inches lemongrass stalk and bulb
1½ teaspoons laos root
1 salam leaf
1 cup coconut milk, made from 1 cup each freshly grated coconut and water (page 42)
2 tablespoons dark-brown or palm sugar
Dash of salt

Heat a wok or skillet and coat with the oil. Add shallots and garlic, and sauté for 1 minute, then remove from wok (leaving the oil) and set aside. Add tempeh to wok and sauté for 3 to 4 minutes, then remove and set aside. Add green chili to wok and sauté for 30 seconds, then add sautéed shallots and garlic together with taucho and water; cook for 30 seconds. Add remaining ingredients plus the tempeh, bring to a boil, and simmer, uncovered, stirring occasionally, for 10 minutes, or until a thick sauce is formed. Remove lemongrass, laos root, and salam leaf before serving.

## Sweet & Sour Tempeh
### *(Tempeh Asam Manis)*  SERVES 1 OR 2

1 red chili, seeded and parboiled for 4 minutes
½ teaspoon laos root
2 shallots
2 tablespoons dark-brown or palm sugar
¼ teaspoon terasi
½ clove of garlic
¼ to ½ teaspoon salt
1 teaspoon fresh or powdered turmeric
2 tablespoons (coconut) oil
2 tablespoons water
¾ cup coconut milk, made from ¾ cup each freshly grated coconut and water (page 42)
2¼ ounces (60 grams) tempeh, cut into 3 slices 2½ by 1¼ by ⅜ inch

Combine the first eight ingredients in a mortar and grind until smooth. Heat a wok or skillet and coat with the oil. Add contents of mortar and sauté for 30 seconds. Add water and coconut milk, mixing well. Add tempeh and simmer uncovered for about 20 minutes, stirring occasionally, until all of the liquid has evaporated. Serve over rice.

*Laos root & chilies*

## Tempeh with Rujak Spice
### Sauce *(Tempeh Bumbu Rujak)*  SERVES 1 OR 2

Rujak is the name of one of Indonesia's most famous spice combinations, which is also popular with tofu or with grilled chicken; it is often used in dishes served on ritual occasions. The mixture is available in the West in ready-mixed, packaged form. Also known as Rujak Tempeh or Bumbu Rujak Champur, a popular version from Surabaya features peté beans.

1 shallot
½ clove of garlic
½ to 1 red chili, seeded if desired
1 candlenut
½ teaspoon salt
⅛ teaspoon fresh or powdered turmeric
¼ teaspoon laos root, firmly bruised
2½ inches of lemongrass stalk and bulb, firmly bruised
1 salam leaf
1 tablespoon oil
1 tablespoon dark-brown or palm sugar
2½ ounces (70 grams) tempeh, cut into slices 1 by 1½ by ⅜ inch
¼ cup water
⅔ cup coconut milk made from ⅔ cup each freshly grated coconut and water (page 42)

Combine the first six ingredients in a mortar and grind until smooth. Add laos root, lemongrass, and salam leaf, mixing well. Heat a wok or skillet and coat with the oil. Add spice mixture and sauté for 1 minute. Add brown sugar and sauté for 30 seconds. Add tempeh and water, mix well, and simmer, stirring constantly, for 2 to 3 minutes. Slowly stir in coconut milk, then bring to a boil and simmer, stirring occasionally, for 20 minutes, or until only a small amount of liquid remains. Remove laos root, lemongrass, and salam leaf, and discard. Serve over rice.

*Palm sugar*

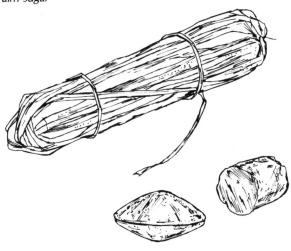

## Overripe Tempeh in Coconut-Milk
### Sauce *(Besengek Tempeh)*  SERVES 2

The various types of besengek are prepared by simmering a basic ingredient in spiced coconut milk; steak is sometimes used in place of tempeh.

2 tablespoons oil
2 shallots, thinly sliced
1 clove garlic, thinly sliced
1 red chili (2½ to 3 inches long), thinly sliced
4 ounces (115 grams) tempeh, cut into pieces ¾ by ½ by ⅜ inch
2 to 2¼ ounces (60 grams) overripe tempeh, cut into pieces ¾ by ½ by ⅜ inch
1 teaspoon laos, crushed
2 salam leaves
¾ to 1 teaspoon salt
1 teaspoon dark-brown or palm sugar
½ cup water
1 cup coconut milk, made from 1 cup each freshly grated coconut and water (page 42)

Heat a wok or skillet and coat with the oil. Add shallots and garlic, and sauté for 1 to 2 minutes. Add chili and sauté for 30 seconds. Add the next seven ingredients, bring to a boil, and simmer, covered, for 5 minutes. Add coconut milk, return to the boil, and simmer, uncovered, stirring occasionally, for 10 minutes. Remove laos and salam leaves. Serve over (or with) rice.

### Okara Tempeh Cubes & Melinjo Leaves in Coconut Milk
#### (Oblok-Oblok Onchom)

SERVES 2

1 tablespoon (coconut) oil
1 shallot, thinly sliced
½ clove of garlic, thinly sliced
½ red chili, thinly sliced
1 fiery dwarf green chili, thinly sliced and, if desired, seeded
1 salam leaf
1 teaspoon laos root, lightly bruised
2½ ounces (70 grams) okara tempeh, peanut-presscake tempeh, or onchom, cut into ¾-inch cubes
1 cup coconut milk, made from 1 cup each freshly grated coconut and water (page 42)
½ teaspoon salt
1½ teaspoons dark-brown or palm sugar
⅓ cup (20) melinjo leaves

Heat a wok or skillet and coat with the oil. Add shallot and garlic, and sauté for 1 minute. Add both types of chilies and sauté for 15 seconds. Add salam leaf and laos root, and sauté for 10 seconds. Add tempeh or onchom cubes and sauté for 10 seconds. Slowly mix in ¼ cup coconut milk. Then mix in the last three ingredients plus ¼ cup more coconut milk; cook for 15 seconds. Stir in the remaining ½ cup coconut milk, cover, and simmer, stirring every three minutes, for 9 to 10 minutes. Remove lid and simmer for 10 to 12 minutes more, or until ¼ to ½ cup of liquid remains. Remove salam and laos. Serve over (or with) rice.

### Overripe Tempeh & Coconut-Milk Sauce (Oblok-Oblok Tempeh Busuk or Bungkil)

SERVES 6 TO 8

A favorite in the Banyumas and Purwokerto areas of Central Java, this recipe can feature either overripe soy or peanut-presscake tempeh. Some cookbooks describe it as a stew or gravy, and it is a close relative of Besengek Tempeh. The well-known mixture of spices called *oblok* [2] is available in the West as a prepackaged mixture.

2 tablespoons oil
2 shallots, thinly sliced
1 red chili, thinly sliced
½ cup melingo leaves
7 ounces (200 grams) overripe soy or peanut-presscake tempeh (very soft), cut into 1-inch cubes
1 cup coconut milk, made from 1 cup each freshly grated coconut and water (page 42)
1 teaspoon dark-brown or palm sugar
½ teaspoon laos root, bruised until partially crushed
3 tablespoons peté beans, minced; or substitute fresh green soybeans or baby limas
1 teaspoon salt

Heat a wok or skillet and coat with the oil. Add shallots and sauté for 30 seconds. Add the next four ingredients, bring to a boil, and simmer for 1 minute. Add the last four ingredients and simmer, uncovered, stirring occasionally, for 3 to 4 minutes. Remove laos root. Serve over rice or vegetables.

VARIATION

* **Tempeh Telur:** Prepare as above but add diced hard-boiled eggs together with the tempeh and use different spices if desired.

### Tempeh & Tofu in Coconut-Milk Sauce (Tempeh Masak Petis)

SERVES 3

2¼ ounces (65 grams) tempeh, cut into slices 1¼ by 1¼ by ⅜ inch, dipped into tamarind water (page 44) and salt, then deep-fried to make Tempeh Goreng (page 76)
2 to 2¼ ounces (60 grams) tofu, cut into pieces 1¼ by 1¼ by ⅜ inch, marinated in tamarind water and salt, then deep-fried
3 eggs, hard-boiled, shelled, and deep-fried until golden brown
¼ cup oil
3 shallots, thinly sliced
2 cloves of garlic, thinly sliced
1 green chili (2½ inches long), thinly sliced
2 salam leaves
1 teaspoon laos root, bruised
2 teaspoons petis (shrimp paste)
1½ tablespoons dark-brown or palm sugar
1¾ cups coconut milk, made from 1¾ cups each freshly grated coconut and water (page 42)

Deep-fry the tempeh, tofu, and hard-boiled eggs, then set aside. Heat a wok or skillet and coat with the oil. Add the next three ingredients and sauté for 1 minute. Add the remaining ingredients together with the fried tempeh, tofu, and eggs. Bring to a boil and simmer, uncovered, stirring occasionally, for about 13 minutes, or until a thick sauce remains. Remove laos root. Serve over rice.

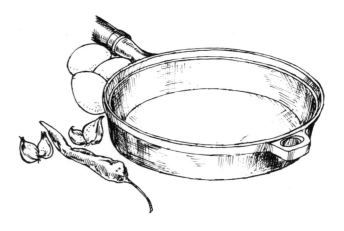

### Tempeh in Coconut-Milk Sauce (Terik Tempeh)

SERVES 1 OR 2

Indonesia's sixth-best-known way of serving tempeh, this recipe may call for either regular or okara tempeh. The combination of coconut milk and palm sugar and the lack of chilies add up to a rich and mellow flavor. Terik Tempeh is closely related to the popular Tempeh Bachem except that the former contains no coriander and is not deep-fried after having been simmered in seasoned coconut milk.

½ cup fresh coconut milk, made from ½ cup grated coconut and ½ cup water (page 42)
½ teaspoon coriander seeds
1 shallot
½ clove of garlic
½ candlenut
½ teaspoon laos root
¼ teaspoon tamarind paste
2¼ ounces (65 grams) tempeh (soy or okara), cut into 5 slices 1¼ by 2 by ⅜ inch
1 teaspoon dark-brown or palm sugar
¼ to ½ teaspoon salt
1 salam leaf

Prepare coconut milk and set aside. Combine the next six ingredients in a mortar and grind until smooth. Combine coconut milk and contents of mortar in a saucepan, mixing well, then add the remaining ingredients. Bring to a boil, and simmer, covered, stirring occasionally, for about 15 minutes, or until a thick sauce remains. Remove salam leaf and serve as is or over rice.

### Tempeh and Tofu Chunks in Coconut-Milk Sauce (Terik Tahu dan Tempeh)

SERVES 2

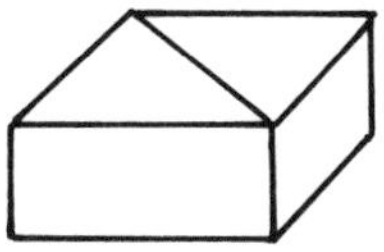

3 shallots
½ clove of garlic
1½ candlenuts
1 teaspoon coriander
1 to 1¼ teaspoons salt
1 teaspoon laos root, mashed
1 salam leaf
5 ounces (140 grams) tempeh, cut into pieces 1¼ by 1¼ by ¾ inch
5¼ ounces tofu, cut into 6 pieces 1½ inches square by ¾ inch thick, then each of these pieces cut diagonally into halves
⅔ cup water
1 cup coconut milk, made from 1 cup each freshly grated coconut and water (page 42)

Combine the first five ingredients in a mortar and grind until smooth. Combine the next five ingredients in a wok or skillet, add contents of mortar, mixing well, and bring to a boil. Simmer, uncovered, for 5 minutes, then stir in

coconut milk, return to the boil, and simmer for 4 to 5 minutes more. Remove laos root and salam leaf. Serve over rice.

### Overripe Tempeh Gravy (Orem-Orem)

SERVES 3 OR 4

3 shallots
½ clove of garlic
1 to 1¼ teaspoons salt
¾ teaspoon coriander seeds
½ teaspoon kenchur root
1 tablespoon oil
1 teaspoon laos root, bruised until partially mashed
¼ cup chopped green onions (1¼-inch lengths)
1 green chili (2½ inches long), thinly sliced
5¼ ounces (150 grams) overripe tempeh
1 cup coconut milk, made from 1 cup each freshly grated coconut and water (page 42)

Combine the first five ingredients in a mortar and grind until smooth. Heat a wok or skillet and coat with the oil. Add contents of mortar and sauté for 30 seconds. Add the remaining ingredients, bring to a boil, and simmer, uncovered, for 4 minutes. (If tempeh is firm, mash during simmering and mix into gravy.) Remove laos root. Serve over rice or vegetables.

### Tempeh in Rich Gravy (Brongkos Tempeh)

SERVES 2 OR 3

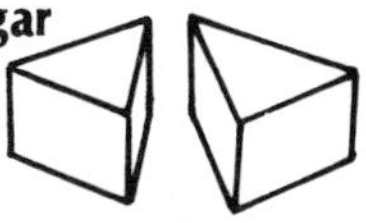

3 shallots
1 clove of garlic
½ red chili (1½ inches)
1½ teaspoons fresh or powdered turmeric
½ teaspoon kenchur root
1½ teaspoons dark-brown or palm sugar
1¼ teaspoons salt
1 kluwak
2 tablespoons oil
4½ ounces (130 grams) tempeh (fresh or overripe), cut into triangles ¾ inch on a side and ½ inch thick
1½ cups coconut milk, made from 1½ cups each freshly grated coconut and water (page 42)
6 green beans (6 inches long), cut diagonally into thin strips

Combine the first eight ingredients in a mortar and grind until smooth. Heat a wok or skillet and coat with the oil. Add contents of mortar and sauté for 30 seconds. Add tempeh and coconut milk, bring to a boil, and simmer for 1 minute. Add green beans and simmer, uncovered, stirring occasionally, for 5 minutes. Serve over rice or vegetables. For variety, substitute ½ to ¾ cup slivered carrots or cabbage for the green beans.

### Tempeh Cooked with Melinjo Leaves (Oseng-Oseng Tempeh)   SERVES 1 OR 2

In Indonesian, *oseng-oseng* means "a mixture of things." There are a great many variations on this basic motif, the more popular of which feature bean sprouts (taugé), mustard greens (daun sawi), fern fronds (daun pakis) or various other vegetables used with the tempeh. This recipe is closely related to Sambal Goreng Tempeh, except that the former contains no coconut milk.

1 tablespoon oil
2 shallots, thinly sliced
½ clove of garlic, thinly sliced
½ to 1 green chili, thinly sliced
2½ ounces (70 grams) tempeh, cut into slices ½ by 1¼ by ¼ inch
1 salam leaf
¼ teaspoon laos root, partially mashed
6 tablespoons water
1¾ ounces (50 grams) melinjo leaves
2 tablespoons sweet Indonesian soy sauce (kechap)
⅛ teaspoon salt

Heat a wok or skillet and coat with the oil. Add shallots and garlic, and sauté for 1 minute. Add chili and sauté for 1 minute more. Add the next five ingredients, mixing well, and simmer for 2 minutes. Add soy sauce and simmer for 3 minutes, then season with salt and simmer for 2 minutes more. Remove salam leaf and serve over rice.

VARIATIONS

* **Tempeh Cooked with Bean Sprouts** (Oseng-Oseng Taugé): Sauté in 2 tablespoons coconut oil: 10½ ounces washed mung-bean sprouts, 5 sliced shallots, 2 cloves minced garlic, and 5 sliced red chilies. Add 3 ounces sliced tempeh, 1 tablespoon brown sugar, ½ to 1 teaspoon salt, and ¾ cup water. Simmer uncovered until about ¼ cup water remains.

* **Tempeh Cooked with Mustard Greens** (Oseng-Oseng Sawi): Fry in 2 tablespoons oil: 3 sliced shallots and 2 cloves minced garlic. When lightly browned, add 5 minced chilies, 2 small pieces of laos, 2 salam leaves, 1 tablespoon brown sugar, and 1 to 3 teaspoons salt. After several minutes, add 3 to 4 ounces sliced tempeh, 1 pound washed and sliced mustard greens (*Brassica rugosa*), and 2 cups water. Simmer for 15 minutes.

* **Tempeh Cooked with Fern Fronds** (Oseng-Oseng Daun Pakis): Sun-dry 10 fern fronds until withered, then cut into 1½-inch lengths. Sauté in 6 tablespoons coconut oil: 5 sliced shallots, 2 sliced cloves of garlic, 10 minced red chilies, 1 laos root, and 5 salam leaves. Add ½ tablespoon brown sugar, ¾ teaspoon salt, 1 ounce mashed overripe tempeh, ½ pound washed bean sprouts, the fern fronds, 1½ cups water, and ½ cup peeled and sliced peté beans. Bring to a boil and simmer for 15 minutes.

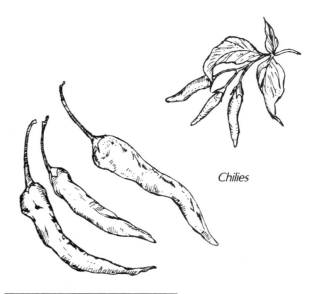

Chilies

### Tempeh Slices in Hot Sauce (Tempeh Bumbu Bali)   SERVES 3

In Indonesian, the word *bumbu* is a general term referring to seasonings of all types (the word *rempah-rempah* is used to refer specifically to spices). Bumbu Bali is a combination of spices and other seasonings that was developed in East Java in imitation of the more elaborate and strongly flavored *bumbu megenep,* the most famous seasoning mixture in Bali. The latter is used in so many Balinese dishes that some foreign cooks have come to suspect that it may be endowed with an almost ritualistic importance. The original Balinese product also contains sesame, kenchur, turmeric, pepper, gingerroot, and a chili-like seasoning called *tabia bun*. Both combinations are now sold in ready-mixed, packaged form.

4¼ to 4½ ounces (125 grams) tempeh, cut into slices 1¼ by 1¼ by ⅜ inch, dipped into tamarind water (page 44) and salt, then deep fried to make Tempeh Goreng (page 76)
3 eggs, hard-boiled, shelled, and deep fried until golden brown
¼ cup oil
3 or 4 red chilies (each 2¾ inches long), seeded and ground
3 cloves of garlic, thinly sliced
4 or 5 shallots, thinly sliced
2 teaspoons laos root
1 teaspoon white sugar
2 tablespoons sweet Indonesian soy sauce (kechap)
½ teaspoon salt
2 salam leaves
3¼ inches lemongrass stalk and bulb

Have the tempeh and eggs ready. Heat a wok or skillet and coat with the oil. Add chilies and sauté for 2 minutes. Add all the remaining ingredients plus tempeh and eggs, and sauté for 3 to 4 minutes. Remove salam leaves and lemongrass. Serve over rice.

## Tempeh Slices in Tomato Hot Sauce (Tempeh Lado)

SERVES 2

10 slices of tempeh, 3½ by 1½ by ⅛ inch
5 red chilies, seeded
1 small tomato, sliced
¼ cup (coconut) oil
1½ tablespoons tomato ketchup
2 teaspoons dark-brown or palm sugar
⅛ teaspoon salt
Dash of pepper
3 to 4 tablespoons water

Combine 4 tempeh slices with the chilies and tomato in a mortar and grind together until smooth. Heat a wok and coat with the oil. Add contents of the mortar plus the next four ingredients, mixing well, then add the remaining tempeh slices and sauté for 2 minutes. Add the water, mixing lightly, and simmer for 2 minutes. Serve tempeh slices hot over rice, with the sauce scooped over the top.

## Tempeh in Coconut-Milk Sauce (Tempeh Podomoro)

SERVES 2

A specialty of Yogyakarta and Central Java, Podomoro also calls for the use of ground beef instead of tempeh.

¼ teaspoon (18) coriander seeds
1 clove of garlic
½ red chili (1½ inches), thinly sliced and seeded
½ teaspoon salt
1 round of laos root, ⅛ inch thick and ½ inch in diameter
1 cup coconut milk, made from 1 cup each freshly grated coconut and water (page 42)
1 tablespoon oil
1 shallot, thinly sliced
2½ ounces (70 grams) tempeh, cut into irregular 1½-inch-diameter thin slices
1 green chili (4 inches), thinly sliced and seeded
1 salam leaf
1 to 2 teaspoons dark-brown or palm sugar

Combine the first four ingredients in a mortar and grind until smooth. Add laos root and bruise firmly, then transfer contents of mortar to a bowl and set aside. Pour ¼ cup coconut milk into mortar and set aside. Heat a wok and coat with the oil. Add shallot and sauté for 30 seconds. Add ground spices from bowl and sauté for 30 seconds. Swish coconut milk around inside of mortar to wash it, then add to wok together with the tempeh, green chili, salam leaf, and sugar; sauté for 30 seconds. Stir in remaining ¾ cup coconut milk and simmer uncovered for 10 minutes. Stir well, then simmer for 3 minutes more, or until about ½ cup of liquid remains. Remove laos root and salam leaf. Serve over rice.

## Tempeh & Tofu Cubes in Tangy Turmeric Sauce (Achar Tahu Tempeh)

SERVES 2

The term *achar* refers to an Indonesian-style relish, vinegared salad, or pickled preparation. Popular ingredients include mixed vegetables, mung-bean sprouts, or gherkins (immature cucumbers or small prickly fruits, *Cucumis anguria*, used for pickling).

4 ounces (110 grams) tempeh, cut into pieces 1¼ by 1¾ by ½ inch
2½ ounces (70 grams) tofu, cut into pieces 1¼ by 1¼ by ½ inch
½ teaspoon salt, dissolved in ⅔ cup water
2 shallots
2 candlenuts
½ clove of garlic
½ teaspoon dark-brown or palm sugar
½ teaspoon fresh or powdered turmeric
½ teaspoon salt
¼ cup (coconut) oil
1 red chili (2¾ inches long), cut diagonally into thin slices and seeded
1½ teaspoons vinegar
½ cup water

Combine the tempeh, tofu, and salt water in a bowl and allow to stand for 10 minutes, then drain well. Meanwhile, combine the next six ingredients in a mortar and grind until smooth. Heat a wok or skillet and coat with the oil. Add the tempeh and fry for 3 to 4 minutes, turning once, until golden brown on each side, then remove from wok. Fry tofu in the same way and remove. Now add contents of the mortar plus the chili to the wok and fry for 30 seconds. Add the last two ingredients and sauté for 15 seconds. Mix in fried tempeh and tofu and sauté for 1 minute more. Serve over rice.

*Shallots*

## Tempeh & Vegetable Sauté
### (Tumis Tempeh)

SERVES 2

The word *tumis* means "to stir-fry or sauté," a cooking technique widely used in Indonesia, where the wok is the most popular cooking pan. There are various *tumis* recipes; all feature vegetables and a number contain tempeh.

¼ cup oil
3 to 4 shallots, thinly sliced
1 clove of garlic, thinly sliced
1 red chili (2½ inches long), thinly sliced
1 green chili (2½ inches long), thinly sliced
1 teaspoon laos root
½ teaspoon tamarind paste
¾ to 1 teaspoon salt
4 ounces (115 grams) tempeh (regular or overripe), cut into pieces ⅜ by 1¼ by ¼ inch
1 ounce (30 grams) melinjo leaves, chopped (or substitute spinach or kangkung leaves)

Heat a wok or skillet and coat with the oil. Add shallots and garlic, and sauté for 1 to 2 minutes. Add the next five ingredients and sauté for 1 minute. Add tempeh and sauté for 3 minutes. Add melinjo leaves and sauté for 1 minute more. Cover and cook for 1 to 2 minutes. Serve hot as a side dish or over rice.

VARIATION

* **Udang Tempeh Lombok** (Spicy Tempeh & Coconut Stew): Simmer together: tempeh, coconut milk, crushed garlic, turmeric, candlenuts, shallots, salt, tiny shrimp (dried or fresh), sugar, chili powder and other spices. Simmer to stew consistency.

## Tempeh & Vegetable Sauté
### (Tumis Kangkung Tempeh)

SERVES 2

2 tablespoons oil
2 shallots, thinly sliced
1 clove of garlic, thinly sliced
1 or 2 green chilies, cut into thin diagonal slices
1 teaspoon gingerroot, slivered
1 bay leaf
1 teaspoon laos root
A 5-inch stalk and bulb of lemongrass (optional)
2 small tomatoes, cut into thin wedges
¾ to 1 teaspoon salt
4 ounces (115 grams) tempeh (soy or okara), cut into pieces ¾ by ½ by ⅜ inch
2¾ ounces (80 grams) kangkung (or substitute spinach), separated into stems and leaves
1 teaspoon dark-brown or palm sugar

Heat a wok or skillet and coat with the oil. Add shallots and garlic, and sauté for 2 minutes. Add the next five ingredients and sauté for 30 seconds. Add tomatoes and salt, and sauté for 1 minute. Add tempeh and sauté for 2 minutes more. Add kangkung stems and sauté for 1 minute. Cover and cook for 1 minute, then stir, re-cover, and cook for 1 minute more. Add kangkung leaves and sugar, and sauté for 30 seconds, then cover and simmer for 30 seconds. Uncover, stir, and cook for 2 minutes, then remove from heat. Remove bay leaf, laos root, and lemongrass. Serve over rice.

## Crumbly Tempeh Sauté
### (Tumis Tempeh Gembus or Tumis Onchom)

SERVES 1 OR 2

1 tablespoon (coconut) oil
¼ teaspoon laos root
1 shallot, thinly sliced
½ clove of garlic, thinly sliced
1 green chili, cut into thin diagonal slices
2¼ ounces (65 grams) okara tempeh or onchom, diced into ½-inch cubes
½ teaspoon dark-brown or palm sugar
½ teaspoon salt
½ cup hot water

Heat a wok or skillet and coat with the oil. Add the next four ingredients and sauté for 1 minute. Add the next three ingredients and sauté for 1 minute more. Stir in ¼ cup hot water and simmer for 1 minute, then stir in the remaining water, bring to a boil, and simmer for 3 minutes, or until all the liquid has evaporated. Serve over rice.

*Salam leaf*

## Overripe Tempeh in Thin Sauce (Pindang Tempeh Busuk)

SERVES 2

4 ounces (115 grams) overripe tempeh, cut into pieces ¾ by ½ by ½ inch
2 shallots, thinly sliced
1 clove of garlic, thinly sliced
1 red pepper (3 to 3¼ inches long), cut diagonally into thin slices and seeded
½ teaspoon tamarind paste or ½ tomato
1 salam leaf
1 teaspoon laos root, thinly sliced
¾ to 1 teaspoon salt
2 cups soup stock or water

Combine all ingredients in a saucepan and bring to a boil. Simmer, covered, for 10 minutes. Remove salam leaf. Serve hot.

## Steamed Leaf-Wrapped Tempeh Preparations

Each of the following is most colorful and flavorful when wrapped in fresh green leaves, especially banana leaves, which impart a prized flavor. Yet fairly good results can also be obtained using aluminum foil.

### Tempeh & Coconut Steamed in Banana Leaves (Botok Tempeh)

SERVES 2

1 shallot
½ clove of garlic
2 teaspoons dark-brown or palm sugar
½ to 1 teaspoon salt
2½ ounces (70 grams) tempeh
⅓ to ½ cup grated coconut
½ ounce (15 grams) melingo leaves
1 green chili (2½ inches long), thinly sliced
¼ teaspoon laos root, bruised until mashed
1 peté bean (optional)
Banana-leaf wrappers (or substitute aluminum foil): one 9-inch and one 4-inch square
1 salam leaf

Combine the first four ingredients in a mortar and grind until smooth. Add tempeh and mash lightly until cake is broken into individual beans. Add the next four (or five) ingredients, mixing well. Place the 4-inch-square banana leaf at the center of the 9-inch square, then place the salam leaf at the center of the 4-inch square (Fig. 5.7) and top with the mixed ingredients. Roll up both banana leaves from one side and fasten both ends with foodpicks. Place in a preheated steamer and steam for 30 minutes. Unwrap and serve hot on the leaves.

Fig. 5.7: Botok tempeh

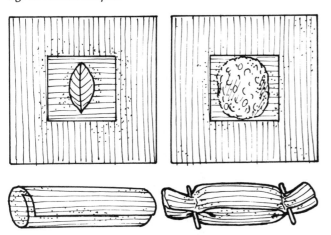

Fig. 5.8: Botok tempeh

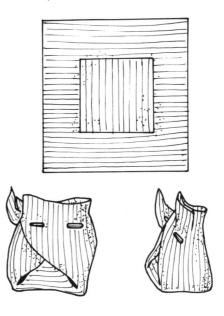

### Sweet Tempeh & Coconut Steamed in Leaf Wrappers (Botok Tempeh 2)

MAKES 13 PACKETS

1¼ cups freshly grated coconut
⅔ cup green leucaena (wild tamarind) beans, or substitute fresh green soybeans or baby limas
1 teaspoon laos root, thoroughly mashed (or grated)
3 shallots, thinly sliced
2 red chilies, thinly sliced (seeded if desired)
3 green chilies, thinly sliced
1½ to 2 teaspoons salt
2 tablespoons dark-brown or palm sugar
½ cup water
6½ ounces (185 grams) tempeh, diced into ⅜-inch cubes (about 4 cups)
Banana-leaf wrappers (or substitute aluminum foil): thirteen 9-inch and thirteen 4-inch squares

Combine the first nine ingredients in a bowl, mixing well, then add tempeh and mix gently but thoroughly. Place one 4-inch banana-leaf square at the center of each 9-inch square. Place ⅓ to ½ cup of the mixed ingredients at the center of each 4-inch square. Fold the leaves as shown above to make a compact packet and seal each with a foodpick. Place in a preheated steamer and steam for 20 minutes. Unwrap and serve over rice.

## Tempeh & Coconut-Milk Custard *(Gadon Tempeh)*

MAKES 4

This recipe is a favorite among Indonesian children. There are no chilies, and the flavor is mellow; the egg adds richness. Although it can be made in a custard cup or bowl, the use of banana leaves is said to produce the finest flavor.

2 shallots
½ clove of garlic
½ teaspoon dark-brown or palm sugar
½ teaspoon salt
¼ teaspoon coriander seeds
⅛ teaspoon white peppercorns
4 ounces (115 grams) tempeh
1 egg, lightly beaten
½ cup coconut milk, made from ½ cup each freshly grated coconut and water (page 42)
Banana-leaf wrappers: four 9-inch squares and four 4-by-9-inch rectangles. Or substitute equal-sized pieces of aluminum foil, or four custard cups or bowls
4 salam leaves

Combine the first six ingredients in a mortar and grind until smooth. Add the tempeh and mash until well mixed. Combine the egg and coconut milk, beating lightly, then add the tempeh-spice mixture, mixing well. Arrange the four banana-leaf squares on a table and place one banana leaf rectangle at the center of each square (Fig. 5.9). Then place one salam leaf at the center of each rectangle. (Or place one salam leaf at the bottom of each of four custard cups.) Put one-quarter of the mixture atop the salam leaf at the center of each wrapper (or in each cup), then fold each wrapper into a packet as shown and seal with a foodpick (or cover cups). Place in a preheated steamer and steam for 30 minutes. If desired, trim top of each packet to give a more elegant appearance before serving.

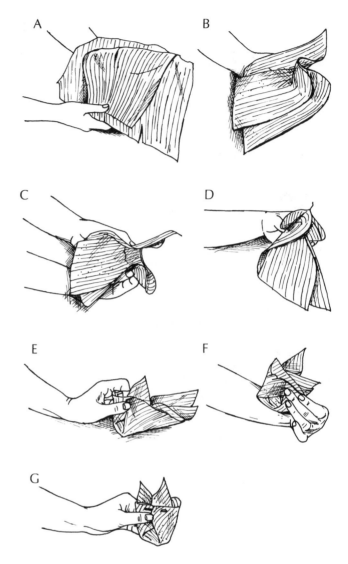

Fig. 5.10: Folding leaf wrappers for gadon tempeh

Fig. 5.9: Gadon tempeh

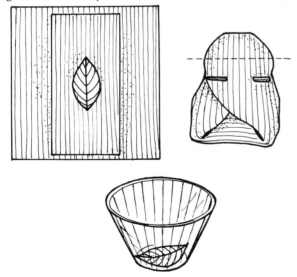

## Spicy Tempeh Steamed in Banana Leaves
*(Pepes Tempeh)*

MAKES 12

The West Javanese call this recipe Papais, while the people in Central and East Java and Bali call it Pepesan. Pepes is a basic type of preparation and the main ingredient may be tofu, onchom, peté seeds, grated coconut, dry or fresh fish, ground beef, etc., instead of tempeh. Learning to make the wrappers is not difficult and opening them at mealtime is like opening presents. The moist-and-steamy, spicy mixture picks up the fragrance of the banana leaves to develop a flavor all its own.

**11 shallots**
**5 red chilies**
**13 candlenuts**
**15 ounces (425 grams) tempeh, cut into pieces 2½ by 1½ by ½ inch**
**¼ cup coconut milk (page 42)**
**¾ to 1 cup kemangi leaves**
**2 teaspoons salt**
**2 teaspoons dark-brown or palm sugar**
**Banana-leaf wrappers (or substitute aluminum foil): twelve 9-inch and twelve 4-inch squares**

Combine the first three ingredients in a mortar and grind until smooth. Add tempeh and mash together well. Add the remaining ingredients, mixing well. Arrange the large square wrappers on a table, place one small wrapper at the center of each, then place about ½ cup of the mixture in a 6-inch-long oval atop each of the small wrappers. Roll up the large and small wrappers from one side to form a 1-inch-diameter cylinder. Fold both ends and fasten with foodpicks (see illustrations below). Place packets in a preheated steamer and steam for 10 minutes, then grill on a charcoal brazier or in an oven broiler for 8 to 10 minutes, or until surface of wrappers is lightly burned. Serve in the leaves.

*Fig. 5.11: Rolling leaf wrappers for pepes tempeh*

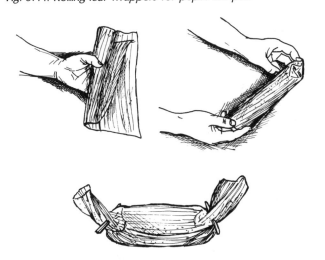

*Fig. 5.12: Pepes tempeh on broiler and packets ready to serve*

VARIATIONS

**\* Ungrilled Pepes:** Wrap in banana-leaf packets as shown in Fig. 5.11, then place in a preheated steamer and steam for 30 minutes.

**\* Balinese Pepesan:** Use as spices: kenchur root, laos root, red chilies, turmeric, salt, plus a little belimbing (star fruit). Fish is also used in some areas. Mix with tempeh, roll in banana leaves, and steam or broil for 15 minutes.

*Fig. 5.13: Folding leaf wrappers for Balinese pepesan*

**\* Leaf-Wrapped Steamed Overripe Tempeh & Coconut** *(Gembrot Sembukan)*. Grind together in a mortar: 3 cloves minced garlic, 3 large chilies, 3 fiery dwarf chilies, 1 laos root, ½ teaspoon coriander, 1 teaspoon salt, and 3½ ounces (100 grams) overripe tempeh. Mix in ¾ cup fresh grated coconut and 1 thinly sliced sembukan leaf *(Saprosma arboreum)*. Wrap ½-cup quantities in banana leaves (or aluminum foil) and steam for 30 minutes, then broil briefly until dry.

### Crumbly Okara Tempeh Steamed in Banana Leaves (Pepes Tempeh Gembus or Pepes Onchom)

SERVES 1 OR 2

A favorite at West Javanese restaurants, this recipe may feature either okara tempeh or peanut onchom as the main ingredient.

1 shallot
1 clove of garlic
1 teaspoon dark-brown or palm sugar
½ teaspoon salt
2½ ounces (70 grams) okara tempeh or onchom
⅓ cup (about 20) kemangi leaves
1 ten-inch square of banana leaf
1 salam leaf

Combine the first four ingredients in a mortar and grind until smooth. Add tempeh, mashing well with other ingredients, then mix in kemangi leaves. Place banana-leaf wrapper on table and set salam leaf at its center. Spoon contents of mortar in a 6-inch-long oval at center of salam leaf, then roll up leaf from one side to form a 1-inch-diameter cylinder. Fold both ends and fasten with food-picks (see figure 5.11). Place packet in a dry wok, cover with a lid, and cook over low heat for 14 minutes. Turn and cook for 7 minutes more. Unwrap and serve on the leaf.

## Tempeh in Sayurs, Soups, Stews, and Curries

The Indonesian term *sayur* may be translated in either of two ways, depending on the context: as "vegetable" or "vegetables" or as "soup" or "stew." The latter usage refers to a wide variety of Indonesian vegetable dishes with consistencies ranging from that of a sauce to a thick soup. They are all basically different from Western soups, however, insofar as they are usually served over or with rice and usually (but not always) employ coconut milk instead of soup stock as the liquid in which the vegetables, tempeh, and other main ingredients are simmered. Soups, which do not figure as prominently in the Indonesian menu as they do in the West, are generally prepared only by foreigners, especially resident Europeans and Chinese.

In Indonesia, there are actually two types of souplike preparations: sayurs and sotos. Sotos are more substantial, somewhat like stews, and generally consist of chicken or beef cooked in broth; tempeh is not generally used in sotos but is widely used in sayurs, together with vegetables such as cabbage, green beans, cauliflower, peas, leeks, eggplants, young jackfruit, carrots, green peppers, onions, celery, and a host of others.

Often grouped together with sayurs and sotos because of their somewhat liquid consistency are the curry-like preparations, opor, karé, and gulai. Indonesian curries are relatively mild.

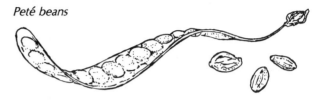

*Peté beans*

### Overripe Tempeh & Vegetables in Coconut-Milk Soup (Sayur Lodeh)

SERVES 2

Also called simply Lodeh or Sayur Tempeh Santan, this spicy concoction is Indonesia's fourth-most-popular way of serving tempeh, and the *most* popular way of serving overripe tempeh, especially in Central and East Java. The word *sayur* means "vegetable" and *lodeh* is a general term referring to numerous souplike preparations containing a relatively high proportion of liquid. Okara tempeh is also widely used in Sayur Lodeh, and taucho (Indonesian soybean miso) is a popular seasoning substitute for salt, especially in West Java.

2 tablespoons oil
2 shallots, thinly sliced
1 teaspoon laos root
1 clove of garlic, thinly sliced
1 red chili (2½ to 3 inches long), thinly sliced and seeded
5 peté beans, chopped, or substitute fresh green soybeans or baby limas
3½ ounces (100 grams) tempeh (overripe, regular, or okara), cut into ¾-inch cubes
1½ cups coconut milk, made with 1½ cups each freshly grated coconut and water (page 42)
1 salam leaf
1 teaspoon dark-brown or palm sugar
½ teaspoon salt or 2 teaspoons taucho
½ cup melinjo leaves
½ cup string beans (1¼-inch lengths)

Heat a wok or skillet and coat with the oil. Add the next four ingredients and sauté for 1 minute. Add the peté beans and tempeh, mixing well, and sauté for 1 minute. Add the next four ingredients, bring to a boil, and simmer for 4 minutes. Add melinjo leaves and string beans, and simmer for 5 minutes more. Remove salam leaf. Serve hot over rice.

VARIATIONS

**\*Jangan Lodeh Lembayung:** Cook regular or okara tempeh together with French bean leaves (daun kachang pajang, *Phaseolus vulgaris*). Popular in Yogyakarta, and Central and East Java.
**\*Tempeh Bumbu Lodeh:** Prepare as Sayur Lodeh but simmer until it attains the consistency of a sauce.

## Tempeh & Tofu Stewed in Coconut-Milk Sauce
### (Opor Tempeh)

SERVES 3 TO 4

Opor, which consists of a main ingredient such as tempeh and/or tofu cooked in spiced coconut milk, is usually classified as a type of curry. The delicious sauce generally contains ground candlenuts, which add a hefty flavor and texture. Also called Tempeh Bumbu Opor, this preparation is a close relative of Terik Tempeh, except that the latter contains less liquid. In Indonesia, chicken opors and pork opors are also popular.

4¼ to 4½ ounces (125 grams) tempeh (fresh or over-ripe), cut into slices 1¼ by 1¼ by ⅜ inch, dipped into tamarind water (page 44) and salt, and deep-fried to make Tempeh Goreng (page 76)
6 pieces tofu, each 1¼ by 1¼ by ¾ inch, deep-fried
½ teaspoon coriander seeds
5 shallots
3 candlenuts
1½ cloves of garlic
½ teaspoon white pepper
3 to 4 tablespoons oil
½ to ¾ teaspoon salt
1 teaspoon tamarind paste
1 tablespoon dark-brown or palm sugar
1½ cups coconut milk, made from 1½ cups each fresh-ly grated coconut and water (page 42)
2 salam leaves
1 teaspoon laos root, bruised
3½ inches of lemongrass stalk and bulb, bruised

Have the tempeh and tofu ready. Combine the next five ingredients in a mortar and grind until smooth. Heat the oil in a wok or skillet, add contents of mortar, and sauté for 1 minute, then transfer to a small pot or saucepan. Add all the remaining ingredients together with the deep-fried tempeh and tofu. Bring to a boil and simmer, un-covered, for 5 minutes. Remove salam leaves, laos root, and lemongrass. Serve over rice. Also delicious refrigerat-ed overnight, then served cold or reheated the next day. For variety, add sliced hard-boiled (deep-fried or regular) eggs together with the tempeh. Popular served with Lon-tong (page 47).

## Tempeh Curried in Coconut Milk *(Karé Tempeh)*

SERVES 1 TO 2

2 shallots
¼ teaspoon coriander seeds
1 teaspoon fresh or powdered turmeric
½ clove of garlic
1 candlenut
½ teaspoon salt
¼ teaspoon tamarind paste
½ teaspoon dark-brown or palm sugar
1 teaspoon laos root
1 tablespoon oil

2 very small potatoes, peeled and diced
4 ounces (115 grams) tempeh, cut into pieces ¾ by 1½ by ½ inch
½ cup chopped cabbage
1½ cups coconut milk, made from 1½ cups each fresh-ly grated coconut and water (page 42)

Combine the first nine ingredients in a mortar and grind until smooth. Heat a wok or skillet and coat with the oil. Add contents of mortar and sauté for 1 minute. Add po-tatoes and tempeh, and sauté for 1 minute more. Mix in cabbage and coconut milk, bring to a boil, and simmer, covered, stirring occasionally, for 20 minutes, or until po-tatoes are tender. Serve over rice.

## Mild Curried Tempeh
### (Nyomok Tempeh)

SERVES 2 OR 3

A close relative of both Gulai and Curry, this recipe also is enriched with the flavor of coconut milk.

4 shallots
1 small kenchur root (1½ inches long)
½ teaspoon coriander seeds
½ teaspoon salt
1 clove of garlic
2 tablespoons oil
1 green chili (2¾ inches long), uncut
2 salam leaves
½ teaspoon laos, bruised until partially mashed
4¼ ounces (120 grams) tempeh, cut into ½-inch cubes
1 cup coconut milk, made from 1 cup each freshly grat-ed coconut and water (page 42)

Combine the first five ingredients in a mortar and grind until smooth. Heat a wok or skillet and coat with the oil. Add contents of mortar and sauté for 30 seconds. Add the next four ingredients and sauté for 1 minute. Add coconut milk, bring to a boil, and simmer, uncovered for 4 minutes. Remove salam leaves and laos. Serve over rice.

## Spicy Curried Tempeh
### (Gulai Tempeh)

SERVES 2

In Indonesia, Gulai (also called *Gulé*) and Curry are first cousins; Curry is generally considered a party dish for special occasions, while Gulai is its everyday equivalent. All real Curries and Gulais have at least one thing in common: they are cooked in coconut milk or coconut cream, which blends and civilizes the most fiery, most pungent collections of spices and chilies into velvety rich, cosmopolitan flavors that somehow add up to much more than the sum of their parts. Gulai, however, is often made with the thinner squeezings of the coconut milk and with fewer spices than its noble relative. Found in various forms throughout Indonesia, Gulai traces its origins to Arabia and India. The most famous Gulai comes from Padang, the culinary capital of West Sumatra, famed and feared for the blazing severity of its chili seasonings. West Javanese recipes also use lots of chilies, often accompanied by tomatoes. Central Javanese recipes like this one are milder, with a touch of the richness of palm sugar and no chilies.

1¾ ounces (50 grams) of any one of the following, stems removed and chopped: cassava leaves, melinjo leaves, cabbage, kale, or spinach
1 clove of garlic
½ teaspoon kenchur root
1½ teaspoons dark-brown or palm sugar
½ teaspoon fresh or powdered turmeric
⅛ teaspoon white peppercorns
1¼ to 1½ teaspoons salt
2 cups coconut milk, made from 2 cups each freshly grated coconut and water (page 42)
2 cloves
4¼ ounces (120 grams) tempeh, diced into ¾-inch cubes
3 to 4 tablespoons Deep-Fried Onion Flakes (page 44)

Bring 2 cups water to a boil in a saucepan, drop in leaves, and return to the boil. Simmer, uncovered, for 15 to 30 minutes, then drain well, chop, and set aside.

Combine the next six ingredients in a mortar and grind until smooth, then combine contents of mortar in a saucepan or pot with the coconut milk, cloves, tempeh, and cassava leaves. Bring to a boil and simmer, covered, for 12 minutes. Remove cloves. Serve over rice, topped with a sprinkling of the onion flakes.

Prepared a day or two in advance and refrigerated, Gulais and Curries generally develop depth and richness of flavor.

## Tempeh Cubes in Thin Dark
### Broth (Semur Tempeh)

SERVES 2

Indonesian Semurs are closely related to Western-style stews; there are many different varieties, and those which do not contain tempeh as the basic ingredient generally feature some type of tofu, meat, or poultry.

2 tablespoons oil
3 shallots, thinly sliced
½ clove of garlic, thinly sliced
⅔ cup water
1½ tablespoons sweet Indonesian soy sauce (kechap)
⅛ to ¼ teaspoon white pepper
Dash of nutmeg
¾ to 1 teaspoon salt
5¼ ounces (150 grams) tempeh, cut into pieces 1 by 1 by ½ inch

Heat a wok or skillet and coat with the oil. Add shallots and garlic, and sauté for 1 minute. Add the next five ingredients, mixing well, then add tempeh, bring to a boil, and simmer for 3 minutes. Serve over rice, like a sauce.

In Indonesia, tofu is often substituted for part of the tempeh or used in place of it.

## Tangy Tempeh, Vegetable & Peanut
### Dish (Sayur Asam Tempeh)

SERVES 2

This Sayur is somewhat unusual in that it contains no coconut milk but is known for the richness of its candlenut flavor and the spiciness of its chilies. Also called Sayur Tempeh.

1 shallot
1 red chili (2½ inches long), seeded
2 candlenuts
¾ teaspoon salt
3 cups soup stock or water
3 tablespoons peanuts
¼ cup green beans (1½-inch lengths)
⅓ cup chopped chayote
1 salam leaf
1 teaspoon laos root
⅔ cup melinjo leaves
1 green chili (2 inches long), sliced and seeded
2½ ounces (70 grams) tempeh, cut into ½-inch cubes
1 small tomato, sliced into thin wedges, or ½ teaspoon tamarind paste
1 tablespoon dark-brown or palm sugar

Combine the first four ingredients in a mortar and grind until smooth. Bring soup stock to a boil in a saucepan. Add peanuts and cook for 4 minutes, or until soft. Add green beans and cook for 3 minutes, then add chayote and cook for 4 minutes more. Add the next five ingredients plus the contents of the mortar, bring to a boil, and simmer for 2 minutes. Add the last two ingredients and simmer for 1 minute. Remove salam leaf and laos root. Serve as a soup.

## Jackfruit & Tempeh Stewed in Spiced Coconut Milk
### (Gudeg Tempeh)

SERVES 7 TO 10

This is a dish the Central Javanese consider their very own. Its main ingredient, the jackfruit, when eaten raw has a firm, yellow, melonlike flesh that tastes something like a mild cheese. In Gudeg, cooked with coconut milk, overripe tempeh, palm sugar, coriander, and various other spices, the jackfruit slices develop a rather delicate meaty flavor, while the tempeh lends its unique richness of texture and aroma. Purists always cook Gudeg in an earthenware pot, claiming that enamel and metal fail to bring out the proper "warmth." Hard-boiled eggs are used in some versions, a single bitter peté bean in others. Ordinarily Gudeg is a side dish, accompanying rice but never eaten with it in the same mouthful. On special occasions, however, Indonesians serve Gudeg Rice (Nasi Gudeg), also called "Complete Gudeg," in which the Gudeg is combined with rice and served topped with any or all of the following delicacies: Tempeh Cutlets (Tempe Bachem), Tofu Cutlets (Tahu Bachem), Fried Whole Egg in Coconut Milk (Sambal Goreng Telur), and Chicken Stewed in Coconut Milk (Opor Ayam). The combination makes a complete meal. Another Gudeg preparation, in which more coconut milk is used to give a more souplike consistency, is called Sayur Gudeg.

2¼ pounds young or unripe jackfruit, peeled, washed, and diced
10 shallots, minced
3 cloves of garlic, minced
5 candlenuts
1 teaspoon coriander
½ to 1 teaspoon salt
2 teaspoons laos root (1 piece)
2 salam leaves
2¼ cups rich coconut milk (page 43)
4 ounces (115 grams) (overripe) tempeh, diced
4 ounces (115 grams) Tempeh Cutlets (page 76) (optional)

Combine jackfruit with water to cover in a pot, bring to a boil, and simmer for about 1 hour, or until soft. Discard cooking liquid and grind jackfruit in a mortar. Grind together the next five ingredients in a separate mortar, then combine them in a pot with the jackfruit, laos root, salam leaves, coconut milk, and overripe tempeh. Bring to a boil and simmer, covered, for 20 to 30 minutes, or until thick. Serve topped, if desired, with the Tempeh (or Tofu) Cutlets.

*Indonesian spices*

## Tempeh Simmered in Clear Soup
### (Sayur Bening Tempeh)

SERVES 2

4 ounces (115 grams) tempeh, cut into pieces ⅜ by ½ by ⅜ inch
2 shallots, thinly sliced
1 red chili (3 to 3¼ inches long), cut diagonally into thin slices (optional)
1 salam leaf
1 teaspoon laos, crushed
¾ teaspoon salt
2 cups soup stock or water

Combine all ingredients in a sauce pan and bring to a boil. Simmer, covered, for 10 minutes. Remove salam leaf. Serve hot.

## Tempeh & Kangkung in Coconut-Milk Soup (Sayur Kangkung Tempeh)

SERVES 4

2 tablespoons oil
1 medium onion, chopped
3 cloves of garlic, chopped
2 large green chilies, thinly sliced
6 ounces (170 grams) tempeh, cut into ½-inch cubes
4 candlenuts, crushed
1½ teaspoons terasi
1 teaspoon ground coriander
¼ teaspoon laos powder or crushed root
1 tablespoon dark-brown or palm sugar
2 salam leaves
½ to ¾ teaspoon salt
4 cups soup stock or water
1 pound kangkung, well washed; or substitute spinach
1 cup coconut milk, made from 1 cup each freshly grated coconut and water (page 42)

Heat a wok or large skillet and coat with the oil. Add consecutively the next four ingredients, sautéing for about 2 minutes after each addition, then mix in the candlenuts and the next six ingredients. Cook for 2 minutes more and remove from heat. Bring the stock to a boil in a pot. Add kangkung or spinach, return to the boil, and simmer for 4 minutes, or until about half done. Add the contents of the wok, return to the boil, and simmer for 2 minutes. Stir in the coconut milk and simmer for 3 minutes more. Serve over rice.

A close relative of the above is Sayur Bobor Kangkung Tempeh, prepared by adding a small piece of ground kenchur root to the ingredients above.

## Broiled Tempeh Preparations

A barbecue, charcoal brazier, oven broiler, or campfire may provide the center of fire and conviviality for these Indonesian favorites that are as ancient as they are tasty.

### Tempeh Shish Kebab with Sauce (Saté Tempeh)

MAKES 6

One of Indonesia's most famous and beloved recipes, saté (also spelled *satay* in the West) consists of delicious bite-size pieces of skewered and marinated meat or tempeh basted with a sauce, grilled over a bed of live coals in a brazier, and then dipped in the sauce just before eating. Although saté originated centuries ago in Java, it is now known as the king of snacks throughout almost all of Southeast Asia. Even in Japan, the famous chicken shish kebab called Yakitori, grilled over long braziers in portable roadside stands, is almost identical to Indonesia's Chicken Saté (Saté Ayam).

Saté comes in almost endless variety; it seems that virtually every town, if not every chef, cherishes a particular recipe for saté sauce and a particular way of grilling. In some districts saté making is a family business which has developed into a fine art. The details of the sauce recipes are jealously guarded and handed down from father to son. In many areas, the most widely used ingredient for skewering is chicken, but tempeh, shrimp or prawns, beef, goat, carabao (water buffalo), turtle, fish, and (in non-Moslem areas) pork are also used. Saté Padang, featuring skewered braised beef with spiced coconut-milk sauce, is considered a great delicacy. The two most widely used types of sauces are spicy hot sauces that contain ground chilies, garlic, shallots, laos, a squeeze of lime juice, and soy sauce, and milder though still somewhat spicy peanut sauces. Mixtures of the two types are also seen.

In Java, the itinerant saté vendor carries all of his ingredients and grilling paraphernalia attached to a handsome portable stand (Fig. 5.15). When the sun goes down (or occasionally during the day), he sets up shop along any busy thoroughfare; favorite spots include railway stations, bus terminals, marketplaces, street corners, and holiday resorts. After firing up his charcoal brazier, he sets out flickering candles or gas lanterns that invite passersby to sit down for an inexpensive treat. Fanning coals with a plaited palm-frond fan or deftly threading his morsels on slender bamboo skewers, he is a vital part of the charm and fragrance of Java. Itinerant saté vendors, ringing their bells and shuffling through the streets, seem to stay up long after everyone else has gone to bed.

Saté is commonly served with Lontong or Ketupat, two varieties of compressed boiled rice (see page 47), usually as a snack. For an Indonesian-style main course, it is usually served with steamed rice, but occasionally with Fried Rice (Nasi Goreng), Yellow Rice (Nasi Kuning) (see Rice Dishes), or Gado-Gado (page 46 or 82). In the West, it is ideal at barbecues. Bamboo skewers seem to work best, since the metal ones retain heat longer and the diner may burn his/her lips. Saté is now one of the most popular dishes at Indonesian restaurants in North America and Europe.

**1 shallot**
**½ clove of garlic**
**2 teaspoons lemon or lime juice**
**¼ teaspoon coriander seeds**
**½ teaspoon salt**
**½ teaspoon dark-brown or palm sugar**
**1 teaspoon laos root**
**1½ tablespoons sweet Indonesian soy sauce (kechap)**
**2 tablespoons water**
**4 ounces (115 grams) tempeh, cut into eighteen ¾-inch cubes, then the upper and lower surfaces scored lightly to aid sauce absorption**
**6 slender (bamboo) skewers, each 6 to 10 inches long**
**½ cup Peanut Sauce, Sweet Indonesian Soy Sauce (kechap), or Sambal Sauce (pages 49, 48, and 99)**

Prepare a charcoal fire in a small brazier. Combine the first seven ingredients in a mortar and grind until smooth, then stir in the sweet soy sauce and water. Dip tempeh cubes into the spice mixture, impale 3 cubes on each skewer, and dip again in spices. Place skewers over brazier and grill for a total of 7 to 10 minutes, dipping them into the spice mixture and turning twice during broiling. Serve sizzling hot, accompanied by the sauce for dipping; or serve without sauce if the dish contains sufficient salt seasoning.

For variety, deep-fry the tempeh instead of broiling.

*Fig. 5.14: Saté tempeh on broiler*

*Fig. 5.15: Saté street vendor in Java*

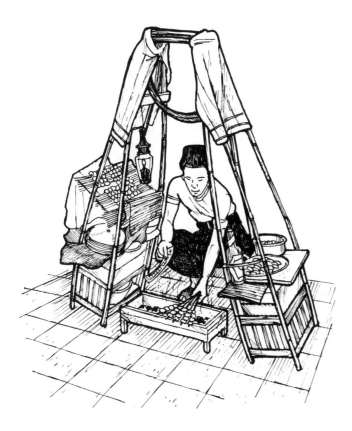

Place sprouts in a saucepan, douse with boiling water to cover, and allow to stand for 3 minutes, then drain and set aside. Combine the next five ingredients in a mortar and grind until smooth. Heat a wok and coat with the oil. Add scallions and sauté for 45 seconds. Add salam leaf and contents of mortar, and sauté for 1 minute more. Add taucho, water, sugar, and sweet soy sauce. Bring to a boil and simmer, uncovered, stirring constantly, for 4 minutes, or until the mixture has a soft, moist consistency like that of a Sloppy Joe or Spaghetti Sauce. Remove salam leaf. Arrange sprouts in a serving bowl and top with the sauce, seasoned with a sprinkling of the vinegar. Serve with Ketupat (page 47).

This sauce is also often served over a combination of mung-bean sprouts, noodles, and fried tofu. A popular close relative, which contains no taucho, is Laksa (Bean Sprouts with Onchom Sauce); sprouts are served topped with the sauce accompanied by Ketupat.

### Tempeh Shish Kebab with Sweet Sauce *(Saté Manis Tempeh)* MAKES 10

½ teaspoon coriander seeds
2 shallots
1 clove of garlic
½ teaspoon salt
3 to 4 tablespoons sweet Indonesian soy sauce (kechap)
7 ounces (200 grams) tempeh, cut into pieces ¾ by ¾ by ½ inch
10 slender (bamboo) skewers, each 6 to 10 inches long
Dipping Sauce
    2 tablespoons sweet Indonesian soy sauce
    1 tablespoon grated laos root
    1 teaspoon lime juice
    1 or 2 dwarf chilies
    ½ teaspoon grated gingerroot (optional)
    2 shallots, thinly sliced

Combine the first four ingredients in a mortar and grind until smooth, then stir in the sweet soy sauce. Add tempeh and marinate overnight. Prepare a charcoal fire in a small brazier. Mix ingredients for dipping sauce. Skewer cubes and grill for 7 to 8 minutes, or until nicely browned. Serve accompanied by the sauce for dipping.

*Fig. 5.16: Saté manis tempeh*

### Bean Sprouts with Tempeh & Taucho Sauce *(Taugé Goreng)* SERVES 1 OR 2

Probably the most popular way of serving okara tempeh and onchom in West Java, this dish is especially popular in the cities of Bogor and Jakarta. It is widely prepared by itinerant street vendors, and sold at prices that all can afford.

3½ cups (200 grams) mung-bean sprouts
½ red chili
2 shallots
1 clove of garlic
½ to 1 teaspoon salt
2½ ounces (70 grams) okara tempeh or onchom
1½ to 2 tablespoons (coconut) oil
5 inches of scallion or green onion stalk and bulb, sliced into thin rounds
1 salam leaf
2 tablespoons taucho (Indonesian soybean miso)
½ cup water
2 teaspoons dark-brown or palm sugar
1 tablespoon sweet Indonesian soy sauce (kechap)
1 teaspoon vinegar

### Grilled Tempeh with Kemangi in Coconut-Milk Sauce
*(Penchok, Pechak, or Pepechak Tempeh)*

SERVES 2 OR 3

This is Indonesia's most popular way of serving peanut-presscake tempeh. The grilled tempeh becomes the basis for a sauce with a consistency like that of a Sloppy Joe or Spaghetti Sauce.

½ **clove of garlic**
1 **red chili (2½ inches long), seeded**
½ **teaspoon kenchur root**
½ **teaspoon salt**
2¼ **ounces (65 grams) tempeh (soy or peanut press-cake)**
3 to 4 **tablespoons kemangi leaves**
½ **cup coconut milk, made with ½ cup each freshly grated coconut and** *hot* **water (page 42)**

Light a charcoal fire in a brazier. Combine the first four ingredients in a mortar and grind until smooth. Grill tempeh over coals until nicely speckled, then add to mortar and mash together with other ingredients. Combine contents of mortar in a bowl with kemangi leaves and hot coconut milk, mixing well. Serve immediately over rice.

### Broiled Okara Tempeh in Thin Sauce *(Gejos Onchom)*

SERVES 1 OR 2

2¼ **ounces (65 grams) okara tempeh or onchom, cut into 1 or 2 slices**
1 **shallot, unpeeled**
½ **clove of garlic, unpeeled**
½ **teaspoon dark-brown or palm sugar**
¼ **teaspoon salt**
½ **cup boiling water**

Place the first three ingredients on a screen over a charcoal brazier; grill the tempeh or onchom for 6 minutes, turning once, and the shallot and garlic for 3 minutes. Combine shallot and garlic in a mortar and grind until smooth. Add sugar and salt, mixing well, then transfer to a serving bowl. Break tempeh or onchom into very small pieces, mix into contents of bowl, then pour on the boiling water and allow to stand, covered, for 5 to 10 minutes before serving.

### Grilled Tempeh *(Tempeh Panggang)*

The verb *panggang* means ''to grill or roast.'' In many Indonesian cookbooks this recipe is listed simply as a preparatory technique. However, in Central Java it is served as a dish in its own right: grilled tempeh is diced and either mixed with or topped with a peanut sauce or sweet Indonesian soy sauce (kechap) mixed with minced chilies and shallots. The method for grilling is described in Chapter 4. The grilled tempeh may also be used as an ingredient in other recipes.

## Tempeh in Sambals

Sambals are fiery condiments that contain hot chilies as their main ingredient, ground together with spices, pickles, grated coconut, and/or other ingredients to attain the consistency of a soft paste or thick sauce, which is eaten especially with rice and curry in and around Indonesia and Malaysia. So essential are sambals to Indonesian cookery that it has been said, ''A meal without rice and a sambal is no meal at all.'' Served somewhat like chutney in amounts of several teaspoonfuls atop the rice and/or vegetables, sambals add both salt seasoning and a zesty wallop to arouse dull palates and sleepy dinner companions.

In Indonesia, sambal making is the test of a young girl's cooking skill. The Javanese, in fact, attach so much importance to sambals that one proverb declares even an ugly girl will find a husband if she can prepare a good sambal. Some sambals seem to have been made from a whole shelf of spices, and they come in many well-known varieties, each with the name of the key ingredient following the word *sambal:* Sambal Taucho is a spiced chili paste with Indonesian soybean miso; Sambal Kachang features peanuts and Sambal Kelapa, grated coconut.

Most Indonesian cookbooks distinguish between sambals and their near relatives sambal gorengs (''fried sambals''). The latter, which we have listed with fried tempeh recipes at the beginning of this chapter, are generally not nearly as hot as other sambals; stir-fried with a main ingredient (such as vegetables, eggs, tempeh, tofu, shrimp, prawns, or meat), they become somewhat more substantial and savory.

In some parts of Indonesia (especially Sumatra) the sambals are so unbelievably hot as to boggle the mind and short-circuit the senses. For Western travelers who bravely partake of local cuisines, the memory of their first tangle with a stark, blazing sambal may remain seared in the mind for a lifetime. In his colorful *Pacific and Southeast Asian Cooking,* Rafael Steinberg describes such an experience vividly: ''And so without thinking I plunged in. With the first bite my mouth caught fire and I could not go on. I felt that all the hottest spices of the marketplace had somehow been forced down my throat in one dose. I had met my match, and not even two glasses of soothing Grief Fruit a Go Go juice could quench the fire.''

Experienced travelers quickly learn that sambals should be spooned in *very* minute quantities onto the side of the plate and then used even more sparingly as a condiment. If you happen to overdose on chilies, do *not* try to put out the fire with ice water, carbonated soft drinks, or beer; they are only temporary palliatives at best, and once they are swallowed the heat will return with twice its original vengeance. Instead try a slice of raw cucumber or banana—you will be amazed and comforted by the results.

The word *sambal* originated in the ancient Malay language and has equivalents in other regional tongues; Sri Lanka's popular sambals, for example, are called *sambola*. Moreover, many Southeast Asian countries have their own versions of the Indonesian sambal known by totally different names. In Vietnam, for example, condiments such as ground chilies, garlic, vinegar, sugar, and a citrus-fruit pulp are added to the famous fish sauce *nuoc mam* to transform it into the exciting *nuoc cham*. In Thailand, the sambal's equivalent is called *nam prik*.

In the recipes that follow, we have reduced the amount of chilies to proportions manageable by a typical Western palate. Ready-made sambals are now sold in the West in small jars. All types should be served in a small open dish with a nonmetallic (wooden, porcelain, or plastic) spoon. Unrefrigerated, most types will keep for weeks.

## Grated Coconut & Overripe Tempeh Sambal (Sambal Jenggot) SERVES 3 TO 6

The word *jenggot* means "beard," but why it is used to refer to this savory sambal, a favorite in Central Java and Yogyakarta, no Indonesian cook has ever been able to tell us.

1 clove of garlic
2 candlenuts
1 red chili (2½ inches long)
½ teaspoon kenchur root
¾ teaspoon salt
2 to 2¼ ounces (60 grams) slightly overripe or overripe tempeh
2¼ ounces (about ⅔ cup) freshly grated coconut

Combine the first five ingredients in a mortar and grind until smooth. Add tempeh and mash together with other ingredients, then stir in grated coconut. Transfer to a cup or bowl (or wrap in banana leaves), cover, and place in a preheated steamer. Steam for 18 minutes. Serve over rice.

## Overripe Tempeh Sambal (Sambal Tepung) SERVES 2 TO 4

In this recipe we have substituted deep-fried tofu for the usual deep-fried cow hide (krechek or krupuk kulit goreng).

2¾ ounces (80 grams) overripe tempeh
3 shallots
¼ teaspoon kenchur root
1 red chili (2½ inches long), seeded
1 clove of garlic
½ to 1 cup (coconut) oil for deep-frying
6 ounces tofu, cut into ½-inch-thick slices, placed between absorbent toweling for 10 minutes to reduce moisture, then cut into ½-inch squares
1 salam leaf
1 lime leaf
½ teaspoon salt
½ teaspoon dark-brown or palm sugar
1½ cups coconut milk, made from 1½ cups each freshly grated coconut and water (page 42)

Combine the first five ingredients in a preheated steamer and steam for 10 minutes. Meanwhile, heat the oil to 375° F. (190° C.) in a wok, skillet, or deep-fryer. Slide in tofu and deep-fry for 3 to 4 minutes, or until crisp and golden brown, then drain and set aside. Transfer steamed ingredients to a mortar and grind until smooth, then combine them in a saucepan with the salam leaf and the next four ingredients. Bring to a boil and simmer for 2 minutes. Add deep-fried tofu and simmer, uncovered, for about 5 minutes more to form a sauce consistency. Remove salam and lime leaves. Serve over rice.

## Spiced Chili Paste with Grilled Tempeh (Sambal Tempeh) SERVES 2

4 ounces (112 grams) soy or okara tempeh
½ red chili
1 fiery dwarf chili
¼ teaspoon kenchur root
¼ clove of garlic
½ teaspoon salt
⅜ cup kemangi leaves, approximately
3 tablespoons boiling water

Grill tempeh on both sides over a charcoal brazier (or in an oven broiler) for about 10 minutes, or until slightly burned, then cut into ½-inch cubes and allow to cool briefly. Combine the next five ingredients in a mortar and grind until smooth. Add tempeh and mash well with other ingredients. Mix in ¼ cup kemangi leaves and mash lightly. Stir in the boiling water to create a pastelike consistency. Serve in a small open dish garnished with the remaining 5 to 6 kemangi leaves and accompanied by a nonmetallic spoon. Use as a topping for rice and/or cooked vegetables (especially cucumbers, cabbage, or spinach).

*Fry the tempeh instead of grilling; substitute terasi for part of the salt.
**Sambal Onchom:** Use okara tempeh or onchom in place of regular tempeh.

## Mashed Overripe Tempeh
### Sambal *(Sambal Tempeh Busuk)*          SERVES 2

Like most sambals, this one is served over plain cooked rice, accompanied by a vegetable. Popular cooked vegetables include leaves of Indonesian amaranth or spinach, cassava, melinjo, or kangkung; or sliced boiled carrots or green beans. The favorite raw vegetables are sliced cucumbers or tomatoes.

2 to 2¼ ounces (60 grams) overripe tempeh
½ teaspoon kenchur root
1 red chili (2½ inches long)
1 fiery dwarf green chili
2 shallots
1 teaspoon dark-brown or palm sugar
½ teaspoon salt

Skewer tempeh and grill over a bed of live coals, turning frequently, for about 7 minutes, or until nicely speckled. Combine the remaining ingredients in a mortar and grind together until smooth. Add tempeh and mash until well mixed.

VARIATIONS

**Pechel or Pechak Tempeh:** Add coconut milk (made with hot water) to the above ingredients and simmer uncovered until most of the liquid has evaporated. (Note: This recipe is fundamentally different from plain Pechel, page 82).
**Sambal Kachang Tempeh:** Add string beans and grated coconut to Pechel Tempeh. Popular in the cities of Solo and Klaten in Central Java.

## Overripe Tempeh Sambal
### *(Sambal Tumpang)*          SERVES 3 TO 4

2½ cups water
3½ ounces (100 grams) mung-bean sprouts
4¼ ounces (130 grams) Indonesian amaranth leaves; or substitute spinach
2 to 2¼ ounces (60 grams) slightly overripe or overripe tempeh, cut into pieces ½ by ½ by ⅜ inch
3½ ounces (100 grams) tempeh, cut into pieces ½ by ½ by ⅜ inch
3 shallots
1 clove of garlic
1 red chili (whole)
4 peté beans, chopped
1 teaspoon laos root
4 inches of lemongrass stalk and bulb
1 lime leaf
¾ cup grated coconut
½ teaspoon kenchur root
½ teaspoon dark-brown or palm sugar
½ teaspoon salt

Bring the water to a boil in a saucepan. Drop in bean sprouts, return to the boil, and simmer for 1 to 2 minutes, then remove from saucepan with a skimmer or tongs. To the same water add the amaranth, return to the boil, and simmer for 4 minutes, then remove as for the sprouts, keeping vegetables separate and reserving 2 cups of the water. Combine this water with both types of tempeh and the next seven ingredients in the saucepan, bring to a boil, and simmer for 7 minutes. Remove shallots, then pour off water through a strainer, leaving solid seasonings and tempeh in saucepan, reserving 1⅓ cups of the water. Mix this water with the grated coconut to make coconut milk (page 42). Combine the kenchur root and shallots in a mortar, and grind until smooth, then add to the ingredients in the saucepan, together with the coconut milk, sugar, and salt. Bring to a boil and simmer, stirring occasionally, for 5 minutes. Remove the laos root, lemongrass, lime leaf, garlic, and chili. Serve over cooked vegetables.

*Fig. 5.17: Tempeh sambal accompaniment for rice*

# 6

# Making Tempeh at Home or in a Community

Making tempeh is quite similar to making yogurt, and just as easy and enjoyable. In both cases you are working with sensitive living organisms that require a warm, clean place to grow. Actually you are farming or gardening, except that the plants are microscopic in size—as are the "weeds," alien microorganisms which grow if the proper conditions are not maintained.

### Homemade Soy Tempeh: The Basic Method

To make soy tempeh you will need soybeans, some starter, and an incubator or other warm place. Using these and other ingredients and utensils that are readily available, you can prepare your own tempeh for about 34 cents a pound. The process requires only about one hour of work plus some 24 hours of incubation. To ensure good results always: (1) see that the proper incubation temperature is maintained; (2) work with well-washed utensils in a clean, dust-free location; and (3) dry beans well before inoculation to inhibit growth of unwanted bacteria, especially in rainy or very humid weather when the relative humidity is above 85 percent.

## Utensils

To prepare tempeh on a small scale, you will need the following common kitchen utensils and an incubator:

A 2- to 2½-gallon cooking pot (the larger the better)
A large colander or strainer, either metal or woven bamboo; the larger the diameter the better
A large spoon (not wooden)

2 standard (12-by-17-inch) baking tins or serving trays, well washed and dried; one lined with 3 or 4 layers of absorbent paper toweling or a very clean dishcloth
A tempeh container: A good container allows the tempeh enough air to grow but not so much that it dries out or sporulates. Transparent plastic or glass containers let you watch interior mold growth. Choose one of the following, listed with our favorites first:

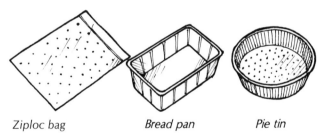

Ziploc bag          Bread pan          Pie tin

* *Polyethylene (plastic) bags:* Two Ziploc storage bags (7 by 8 inches), available at most supermarkets, are ideal. Bags without Ziploc may be sealed by folding the mouth closed and passing it over a candle flame. Polyethylene is very enzyme and heat resistant, insulates well, and has a desirable slight permeability to oxygen.
* *Bread pans, skillet, casserole, or pie plate:* 2 or 3 standard loaf pans, or a 9-inch-square casserole, or a 10-inch-diameter skillet or pie plate. Wash each thoroughly to get rid of all oil or fat, then cover with a tightly-stretched piece of aluminum foil or wax paper, or a loose-fitting lid.

103

* *Aluminum-foil baking pans or pie tins:* 2 or 3 of the rectangular or round reusable or disposable types work well, allowing the bottom to be perforated easily. Or substitute shallow Tupperware or plastic containers. Wash and cover as described above.

For other tempeh containers, see Variations (page 109).

Fig. 6.1: Flowchart for Homemade Soy Tempeh

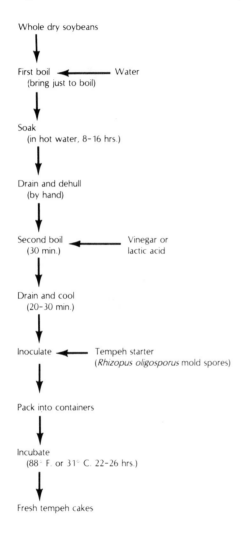

Whole dry soybeans

First boil ◄——— Water
(bring just to boil)

Soak
(in hot water, 8- 16 hrs.)

Drain and dehull
(by hand)

Second boil ◄——— Vinegar or
(30 min.)                lactic acid

Drain and cool
(20-30 min.)

Inoculate ◄——— Tempeh starter
(*Rhizopus oligosporus* mold spores)

Pack into containers

Incubate
(88° F. or 31° C. 22–26 hrs.)

Fresh tempeh cakes

An incubator: The incubator keeps the inoculated soybeans warm as they become tempeh. A good design should be easy and inexpensive to make, require low energy input, and be able to maintain a temperature of about 88° F. (31° C.). If you have an area (such as above a hot-water heater, wood stove, or space heater, or almost anywhere in very warm weather) that is already at this temperature, simply use a large cardboard box as an incubator to keep the tempeh clean. An oven kept at 80° to 90° F. (27-32° C.) by a pilot light or by a 15-watt light bulb attached to a dimmer, also works very well.

Otherwise choose one of the following incubators with its own heat source, listed in order of preference:

**A polystyrene (Styrofoam) cooler and temperature regulator:** This simple and inexpensive incubator consists of five parts: (1) A standard 11-by-17-by-13-inch-deep Styrofoam cooler or ice chest, sold inexpensively at supermarkets. (2) an electric cord with a plug at one end and, on the other, a porcelain socket with a 7.5 watt (ornamental) bulb for use where the surrounding temperature is 65–77° F. (18–25°C.) or a 20-watt bulb for colder environments. (3) A temperature regulator that is simple, inexpensive, and accurate. A simple way of regulating temperature is to put a typical lamp dimmer (sold at hardware stores) between the socket and the bulb, as shown in Figure 6.2. You will soon learn where to set the dimmer dial to keep the temperature at 88° F. (31° C.), the ideal. By far the best way of regulating temperature is to buy a thermostat. One good model is "Thermostat Switch #59–S" made by Brower Mfg. Co., P.O. Box 251, Quincy, IL 62301. Phone 217-222-8561. Used for

*Fig. 6.2: Tempeh incubators*

*With aquarium thermostat*

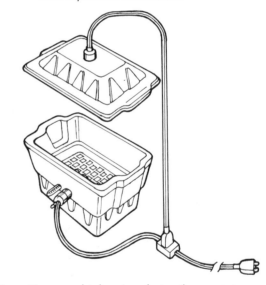

*With Sears/Brower chicken incubator thermostat*

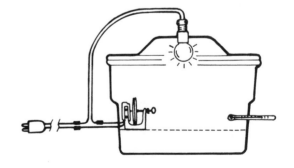

chicken incubators and selling for $10 in 1984, it is also available through the Sears Farm Catalog under Chicken Incubators, and is sold at some feed and supply stores. A bimetallic aquarium thermostat (that can be used out of water) works very well but is hard to find in the U.S. (4) a thermometer that registers in the area of 88° F., as used for aquariums, yogurt, or incubators. (5) A slatted or perforated rack; wire, thick plastic, cardboard, etc.

To assemble: cut a round hole in the cooler lid with a knife, snugly insert the socket, and screw in the bulb (and, if used, the dimmer). Insert the rack 1–5 inches above the bottom of the incubator, making sure it is not too near the bulb. Poke a hole in the middle of one side of the incubator about 2 inches above the rack and insert the thermometer. (If using a thermostat, install and set at 88° F. according to its instructions.)

Alternative designs: In large incubators, insert the bulb near the bottom of one side for more uniform heating. In place of the storebought cooler, try using an old oven or thick wooden box lined with Styrofoam, or use an abandoned refrigerator. Heat with an electric heating pad or with several large jars or hot-water bottles filled with hot water.

Total cost of electricity for incubating one batch (22–26 hours) should be less than ½ cent in temperate environments, less than 2½ cents in very cold environments.

* *Two cardboard boxes:* Choose one about 12 by 16 by 12 inches deep and another about 2 inches larger in all dimensions. Place cotton or polystyrene insulation at the bottom of the larger box, then set the smaller box into the larger and pack the spaces between the walls with insulation, then proceed as for the cooler above.
* *A homemade or storebought yogurt incubator:* The larger the better. Be sure temperature range is between 86° and 99° F. (30–37° C.).
* *A food dehydrator* set at 88° F. (31° C.)

## Ingredients

Each of the basic ingredients for homemade tempeh is now available at reasonable prices in North America. At least one tempeh kit, including tempeh starter, dehulled cracked soybeans, an instruction manual, and an incubation container, is now on the market (see below). The quality of the starter, soybeans, and water has a clear and pronounced effect on the quality and flavor of the tempeh.

**Tempeh Starter or Inoculum** (*Rhizopus* mold spores) and other tempeh-making ingredients are available from:

The Tempeh Lab, P.O. Box 208, Summertown, TN 38483. Phone 615-964-2286. Home or commercial quantities available.

*GEM Cultures,* 30301 Sherwood Rd., Ft. Bragg, CA 95437. Phone 707-964-2922 or 5414. Tempeh, miso, shoyu, and viili starter cultures.

Tempeh starter comes in four basic textures and concentrations listed in the recipe for Homemade Soy Tempeh (below) and described, together with methods for their home preparation in Chapter 7. To preserve the starter's potency, it should be refrigerated (or at least kept in a cool place), preferably together with a desiccant, in sealed plastic bags inside a sealed jar. Do not freeze. For storage details, see Chapter 7.

**Soybeans:** Either whole soybeans or dehulled cracked soybeans (also called full-fat soy grits) can be used to make tempeh; the latter are described in Variation 1. Whole dry soybeans are now available at reasonable prices from most natural- or health-food stores, co-ops, and some supermarkets; order in bulk for substantial savings. In rural areas they may be purchased for pennies a pound at feed stores, farmers' co-ops, grain elevators, or directly from the farmer, but be sure not to get mercury-coated seed beans, required by law to be labeled and dyed.

*Soybean (enlarged)*

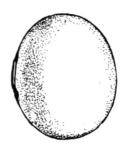

**Water:** Tempeh microorganisms seem to grow well on beans that have been cooked in commercial tap water. However, if the water contains large amounts of chlorine or other "purifying" agents, or if it is "hard" (i.e., has excessive calcium or magnesium salts), the mold may take a few hours longer to form a good mycelium. Of course, the mold grows best when pure, soft spring or well water is used for cooking the soybeans.

**Vinegar or Lactic Acid:** Mixing either of these substances with the soak water or the inoculated beans creates an acidic condition and lowers the pH, thereby making it more difficult for alien bacteria to grow and allowing the use of a smaller amount of starter. In Indonesia, the dehulled beans are soaked overnight in hot water; a slight prefermentation takes place that acidifies the water and beans, so that no acid need be added. In the West these acidic substances may be omitted if a relatively large amount of starter is used, the beans are thoroughly air dried, the humidity is below 70 percent, and the environment is very clean. Some prefer the distilled white vinegar available at any food market. If using lactic acid, use the 85 percent food grade or drug concentrate available from drugstores, laboratories, or chemical supply houses.

*Soybeans in the pods*

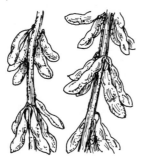

## Homemade Soy Tempeh
*(from whole dry soybeans)*

MAKES ABOUT 1¾
POUNDS (800 GRAMS)

This method is suited for use in North America, Europe, and other regions having a temperate climate. For a method suited to tropical or semitropical climates, requiring no incubator or vinegar, see Variation 4.

**2½ cups (590 cc or 1 pound) whole dry soybeans, washed and drained**
**15½ to 20 cups water**
**1 to 1½ tablespoons (distilled white or regular rice) vinegar or 85 percent lactic acid**
**Tempeh Starter** *(Rhizopus oligosporus* **mold spores)**
  **Choose one:**
  **1 teaspoon extended spore-powder or meal-texture starter**
  **¼ teaspoon spore-powder starter**
  **½ teaspoon meal-texture starter**
  **2 ounces minced fresh tempeh**

1. Combine soybeans and 7½ cups water in the cooking pot and bring just to a boil. Remove from heat, cover, and allow to stand at room temperature for 8 to 16 hours for prefermentation. (Or, to shorten soaking time, simmer for 20 minutes and allow to stand for 2 hours.)

2. Pour off water in pot, then rub beans in pot vigorously between the palms of both hands (or squeeze repeatedly in one hand) for 3 minutes in order to remove hulls (seed coats).

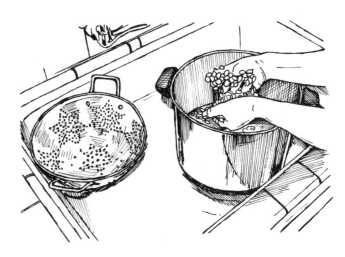

Add water to pot until it is almost full, stir gently in a circle, causing hulls to rise to surface, then pour off water and hulls into a colander or strainer (to catch any stray beans). Refill pot with water, stir, and again pour off water and hulls. Now repeat process of rubbing beans for 1 minute, filling pot, and pouring off hulls three to five times, or until all beans have been dehulled and, in the process, split lengthwise into halves. (Don't worry if a few detached hulls remain mixed with the dehulled cotyledons.) Transfer any beans from colander back into pot; discard hulls, using them as compost or livestock fodder.

3. Add 10 cups (hot) water and 1½ tablespoons vinegar or lactic acid to drained beans in cooking pot. (If you have only a small pot, use 7½ cups water and 1 tablespoon vinegar.) Bring to a boil and cook, uncovered, at an active boil for 30 to 45 minutes. (Do *not* pressure cook.) For best results, skim off excess hulls with a mesh skimmer.

4. Pour beans into large colander set in the sink. Allow to drain dry for several minutes, then shake well to expel as much moisture as possible. Now transfer beans to a baking tin or tray, lined with toweling (or to a large-diameter, shallow colander). Spread them in an even layer, and allow to stand for 20 to 30 minutes, stirring occasionally, until they have cooled to body temperature and their surface is almost dry.

Pat the surface of bean layer with toweling (or fan it) to remove excess moisture, then transfer beans to another well-washed baking tin, large mixing bowl, or plastic bag. Sprinkle tempeh starter evenly over beans and mix for about 2 minutes with a large spoon to distribute starter evenly. (Or cover mixing bowl with a plate or seal mouth of bag and shake vigorously to mix.)

Wash hands well, then prepare and fill the incubation containers you have chosen:

*Polyethylene (plastic) bags:* Lay two bags congruently on top of one another atop a clean wooden board or four thicknesses of clean, soft cloth. Using a sharp ice-pick, slender clean nail, or fat sewing needle (about 0.6 milimeters in diameter) poke holes simultaneously through both bags in a grid pattern at intervals of ½ to ¾ inch; this will let in oxygen, allowing the mold to breathe.

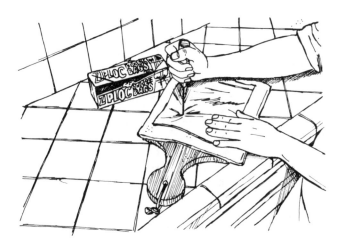

Spoon beans into bags so that each is about half full.

Seal mouth with Ziploc or candle flame, then place bag on a flat surface and press upper surface of bag with the palm of the hand or a flat board or spatula to distribute beans in a uniform layer ½ to 1 inch thick.

*Baking pans, skillet, etc.:* Spoon inoculated beans to a depth of ½ to 1 inch into one or more well-washed containers. Level surface of beans and pack firmly, then cover each container with a sheet of aluminum foil to prevent both drying and contamination. (There should be a space of ½ to 2 inches between beans and foil.) Perforate foil at intervals as described above.

*Aluminum-foil baking pans or pie tins:* Perforate the bottom of each container as for the polyethylene bags, then proceed to fill, pack, and cover as described above. Shallow (½ to ¾ inch deep) containers should be filled to the brim and covered tightly with perforated plastic wrap.

5. Place tempeh containers in incubator. (Set perforated bags on a rack to aid air circulation.) If possible, place incubator in a fairly warm location; in cold weather, cover incubator with blankets. Incubate tempeh at 86° to 88° F. (30–31° C.) for 22 to 28 hours. (If temperature drops to 80°F. (27° C.), incubation may take up to 50 hours.)

Check temperature occasionally and adjust accordingly; if using a dimmer, calibrate dial. Remember, the main cause of tempeh failures is overheating. If you are not using a thermostat, about three-fourths of the way through the incubation, when tempeh starts abundant production of its own heat and condensation begins to form, turn down the temperature several degrees, or turn heat source off. (Or move an unheated incubator to a cooler place.) If mycelium on the bottom of tempeh in plastic bags is underdeveloped or moist, turn over bags and place end that was nearest heat source away from it. Toward the end of the incubation, check tempeh every two hours, since the mold grows very rapidly. Finally examine tempeh (top and bottom) for quality:

**Good Tempeh:** Beans are bound into a firm, compact cake by a dense, uniform, white mycelium, which should permeate the entire cake. In bread pans or aluminum-foil containers, beans are barely visible under a cottony cover resembling the nap on a new tennis ball. Tempeh has a pleasant, clean, subtly sweet or mushroomy aroma. The entire cake can be lifted up as a single, cohesive cake, and thinly sliced pieces should hold together well without crumbling. Tempeh with a gray mycelium or black sporulation near pinholes or edges has been incubated a little too long, but is fine if the aroma is fresh without a strong ammonia odor. Overripe tempeh (which is popular in Indonesia) has been incubated much too long. Although it has a tan to light-gray surface and a slight ammonia aroma, it is nice to use as a seasoning agent, tasting like Camembert cheese. Cook good tempeh (do not scrape or cut off mold) as described in Chapters 4 and 5. Crumbly tempeh having a weak mycelium may be used in salad dressings or crumbly condiments. Tempeh which is not used immediately should be refrigerated (it will keep for 4 to 5 days), frozen (it will keep indefinitely), or stored in a cool place. Do not stack tempeh cakes during refrigeration lest the live mold cause overheating.

*Fig. 6.3: Good soy tempeh*

**Unfinished Tempeh:** Beans are bound together only loosely by a sparse white mycelium. Thinly sliced pieces crumble easily. Incubate longer, unless it has gone more than 8 hours past the recommended time, in which case discard.

**Inedible Tempeh:** Beans are foul or rotten smelling like strong ammonia or alcohol, indicating the development of undesirable bacteria due to excess moisture or overheating. Tempeh cake is wet, slimy, mushy, or sticky, with a collapsed structure that is limp when bent. Color is tan to brown. Mold grows in sparse patches, or there may be small colonies of alien microorganisms different in color or form from the predominant white mold. Discard bad portions and try again.

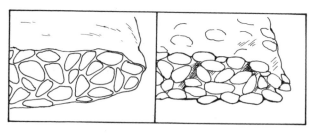

Good tempeh          Bad tempeh

## Troubleshooting

If your tempeh did not turn out properly, find the problem and correct it next time:

*Tempeh is too wet:* Cakes may have been too near the heat source, incubator temperature may have been too high or low, environmental humidity may have been too high (above 85%), beans may have been too wet (i.e., not drained and dried sufficiently before packaging), perforations may have been too small or too far apart, or unclean utensils may have been used. Try opening incubator lid ½ inch for air circulation.

*Mold is sparse and does not bind beans tightly:* Your starter may have lost its potency; increase amount of old starter by 50 percent, incubate longer, or order a new batch and keep it well sealed with a desiccant in the refrigerator.

*Tempeh surface contains black spots or patches:* This sporulation occurs when the tempeh mold receives too much oxygen or is incubated for too long or at too high a temperature.

*Tempeh smells strongly of ammonia:* The tempeh got too hot or was incubated too long. See also reasons for tempeh becoming too wet.

*Mold grew abundantly in some places but was sparse in others:* The sparse places may have been too near the heat source or lacked proper aeration. Or the starter may not have been mixed uniformly with the beans.

**1. Using dehulled cracked soybeans:** These are made by running whole dry beans through a loosely set mill (a Corona hand mill works well) which cracks them into several pieces, after which the hulls (seed coats) may be winnowed, blown, or aspirated off, as with a fan or hair dryer, or skimmed off during cooking. The best quality are splits, those cracked into two cotyledons or halves; however, those cracked into four or five pieces are also acceptable. Using these beans has the advantage of substantially reducing the time and work involved in making tempeh, since they are already dehulled and have a shorter cooking and soaking time. Yet they have several basic disadvantages which we and numerous others feel outweigh their advantages; they lose much more protein and other soluble nutrients in the soaking and cooking water than whole soybeans, are less readily available and more expensive, and yield tempeh with a granular texture which some people find less appealing. In North America split soybeans may be purchased from The Farm (see above) or from better natural- or health-food stores. Cracked beans with the hulls mixed in may also be used.

Soy grits are, by definition, any particles which will not pass through a 100-mesh screen (i.e., one with 100 openings per linear inch). It is best not to use finely cracked grits or hexane-extracted defatted grits or flakes.

Substitute 2½ cups (1 pound) dehulled cracked soybeans for the whole dry soybeans in the basic recipe. Combine beans with 8 cups water and 1½ tablespoons vinegar in the cooking pot, bring to a boil, and cook at a lightly rolling boil for 45 to 90 minutes. (Or soak beans for 2 hours in water and boil or steam for 30 to 60 minutes.) Use a slotted spoon or mesh skimmer to skim off any hulls and foam that float to the surface during cooking. Then proceed from Step 4 above as for whole soybeans.

**2. Alternative tempeh containers:** Each of these yields tempeh with a unique form.

*Sausage-shaped or cylindrical containers:* 4 or 6 polyethylene bags each 1½ inches in diameter and 5 to 7 inches long. Or several sheets of aluminum foil, plastic wrap, or wax paper, each 7 by 11 inches to wrap tempeh. If using bags, perforate, fill and seal as for polyethylene bags (page 107). If using sheets, perforate at ½-inch intervals, arrange about 1 cup inoculated beans in a row near one edge of sheet, then roll up sheet around beans to form a cylinder not more than 1½ inches in diameter. Twist ends closed and, if necessary, seal seam with tape.

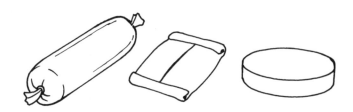

*   *Aluminum-foil packets:* 15 pieces of aluminum foil, each 7 by 9 inches. Divide beans into 15 equal portions (⅓ to ½ cup each), placing one portion at the center of each piece of foil. Fold over sides, then ends of foil, sealing tightly, to form a 4-by-3-by-½-inch packet. Perforation is unnecessary.

*   *Petri dishes or jar lids:* 10 to 12 dishes or lids, each 3¼ to 4½ inches in diameter and ½ to ¾ inch deep, for making tempeh burgers or patties. Pack each dish or lid full of beans, then press surface firmly so it is level with rim. Cover dishes with their unperforated lids; cover lids tightly with plastic wrap or aluminum foil perforated at ½-inch intervals and secured in place with rubber bands.

**3. Using frozen inoculated soybeans:** If you wish to prepare more inoculated soybeans than your incubator can hold at one time, simply pack the inoculated beans into unperforated plastic bags, seal, and freeze them. When the incubator is available, thaw beans in bag, perforate bag with a clean needle, etc., wipe moisture from surface of bag with clean (paper) towels, and incubate at 88° F. (31° C.) for 22 to 26 hours. Freezing does relatively little to retard mold growth.

**4. Homemade tempeh in tropical climates:** Soak beans overnight in water, bring to a boil, and simmer for 15 minutes. Add water until lukewarm, then dehull by hand. Add 10 cups water to drained beans (omit vinegar or lactic acid) and allow to stand for 24 hours while prefermentation acidifies soak water. Bring to a boil in soak water and cook at an active boil for 30 minutes. Then proceed from Step 4 in the basic method, but incubate beans in a clean place on a perforated rack preferably 6 to 8 feet above floor level (where it is warmer) at the natural temperature of the environment (about 77° F. or 25° C. for 48 to 52 hours); cover with a heavy cloth for the first 12 hours of fermentation.

**5. Steaming instead of boiling:** Steaming is a good way to cook grits (and presscake residues) to avoid loss of nutrients in the cooking water and, for softer beans, to prevent them from getting mushy. Wash 2½ cups grits well and pour off any loose hulls. Combine with 6 to 8 cups water and 1 tablespoon vinegar; soak overnight. Drain grits, place in a preheated steamer (wrap in a dishcloth if desired), and steam, covered, for 45 minutes. Whole beans may also be steamed for the same time, but they must first be parboiled to aid removal of the hulls by hand.

**6. Presalted tempeh:** This method yields a preseasoned tempeh which can be deep-fried as is; adding salt increases the incubation time but does not prevent good mold growth. Just before inoculation, mix ½ teaspoon salt with the drained, cooled soybeans. Then mix in starter and proceed as for soy tempeh but incubate for 30 to 36 hours. Slice and fry.

**7. Preseasoned coriander & garlic tempeh:** This basic concept suggests many exciting possibilities for making ready-to-cook tempeh varieties. Just before inoculation, mix with the soybeans: ½ teaspoon salt, ¼ teaspoon ground coriander, and ⅛ teaspoon garlic powder. (Or try substituting comparable amounts of other favorite ingredients, such as minced onions, crushed garlic, curry powder, etc.) Mix in starter and proceed as for soy tempeh, but incubate for 28 to 34 hours. Slice and fry.

**8. One-bean-thick tempeh:** This thin product, which eliminates the need for slicing and the occasional problem of crumbling during cutting, works like a slice of cheese in sandwiches. Use only ⅔ cup dry soybeans. After precooking and dehulling, simmer in 6 cups water with 2 teaspoons vinegar for 45 minutes. Inoculate with ¼ teaspoon starter then pack in a single layer into 2 perforated polyethylene bags. Incubate as for soy tempeh. Serve fried.

**9. Using black soybeans:** This variety of beans, very popular in East Asia, has a jet-black seed coat and white interior, and yields a delicious tempeh. Prepare as for regular soy tempeh but boil for only 30 minutes (step 3) and incubate for slightly longer.

**10. High-fiber high-yield tempeh:** After dehulling, mix from 25 to 100 percent of the hulls back in with the dehulled beans. Proceed to simmer as for regular soy tempeh.

**11. Overripe tempeh:** Prepare as for regular soy tempeh but incubate at 90° F. (32° C.) for 42 to 50 hours. Although mycelium may remain very white and cake stay firm, inner tempeh will develop overripe flavor and aroma. For recipes, see Index.

### Other Legume Tempehs

Legumes that make the best tempeh have a fairly high protein content and low starch content, and are large enough to dehull easily by hand. The following are listed in order of our flavor preference and the availability of the legumes. Small beans with a high carbohydrate content (such as mung, azuki, and lentils) do not work well; after cooking they become mushy. Larger beans with a high starch content (such as broad beans) can be prevented from turning into a mush when the starch grains are broken by boiling them in a 0.1 percent calcium chloride solution (i.e., 1 gram $CaCl_2$ mixed with 4¼ cups water). Cooking time for beans when making tempeh is roughly one-fifth the time required when cooking the beans for direct food use. If beans must be cracked, the coarser the better. In each recipe use the same quantities of ingredients as for Homemade Soy Tempeh.

**Peanut Tempeh:** Although peanuts are hard to dehull, the delicious tempeh is worth the work. Prepare as for Soy Tempeh but soak raw peanuts for 12 to 16 hours after first boil. Split peanuts into halves (or chop) after dehulling, steam in a cloth on a rack over 1 to 2 cups water at 15 pounds in a pressure cooker for 10 minutes, then allow pressure to come down naturally. Mix steamed peanuts with steaming water (1 to 2 cups) and vinegar, drain well, and proceed to inoculate and incubate as for soy tempeh. Serve thinly sliced as Seasoned Crisp Tempeh (page 54).

**Peanut and Soy Tempeh:** Bring 1¼ cups soybeans just to a boil in 4 cups water, cover, and allow to stand overnight. Do the same for 1¼ cups raw peanuts. Dehull soybeans and peanuts separately by hand. Then combine the two legumes and proceed from Step 3 as for Homemade Soy Tempeh.

**Garbanzo-Bean (Chickpea) Tempeh:** Prepare as for Soy Tempeh, but incubate slightly longer.

**Pink-Bean, Pinto-Bean, Red-Kidney-Bean, California-Blackeye-Pea, or Black-Bean Tempeh:** These beans, listed with our tempeh favorites first, are available at many U.S. supermarkets and natural-food stores. Prepare as for Homemade Soy Tempeh, but, after dehulling, bring to a boil and simmer for only 10 minutes for pink, pinto, and black beans, 5 minutes for red kidney beans, and 30 seconds for California blackeye peas. Then proceed as for Soy Tempeh. Deep-fried, the tempeh is slightly crunchy, creamy, and sweet, with an occasional heady, faintly alcoholic aroma.

**Great-Northern-Bean or Navy-Bean Tempeh:** Combine 2½ cups beans with 8 cups water, bring to a boil, remove from heat, and allow to stand for 1½ hours for great northerns, 1 hour for navys. Drain, add plenty of cool water, and allow to stand until beans are cool and firm, then drain again and dehull by hand. Combine dehulled beans in a pot with 8 cups water and 1 tablespoon vinegar, bring just to a boil, then remove from heat for great northerns, simmer for 5 minutes for navys. Drain and proceed as for Soy Tempeh.

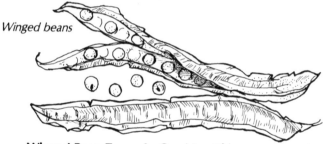

*Winged beans*

**Winged-Bean Tempeh:** Combine 2½ cups winged beans with 1 tablespoon baking soda and 6 cups water. Bring to a boil, simmer covered for 20 minutes, and allow to stand for 16 hours. Dehull by hand, then proceed as for Soy Tempeh, but simmer for only 25 (instead of 45 minutes), and incubate for slightly longer than Soy Tempeh. For variety, dice beans (roughly 4 chunks per bean) before inoculation to give finished tempeh a less crumbly, more sliceable texture; incubate at 88° F. (31° C.) for 48 hours to decrease beany or bitter flavors. Serve as Seasoned Crisp Tempeh (page 54).

**Broad-Bean (Fava-Bean) Tempeh:** Omit first boil. Soak for 18 hours at room temperature, hand dehull, boil for 10 minutes in a 0.1 $CaCl_2$ solution, inoculate with *Rhizopus arrhizus*, and incubate at 88° F. (31° C.) for 24 to 30 hours.

**Cowpea (Blackeyed-Pea) Tempeh:** Omit first boil. Soak for 18 hours at room temperature, hand dehull, boil for 20 minutes, inoculate with either *Rhizopus oligosporus* or *R. oryzae*, and incubate at 88° F. (31° C.) for 24 hours.

**Lupin Tempeh:** Prepare as for Soy Tempeh.

**Tofu Tempeh:** This new variety was developed by the Soybean Collective at the Wolfmoon Bakery in East Lansing, Michigan. The basic idea is closely related to that of China's popular fermented tofu (*toufu-ru, fuyu,* or *sufu*), except that the latter is pickled in a brining liquor to make a cheeselike product. Cut 8 ounces of firm tofu into ½-inch-thick, 2-inch-square pieces; place between triple layers of clean dishtoweling for 20 minutes to reduce moisture content. Sterilize a flat cookie sheet, baking pan, or large plate (at least 12 inches square) with boiling water, invert, and drain dry. Sprinkle ½ teaspoon (extended-meal-texture) tempeh starter over bottom of a cookie sheet, distributing it evenly with a cotton swab or sterilized knife. Arrange tofu pieces atop one-half of the starter, rubbing gently to inoculate the bottom side of each piece, then flip pieces over onto the other half of starter to inoculate the other side. Now use a cotton swab (or one small cube of tofu) to dab the remaining starter on all sides of each tofu piece. Place tofu pieces on a wire rack, place rack in a small open box, cover box with perforated aluminum foil, then place in incubator at 88° F. (31° C.) for 25 to 35 hours, or until tofu is covered with a ¼-inch-thick fragrant white mycelium. Serve as for Shallow-Fried, Pan-Fried, or Seasoned Crisp Tempeh (page 54).

### Homemade Grain (or Seed) & Soy Tempehs

The combination of cereal grains and soybeans yields tempeh which is tasty and has high-quality protein owing to protein complementarity. All tempeh containing cereal grains should be inoculated with a starter (such as *Rhizopus oligosporus* strain NRRL 2710) with weak amylase enzyme activity. Hesseltine, in 1965, found that strains with strong amylase activity break down the grain starch into sugars which are fermented into organic acids that give an undesirable flavor. Fortunately most of the starter sold in North America is derived from NRRL 2710, isolated at the Northern Regional Research Center in Peoria, Illinois. The following tempeh types, each weighing 33 to 35 ounces, are listed with our favorites first.

*Fig. 6.4: Four types of homemade tempeh*

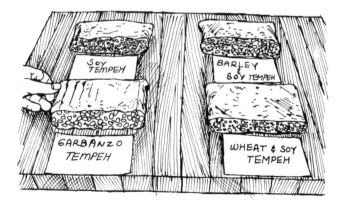

**Millet & Soy Tempeh:** Combine 1½ cups washed soybeans with 6 to 8 cups water, bring just to a boil, remove from heat, cover, and allow to stand overnight. Dehull by hand. To drained beans add 8 to 10 cups fresh water and 1 tablespoon vinegar. Bring to a boil and simmer for 45 minutes. Exactly 12 minutes before the end of cooking (i.e., after 33 minutes), add to the beans 1 cup washed and drained, unsoaked whole millet, return to the boil, and simmer for 12 minutes together with the soy. Proceed as for Soy Tempeh to drain the mixture of cooked grains and soy, cool and dry on paper towels, inoculate with 1 teaspoon tempeh starter, and incubate at 88° F. (31° C.) for about 22 hours.

**Rice & Soy Tempeh:** Prepare as for Millet & Soy Tempeh except substitute brown or white rice (long-grain works best) for the millet. For brown rice, soak long-grain rice in 3 to 4 cups water overnight in a separate container, then drain and add to the simmering soybeans 30 minutes before the end of the 45-minute cook. For white rice, wash long-grain rice and drain but do not soak; add to the simmering soybeans 10 minutes before the end of the 45-minute cook. Proceed as for Soy Tempeh. Do not incubate longer than 24 hours, lest the tempeh develop an overly sweet, alcoholic aroma.

To use cracked white rice: Substitute an equal quantity of very coarsely cracked white rice for the cracked wheat in the recipe for cracked soybeans & cracked wheat, below. Or use ½ cup cracked rice and 1½ cups cracked soybeans.

**Pearl Barley & Soy Tempeh:** Prepare as for Millet & Soy Tempeh except substitute pearl barley for the millet; add rinsed and drained pearl (pearled) barley to the simmering soybeans 20 to 22 minutes before the end of the 45-minute cook. Proceed as for Soy Tempeh. Do not incubate for too long, lest tempeh develop an alcoholic aroma.

**Bulgur Wheat & Soy Tempeh:** Prepare as for Millet & Soy Tempeh but add unsoaked bulgur wheat (often sold in America under the brand name Ala) or couscous to the simmering soybeans 5 minutes before the end of the 45-minute cook.

**Wheat & Soy Tempeh:** The easiest way to prepare Homemade Wheat & Soy Tempeh is to use bulgur wheat, as described above. The key to using whole wheat lies in cracking each kernel coarsely, into only 2 or 3 pieces; otherwise the cooked product will become mushy. Coarse cracked wheat is now sold at many natural-food stores. To make your own, use a very loosely set (Corona) hand mill, burr mill, or meat grinder. (A blender will not work.) Moreover, whole wheat will not work, since the mycelium cannot penetrate each kernel's tough bran layers. Prepare as for Millet & Soy Tempeh adding unsoaked cracked wheat to the simmering soybeans 12 minutes before the end of the 45-minute cook, or steam wheat separately for 10 to 20 minutes, stirring occasionally.

To use cracked soybeans: Combine 1¼ cups dehulled, coarsely cracked soybeans with 5 cups water and 2 teaspoons vinegar, bring to a boil, and simmer for 30 minutes. Meanwhile combine 1¼ cups coarsely cracked wheat in a second pot with 5 cups water and 2 teaspoons vinegar, bring to a boil, and simmer (or steam) for 12 minutes. (Simmer pearled wheat for 30 minutes.) Drain the contents of the two pots separately on a fine mesh strainer or in a colander, not on paper towels lest the wheat stick. (Save the wheat cooking liquid if desired for use in stocks, breads, etc.) When grains have cooled to body temperature, mix, inoculate, and proceed as for Soy Tempeh, but incubate at 88° F. (31° C.) for 20 to 40 hours. Serve as for Soy Tempeh.

For a higher protein content and lower proportion of wheat, use ½ cup cracked wheat cooked in 2 cups water and 1½ cups cracked soybeans cooked in 6 cups water.

**Cracked Oats & Soy Tempeh:** Prepare as for Millet & Soy Tempeh except: substitute coarsely cracked oats (oat groats, available at most natural-food stores) for the millet. Add unsoaked, washed and drained oats to the simmering soybeans 10 minutes before the end of the 45-minute cook. Incubate at 88° F. (31° C.) for about 30 hours.

**Sesame & Soy Tempeh:** Prepare as for Homemade Soy Tempeh except: use only 2 cups soybeans; while the cooked drained beans are cooling, roast ½ cup sesame seeds in a heavy dry skillet until seeds are fragrant and golden brown and just begin to pop. Mix warm seeds into cooling soybeans, wait several minutes, then mix in tempeh starter. Proceed as for Soy Tempeh. The roasted sesame seeds absorb moisture from the soybeans, which seems to improve the quality (especially the firmness) of the tempeh. Sesame is also rich in the sulfur-containing amino acids, thereby complementing soy. And sesame gives the tempeh a nice nutty flavor.

**Sunflower Seed & Soy Tempeh:** Prepare as for Millet & Soy Tempeh except: substitute hulled sunflower seeds for the millet; simmer the (unsoaked) sunflower seeds together with the dehulled soybeans for the full 45 minutes; incubate at 88° F. (31° C.) for 24 to 26 hours.

**Other Legume & Grain Tempehs:** Try substituting other legumes (listed at Other Legume Tempehs, above) for the soy in the above recipes. A combination of 2 parts wheat to 1 part broad (fava) beans has been shown to have a high protein quality (chemical score of 62.1 vs. 63.9 for soy tempeh); however it was found that frying made the wheat kernels hard and the tempeh had a slightly yeasty, sour flavor (Djurtoft and Jensen, 1977).

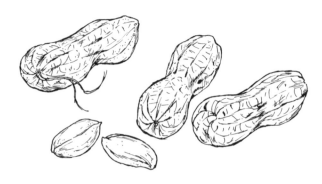

## Homemade Grain Tempehs

Some grains such as millet and long-grain brown rice may be used in their whole, natural state. Most others, however, must be pearled, dehulled, or coarsely cracked (into no more than 2 or 3 pieces); more finely cracked grains will become mushy during cooking, making tempeh preparation impossible. Pearling is a process whereby, without crushing the grain, the bran layers and germ are removed. Grain tempehs generally have a soft, moist texture and a slightly sweet, yeasty, and heady (or occasionally sweet-and-sour) flavor and aroma. To give added firmness, add okara or cracked soybeans. For home cracking into coarse pieces, use a Corona-type hand mill, burr mill, or meat grinder. Soft grain tempehs may spatter during deep-frying; to prevent this, coat with a dusting of flour or a bound breading of flour, egg, and bread crumbs. Most varieties are especially tasty served as Seasoned Crisp Tempeh (page 54), topped with a sprinkling of Worcestershire or soy sauce.

As mentioned under Homemade Grain (or Seed) & Soy Tempehs (above) all grain tempehs should be inoculated with low-amylase *Rhizopus oligosporus* strain NRRL 2710. Yet, even with this starter, corn (fresh or dried) does not make good tempeh since it is too difficult to dehull; both successes and failures are reported for sorghum.

The following are listed with our favorites first.

**Millet Tempeh:** Wash 2 cups whole millet, combine in a cooking pot with 8 to 10 cups water and 1 tablespoon vinegar, bring to a boil, and simmer for 12 minutes. Drain well in a large colander, then dry on paper towels. When grain cools to body temperature, mix well with 1 teaspoon extended-meal-texture tempeh starter and pack to a depth of ½ to ¾ inch into perforated Ziploc bags (or any other containers described in Homemade Soy Tempeh). Incubate at 86° to 88° F. (30° to 31° C.) for about 22 hours. Makes about 2 pounds.

**Rice Tempeh:** Prepare as for Millet Tempeh except: use 2 cups long-grain brown rice; soak rice overnight in excess water, drain, then combine with water and vinegar; bring to a boil and simmer for 30 minutes. Drain on a fine mesh strainer rather than on paper towels (lest it stick to towels). Incubate for about 30 hours. Use a coating when deep-frying.

To use long-grain white rice or converted rice (or coarsely cracked white or brown rice): Do not presoak. Combine with 6 cups water and 1 tablespoon vinegar, bring to a boil, and simmer for 7 to 10 minutes. Drain well, dry on paper towels, then proceed as above.

**Barley Tempeh:** Prepare as for Millet Tempeh except: use 2 cups pearl (pearled) or dehulled barley; soak overnight in excess water, drain, then combine with water and vinegar; bring to a boil and simmer for 15 minutes.

**Bulgur-Wheat Tempeh:** Prepare as for Millet Tempeh except: use 2 cups bulgur wheat (often sold in America under the brand name Ala) or couscous; simmer in 6 cups water for only 5 minutes.

**Noodle or Pasta Tempeh:** The secret of preparing this unusual and tasty tempeh lies in using short lengths of pasta, which allow proper mixing with the starter. Our favorite is made from whole-wheat pasta (eggless noodles) available at most natural-food stores. See that noodles are precut or broken into 1- to 2-inch lengths. Bring 8 cups water to a boil. Add 1½ teaspoons vinegar, then drop in 6 ounces cut noodles, return to the boil, and simmer for 5 minutes. Rinse briefly with cold water to decrease stickiness. Drain well, dry on paper towels, inoculate, and pack into one perforated polyethylene bag. Incubate at 88° F. (31° C.) for 31 hours. Serve as for Soy Tempeh. Make 12 ounces.

**Oat Tempeh:** Prepare as for Millet Tempeh except: use 2 cups coarsely cracked oats (oat groats, dehulled pieces that are larger than grits; available at most natural-food stores); simmer for 10 minutes; incubate for 30 to 36 hours.

**Wheat Tempeh:** Prepare as for Millet Tempeh except: use 2 cups coarsely cracked regular or toasted wheat (the use of toasted cracked wheat gives a savory, somewhat meaty flavor to fried tempeh, whereas regular wheat gives a popcornlike flavor); simmer for 12 minutes or steam for 20 minutes, stirring occasionally. Unlike Soy Tempeh, Wheat Tempeh still possesses a very pleasant flavor and aroma after as much as 43 hours of incubation, although it is ready after only 20 hours. Some difficulty may be encountered in slicing the finished tempeh prior to frying, since the individual kernels are quite firm. Frying may make kernels crunchy.

**Buckwheat Tempeh:** Prepare as for Millet Tempeh except: use 2 cups pearl (pearled or dehulled) buckwheat, available at some natural- or health-food stores; soak overnight in excess water, drain, combine with cooking water and vinegar, bring to a boil, and simmer for 15 minutes.

**Rye Tempeh:** Prepare as for Millet Tempeh except: use 2 cups coarsely cracked, pearled, or polished rye; simmer for 12 minutes.

**Combination Grain Tempehs:** Interesting flavors and textures can be obtained by combining equal parts of various grains; try Millet & Brown-Rice Tempeh. Wheat and Sorghum is also said to be good.

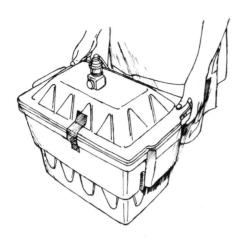

### Homemade Okara Tempeh

Here is a delicious way to use the abundant and nutritious okara (soy pulp) that remains after the preparation of soymilk or tofu. Coarse-textured okara which has been firmly pressed or hand-squeezed to expel its soymilk (reducing the moisture content by 80%) and impart a light crumbly texture, allows the best air circulation during incubation and therefore produces the best tempeh. Fresh okara, remaining from the preparation of homemade soymilk or tofu, regardless of the length of time the soymilk has been cooked, can be used without further cooking to make tempeh; the tempeh mold grows very well on uncooked okara, and subsequent deep-frying will deactivate trypsin inhibitor. Okara purchased commercially or more than one day old should be placed in a (cloth-lined) preheated steamer and steamed for 20 minutes to sterilize it; it should then be re-pressed to expel excess moisture and cooled to body temperature before inoculation. Even fresh okara when steamed develops a slightly better flavor and texture.

**Okara Tempeh:** Measure out 2¼ to 2½ cups firmly packed fresh or presteamed okara; this is just the amount remaining from one batch of tofu or soymilk made from 1½ cups dry soybeans. Spread okara on a large, clean (or sterilized) tray, mix with 1½ teaspoons vinegar, and allow to cool to body temperature. Mix in ½ teaspoon (extended-meal-texture) tempeh starter, then proceed to pack into *one* perforated polyethylene bag (or other tempeh container) to a depth of ½ inch. Proceed as for Homemade Soy Tempeh to incubate at 88° F. (31° C.) for 24 to 30 hours. Served as Seasoned Crisp Tempeh (page 54), this variety has a flavor and texture resembling fish fillets or French-fried potatoes. Makes about 1 pound.

**Other Presscake Tempehs:** Tempeh can also be made from the presscakes of peanuts, mung beans, kapok seeds, and cottonseeds, as well as defatted soy meal. Do not try to use coconut presscake or grated coconut shreds, which may produce a toxic product (tempeh bongkrek).

### Soy (or Grain) & Okara Tempehs

These firm-textured tasty products represent good ways of utilizing leftover okara.

**Soy & Okara Tempeh:** Prepare as for Homemade Soy Tempeh except use only 2 cups whole dry soybeans; after the beans are drained and cooled, mix in ½ to 1 cup fresh or steamed okara. Proceed as for Soy Tempeh.

**Brown Rice & Okara Tempeh:** Wash 1 cup long-grain brown rice and soak overnight in excess water. Drain, then combine in a pot with 6 cups water and 1 tablespoon vinegar. Bring to a boil and simmer for 30 minutes. Drain well, mix in 2 to 2½ cups fresh or steamed okara, and cool to body temperature. Mix in 1 teaspoon (extended-meal-texture) tempeh starter and proceed as for Homemade Soy Tempeh. Incubate for 27 to 28 hours.

**Bulgur Wheat & Okara Tempeh:** Combine 1 cup bulgur wheat with 4 cups water and 1 tablespoon vinegar. Bring to a boil and simmer for 5 to 6 minutes. Then proceed as for Brown Rice & Okara Tempeh, above. Incubation may take somewhat longer.

**Other Combinations:** Try okara with other grain or legume combinations as described above.

*Dry soybeans in pod on plant*

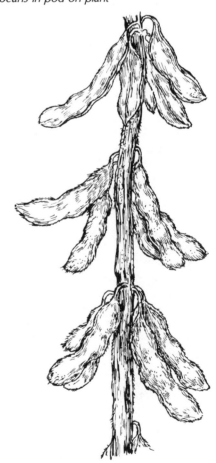

**Community Tempeh**

The process for making tempeh on a community scale in temperate climates is basically the same as that described in Homemade Soy Tempeh, above. However, the larger scale (eight times that of the homemade recipe) requires the use of larger equipment, especially the incubator. The method will be given here in outline, emphasizing the points of difference from the homemade method. A more detailed discussion plus full instructions for setting up a small-scale or middle-level commercial tempeh shop or a modern tempeh factory is given in *Tempeh Production: The Book of Tempeh,* Volume II, available from The Soyfoods Center.

**1. Ingredients:** Use 8 pounds (20 cups) whole dry soybeans to yield 14 pounds of tempeh, enough for 56 four-ounce servings, or 16 fourteen-ounce packages.

**2. Cooking:** Use a 5- to 7-gallon pot set over a hot-water-heater burner or a candy stove (used for commercial candy production).

**3. Dehulling:** Run through a loosely set (Corona) hand mill, or cut into large pieces with a meat grinder having a coarse blade. Float off the hulls in water in the cooking pot. The beans may also be dry dehulled (or dehulled after soaking) and coarsely cracked with a Corona mill, brought to a boil in water, and simmered for 75 minutes.

**4. Cooling and drying:** Drain beans for several minutes in large colanders. To cool, construct a special draining tray having a perforated (Plexiglass or screen) bottom; if desired, dry with an electric fan. Or spread beans on a large piece of absorbent toweling.

**5. Tempeh containers:** Any of the following, listed with our favorites first, may be used.

a. *Polyethylene bags:* The Ziploc type are ideal. They are cheap and easy to use, give consistently good tempeh, eliminate cleanup, disinfection with bleach water, and overheating problems sometimes found when using trays, and they allow the tempeh to be stored (or sold) in the container in which it is incubated, which saves repackaging and is more sanitary. Perforate bags with an icepick or "bed of nails" (page 107 or 128) and seal with a candle (as described in Chapter 8) or with a simple electrical heat sealer. Five-by-7-inch packages cut from oven roasting bags also work well and are highly heat resistant but require sealing both ends.

b. *Trays:* Shallow wooden trays with slatted bottoms (see Chapter 8) lined with perforated plastic sheeting work well, as do compartmented wooden trays (each compartment lined with plastic wrap and having a perforated bottom) as shown below. Aluminum trays (roughly 26 by 21 by 1 inch deep) with bottoms perforated at ½-inch intervals work fairly well when covered with plastic sheeting perforated every 3 inches with a ⅛-inch-diameter hole. Stainless steel may cause problems; since this metal is a poor heat conductor, there is often a buildup of heat at the center of the tray, which spoils the tempeh there.

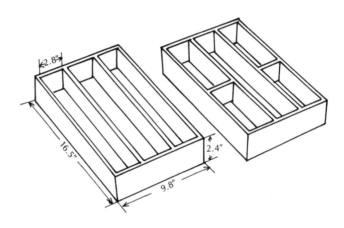

Fig. 6.5. Wooden tempeh incubation tray designs (Source: Martinelli and Hesseltine, 1964)

c. *Patty rounds:* To make patties for Tempeh Burgers, cut 3½-to-4-inch-diameter firm food-grade polyethylene pipe into sections ½ to ¾ inch wide; incubate tempeh in many rounds arranged honeycomb fashion in a single layer between two large trays or between sheets of glass or Plexiglass. As with petri dishes the small size of the containers makes perforation unnecessary.

d. *Tupperware boxes:* Small community shops wishing to use inexpensive, recyclable containers may try this method developed by the Toko Baru shop in Los Angeles. Tupperware containers, each 5 inches square and 1 inch deep, are perforated with 5 holes in each side, 12 holes in the top, and none in the bottom. Each container is filled to the brim with inoculated beans and the top held in place with 4 strong rubber bands, two wrapped in each direction.

**6. Incubators:** Designing and making (or purchasing) a good incubator is probably the most important part of setting up a community shop. The easiest incubator to obtain may be a junked or nonoperative refrigerator or "reach in"; it should be well insulated and have perforated shelves. Mount two 100-watt bulbs or warming trays (or a heating coil in a compartment) near the bottom and attach these to a typical home thermostat. The Toko Baru shop (mentioned above) heats its refrigerator incubator with the hot water saved from cooking the soybeans, placing it in a covered or uncovered pot. Summer incubation is 2 days; winter is 3½ days. Attach absorbent toweling to the incubator's interior ceiling to keep condensation from falling on the tempeh. A small fan may be helpful in providing heat circulation, and an open pan of warm water may be necessary to increase the humidity. A large insulated box the size of a refrigerator will also work well. The figure below shows an upright community-scale incubator developed by a branch of The Farm in Houma, Louisiana. Made from a plastic Thermos cooler 24 by 14 by 14 inches, it contains a thermostat and thermometer at the center of the back wall above a 25-watt light bulb, the heater. There are slats for seven Plexiglass shelves, each ⅛ inch thick with ⅛-inch-diameter holes every ¾ inch to aid air circulation.

*Fig. 6.6: Community tempeh incubator*

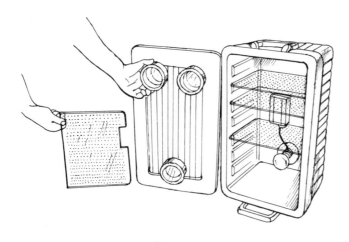

For slightly larger production, make an incubation room 6 by 6 by 6 feet tall. Insulate the walls and line the inside with smooth, nonporous, easy-to-clean (aluminum) panels. On the floor place an electric space heater (1500 watts) and connect it to the Sears thermostat described earlier. In one wall, just above floor level, make a vent, covered with a milk filter. Place a dehumidifier on the floor of the room right in front of the vent and 1 to 2 inches from it. In shops in cold climates, the dehumidifier is used from about October until June to lower the humidity. In the roof of the incubation room, place an exhaust fan, ideally connected to a thermostat so that it turns on when the incubation room temperature reaches

35° C. (95° F.). Even when the fan is off, air circulates in the room by convection, coming in through the low vent and out the exhaust fan. In the room's front wall make a well-insulated door, just large enough to allow the passage of rolling racks. If rolling racks (which are more convenient and easier to clean) are not used, build six shelves into one or more walls, the bottom shelf being 2½ feet above floor level and adjacent shelves spaced about 5 inches apart; the heater is then placed directly below the shelves. The temperature difference between the top and bottom shelves will be only about 2.7° C. (5° F.). The room should be sanitized weekly: put on rubber gloves, dip a sponge in 100% ethanol (ethyl alcohol), and rub down the walls.

Regardless of the incubator design, incubate for the times and at the temperatures and humidities recommended for homemade tempeh. Or increase the temperature and decrease the time (or visa versa) as shown in the following graph.

Fig. 6.7. Incubation time versus temperature for soy tempeh

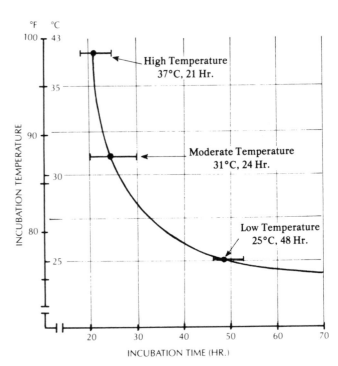

# 7

# Making Tempeh Starter

As with the preparation of yogurt, bread, miso, and most other fermented foods, the preparation of tempeh requires an inoculum or "starter." There are basically three ways of preparing tempeh starter: (1) using sporulated, inoculated soybeans, rice, or other similar substrate; (2) using sporulated tempeh; and (3) using unsporulated fresh tempeh. Only the first method requires that you have some starter to begin with; each has its advantages and disadvantages as described in the introductions to the various methods, below.

The various ready-made starters come in three different degrees of "purity": (1) *Pure-culture* starters, typical of the spore-based types now sold in the West, contain only one strain of mold spores. (2) *Mixed-culture* starters, typical of those made in Indonesian tempeh shops or found in sporulated or unsporulated mycelium, may contain various molds, bacteria, and yeasts, many of which may be beneficial to the fermentation. (3) *Mixed pure-culture* starters, not yet available, are probably the finest type, containing for example pure *Rhizopus oligosporus* mold spores in a balanced combination with a pure culture of *Klebsiella* bacteria, which produce tempeh's vitamin $B_{12}$.

The various starters also come in three different textures and concentrations: (1) *Spore-powder* texture consists of pure spores separated from the substrate and has the highest concentration, i.e., the greatest number of spores or propagules per unit volume or weight. (2) *Meal texture* is made by grinding the sporulated substrate, and thus contains a mixture of spores, substrate, and mycelium. (3) *Extended spore-powder or meal texture* is made by mixing one of the previous two starters with a sterilized, inert extender, such as wheat or rice flour or starch, cornstarch, tapioca, etc. Adding the extender facilitates accurate measurement of the starter. In some cases, to improve storage properties, the sporulated substrate is freeze-dried, ground to a fine meal, then mixed with the extender. To make tempeh from 1 pound of dry soybeans you will need one of the following:

⅛ to ¼ teaspoon spore-powder starter
⅛ to ½ teaspoon meal-texture starter
1 teaspoon extended spore-powder or meal-texture starter
2 ounces minced fresh tempeh

In producing tempeh starter, there are several basic principles that should be observed to obtain best results. First, work with very clean (aseptic or sterile) utensils to minimize contamination. Fortunately *Rhizopus* molds are so hardy that they tend to "crowd out" and dominate many unwanted microorganisms (see Appendix C, Coconut-Presscake Tempeh, and Appendix G, Onchom, in the professional edition), yet it is still extremely important to be rigorously sanitary. Second, keep the incubation temperature between 86° and 90°F. (30-32°C.); maximum sporulation occurs at the latter temperature. Typical incubation times range from 3 to 5 days after inoculation. Third, remember that any starter is most potent just after it has been prepared and most of the loss of potency occurs during the first month of storage. And fourth, where so instructed, carefully control the moisture content of the substrate to obtain optimum sporulation.

Using the methods that follow, it is just as easy to make your own tempeh starter at home as it is to make tempeh. In fact, it's easiest to make both at the same time. Not only will you become self-sufficient, you will also save money and develop a deeper insight into the secret (and fascinating) life of "micro-plants." Tempeh starter is now available by mail order from good sources (see Chapter 6). It is our hope that it will soon be as readily available in food stores and as inexpensive as baker's yeast. Already a number of large American companies (such as Dairyland Food Laboratories in Waukesha, Wisconsin, and Red Star Yeast Company in San Francisco) making similar fermented products have shown an interest in producing tempeh starter once the market becomes somewhat larger; they might market their starter in small foil packets for home use and in bulk for commercial production.

### Storing Tempeh Starter

Whether you purchase ready-made tempeh starter or prepare your own, it is very important that you store it at low temperature and humidity to minimize losses of potency or viability. Where a refrigerator is available, we recommend the following storage method. Put the starter in a clean (unused) polythylene (plastic) bag or wrap in plastic wrap; seal the mouth tightly with a rubber band. Put this bag together with a small packet of desiccant (such as silica gel or calcium chloride) inside a second plastic bag and seal this with a rubber band. Now place these two bags in a jar or canister, put the lid on tightly, and store in the refrigerator, ideally between 41° and 50°F. (5-10°C.). Freeze-dried starter keeps very well at room temperature but is best stored as described above. For best results, starter should not be frozen, since repeated freezing and thawing can rupture and kill some of the cells, especially if their moisture content is high. Yet homemade starter, which contains mold mycelium as well as spores, may be frozen at just below zero to stop mold growth, which temperatures above zero cannot do.

The basic measure of a starter's potency is the percentage of viable spores it contains, i.e., those that will germinate at any given time. Tempeh starters are most potent just after they are freshly made, at which time approximately 69 percent of the spores will germinate. These figures decrease rapidly during the first 4 to 6 weeks of storage as shown in the following figure: Rusmin and Ko, in 1974, found that (1) under conditions of low temperature and humidity, the spore germination rate drops to about 35 percent after 6 weeks and remains virtually unchanged for more than one year; (2) tempeh made from starter stored for one year under these conditions is identical in quality and fermentation time to tempeh made from fresh starter; and (3) even in tropical climates or where refrigeration is not available, the starter can be stored at room temperature (77°F. or 25° C.) with very little loss in potency as long as the relative humidity is kept near zero. Note that starter stored at room temperature without a desiccant would lose all its potency in 12 to 14 weeks.

Starter having decreased potency can still be used in standard amounts to make good-quality tempeh; however, the incubation time must be increased. Or the standard incubation time may be used and the amount of starter increased in proportion to its decrease in potency. The latter method generally gives better results.

### Making Tempeh Starters in the West

The following methods describe household-scale preparations only; for larger scale, see *Tempeh Production*. The word "sterilize" used below actually means "partially sterilize."

#### Sporulated Inoculated Soybean Method

If you have some tempeh starter, this is the quickest and easiest way to make more. A close relative of the hibiscus-leaf method most popular in Indonesia, it can be done as part of the process of making more tempeh; the resulting starter produces tempeh of the finest quality. By being clean and careful, we have had no trouble keeping this starter going for more than 20 generations, each generation yielding more than 15 batches of homemade tempeh.

*Fig. 7.2: Sporulated tempeh for starter in bread pan*

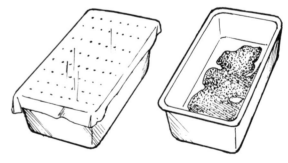

Douse a clean bread pan with plenty of boiling water to sterilize it, then invert pan on a rack to dry. While making soy tempeh (see Chapter 6) set aside about ¼ cup freshly inoculated soybeans; place them in a *single* layer covering the bottom of the sterilized bread pan.

Fig. 7.1. Loss of tempeh starter potency when stored at various temperatures and humidities (Source: Rusmin and Ko, 1974)

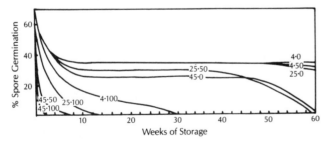

Cover pan with a sheet of aluminum foil, and perforate foil at ¾-inch intervals. Incubate at 86° to 90° F. (30-32° C.) for 48 hours, or until the mold sporulates and the upper surface looks like uniformly dark gray or black cotton. Then remove foil cover and continue to incubate for 24 hours more to dehydrate substrate and increase sporulation. Extract spores, using any one of the following dry or wet methods, listed in order of preference.

**Dry-Strainer Spore Extraction:** This method is quick and easy, and requires only simple utensils that are easily sterilized. Makes 1½ to 2 tablespoons of top-quality, long-lasting spore-powder starter, enough for 18 to 24 batches tempeh.

Sterilize several cups of water by boiling for 5 minutes; pour boiling water over a clean 8-inch-diameter medium-mesh strainer and a metal (soup) spoon, then drain-dry. Place several letter-size sheets of clean, smooth paper on a table or plate. Holding the strainer over the paper, place sporulated starter in the strainer, break into many small pieces with a sterilized spoon, then rub pieces firmly against the bottom of the strainer for 2 to 3 minutes so that spores fall through onto the paper, leaving dry grayish-white beans in the strainer. One-quarter teaspoon of this starter will make 1 batch of Homemade Soy Tempeh (Chapter 6). Wrap unused starter in plastic wrap and refrigerate in a sealed jar. Use within 6 to 12 months; after 1 month increase dosage by 50 percent.

*Fig. 7.3: Dry-strainer spore extraction*

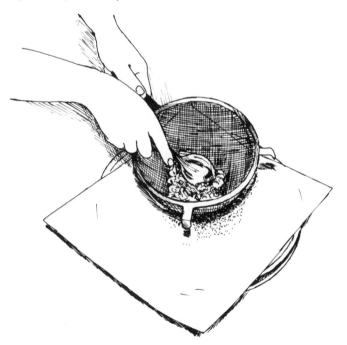

**Wet-Jar Spore Extraction:** This method is easy and requires no drying; the starter is potent, but it must be used within 10 days or it will spoil. Makes about 10 tablespoons of starter liquid, enough for 15 batches of tempeh.

Sterilize 4 cups of water by boiling for 5 minutes. Use 2 cups of this boiling water to wash out two 1-pint screw-top jars. Pour about ¼ cup of the water into the first jar, cover, set in cold water, and allow to cool to room temperature. Douse a knife with some boiling water and use it to cut sporulated substrate into three equal portions. Wrap two of these portions in plastic wrap and freeze (or, for storage of less than 2 weeks, refrigerate) in a sealed jar. Place the remaining sporulated substrate in the ¼ cup water, cover jar, and shake vigorously for 1 to 2 minutes, or until water is dark black. Douse a small strainer with some of the remaining boiling water, cool briefly, then pour the contents of the first jar through the strainer into the second jar so that the black-spore starter solution is in the second jar. Discard the lumps of tempeh left in the strainer (or rewash with more sterile water to extract more spores); 2 teaspoons of this starter will make 1 batch of Homemade Soy Tempeh. (When measuring out starter, pour the liquid into a teaspoon; do not put the teaspoon into the liquid, or the spoon may contaminate the remainder.) Seal jar and refrigerate. Use within 10 days, or while it still smells fresh.

Or, simply leave a small chunk of tempeh in the incubator long enough for it to sporulate and turn black. Then place it in a small jar of water and shake well. Remove the tempeh—if fresh, you can cook it—then refrigerate the liquid in the sealed bottle overnight. When the spores have settled, pour off and discard the clear liquid. Use 1 teaspoon of this spore paste to inoculate 2½ cups dry soybeans.

**Dry-Blender Spore Extraction:** Compared with a strainer, it is more difficult to sterilize most blenders and, after pulverization, to remove all of the starter. Makes about ¼ cup of meal-texture starter, enough for 24 batches of homemade tempeh.

Pour into a blender and over a medium-mesh strainer boiling water that has been boiled for 5 minutes, then invert and drain dry. Transfer sporulated starter to the blender and pulverize for 1 minute, or until it looks like a uniform dark-gray meal or granules. Place several letter-size sheets of clean, smooth paper on a table or plate. Holding the strainer over the paper, transfer all pulverized starter to the strainer and, shaking the strainer, sift onto paper. Discard any coarse particles remaining in the strainer; ½ teaspoon of this starter will make 1 batch of Homemade Soy Tempeh (Chapter 6). Wrap unused portions in plastic wrap and refrigerate in a well-sealed jar.

VARIATION

**\* Sporulated Inoculated Rice Method:** Rinse ¼ cup long-grain white rice twice and drain. Combine rice in a saucepan with ¾ cup water, bring to a boil and simmer, uncovered, for 5 minutes. Drain and cool on paper towels to body temperature. Place rice into a bread pan sterilized (as above) with boiling water, mix in ⅛ teaspoon tempeh starter, cover with perforated foil and proceed to incubate as above. Incubation may take 5 to 7 days. To use, pulverize in a blender.

## Sporulated-Tempeh Method

If you have some fresh or frozen (homemade or store bought) tempeh, you can cut thin slices from the tempeh surface, incubate them until the mold sporulates, then use the spores as a starter for making more tempeh. Although not quite as quck and easy as the Unsporulated-Tempeh Mycelium Method (described below), these spores make better-quality tempeh and can be stored for 6 to 12 months, refrigerated.

Set aside 1 cake of tempeh; the mycelium should be uniformly white or light gray, dense, and fresh smelling, and the tempeh should be firm. Sterilize several cups of water by boiling for 5 minutes, then pour boiling water over a sharp knife and over a small (roughly 8-by-10-inch) wire rack to partially sterilize them; allow both to drain dry. Touching the tempeh surface as little as possible with your hands, use the sterilized knife to shave a number of thin (1/16-inch-thick) strips from the entire tempeh surface and lay these with the cut side down on the sterilized rack. Place the rack in a small box about the size of the rack but having sides that rise about 2 inches above the rack. Cover the box with a sheet of aluminum foil and perforate foil at ¾-inch intervals. Reserve the remainder of the tempeh cake for use as a food; do not try to use the whole thick cake to make starter lest the interior spoil during subsequent incubation. Now place the box containing the tempeh on a rack in an incubator and incubate at 86° to 90° F. (30°-32° C.) for 48 hours. Starter is ready when tempeh slices are covered with a black, richly sporulated mycelium (there should be no brown or pink spores) that has a pleasant aroma (it should not have a strong ammonia or spoiled odor). Proceed to extract spores, using dry or wet methods as described in Sporulated Inoculated Soybean Method, above.

## Unsporulated-Tempeh Mycelium Method

If you have some fresh tempeh, this is the quickest and easiest method to make more. The tempeh is simply minced and mixed with cooked, dehulled soybeans; the mycelium continues its rapid growth without the use of spores. Tempeh made with this starter, however, generally has a slightly weaker mycelium and the incubation time is a little longer than for tempeh made with starter from the sporulated methods described above. And remember that there are always some unwanted bacteria in the original tempeh. If you are careless and/or if the humidity is high, their numbers will increase with each generation until eventually they prevail, preventing the formation of good tempeh. However, by being clean and careful, we have had no trouble keeping this starter going for more than 12 generations of homemade tempeh. Djurtoft, in 1977, working under sterile laboratory conditions, reported propagation of 30 generations of quality soy tempeh. (However, when other legumes were substituted for the soybeans, the method failed for reasons not yet determined.)

Set aside a cake of good-quality homemade or store bought tempeh (preferably fresh, but frozen or refrigerated usually also works). The mycelium should be uniformly white or light gray, dense, and fresh smelling, and the tempeh should be firm. Sterilize several cups of water by boiling for 5 minutes, then pour boiling water over knife. Place tempeh on a sheet of plastic wrap (or other very clean surface), set on a cutting board, and cut off a 2-ounce (1-by-2-by-¾-inch) portion from the best-looking part of the tempeh. Using the knife, mince this as fine as possible on the plastic wrap. Then combine with 2½ cups (1 pound) of dry soybeans that have been dehulled, cooked, drained, and cooled to body temperature. Mix very thoroughly, then pack and incubate at 88° F. (31° C.) for 22 to 26 hours if using fresh tempeh as the starter, or 30 to 32 hours if using frozen tempeh. Proceed as for Homemade Soy Tempeh (Chapter 6).

Wang et al. (1977) report good results from pureeing 1 ounce of good fresh tempeh in a blender with 1 to 2 tablespoons of cool boiled water to make a thin paste, then using this to inoculate 1 pound of dry soybeans that have been cooked, etc.

VARIATION

**The Farm Method:** For a detailed description of this method see *The Farm Vegetarian Cookbook,* revised edition, 1978. Makes enough starter to inoculate 20 pounds of dry soybeans. Into a 100-ml clear glass cough-syrup bottle place 1 tablespoon converted white rice and 1½ teaspoons water. Lay bottle on side and use a spoon handle to smooth rice into an even layer on flat side of bottle. Plug mouth of bottle with plenty of cotton. Pressure-cook bottle on a rack over water (as described above) at 15 pounds pressure for 15 minutes, then allow bottles to cool to body temperature. Using the tip of a paring knife sterilized with alcohol and a burner, transfer "a speck" of dried tempeh starter onto the rice in the bottle, then immediately replug with cotton. Incubate at the temperatures and times described above, until sporulated rice is dark gray or black. Using 5 small steam-sterilized jars and a sterilized spoon, divide the starter among the five jars. Seal mouths of jars, then seal jars in plastic bags, and freeze or refrigerate. To use, add 1 tablespoon cool water to the rice in one jar; mix and shake well to break up rice clumps, then mix contents of jar (rice and black water) into 4 pounds of dry soybeans which have been cooked and cooled to body temperature. Using this method, you can make hundreds of pounds of tempeh from one packet of dried starter.

## Sporulated Rice, Pressure Cooker, and Jar Method

If you have a little tempeh starter, you can use this method to make more. Since the starter is prepared under sterile or almost sterile conditions (especially if you can sterilize your blender), the purity of the original starter will be retained more completely than with the previous methods. Developed by Dr. H. L. Wang at the Northern Regional Research Center in Peoria, Illinois, this is the method used to prepare the (freeze-dried) starter which thousands of people have ordered from the center. Makes about 2½ tablespoons of top-quality meal-texture starter, enough for 15 to 60 batches of tempeh.

The basic utensils required are a 1-pint mason jar (or any 1-pint jar with a piece of twine used instead of the screw-on lid); a pressure cooker; a 4-inch-square sterile bandage pad, or a milk filter (sold at some drugstores), or a ¼ - to ⅛ -inch-thick layer of cotton sandwiched between two layers of gauze (used in place of the jar lid, thus allowing the mold to breathe, while keeping out bacteria and dust); a tempeh incubator (see Chapter 6); and a blender, preferably one with both up- and down-turned blades that allows pulverizing of small quantities of ingredients. You may double the recipe if your pressure cooker is large enough to hold two 1-pint jars; use ¼ cup rice, etc. in each jar.

Combine rice and water in the jar, cover with the sterile pad or milk filter in place of the flat top, then screw on ring or tie with twine. Allow jar to stand at room temperature for 1 hour; shake every 5 or 10 minutes so rice will absorb water uniformly. Place jar upright (on a rack if available) in a pressure cooker, run 2 cups water into cooker, bring to full pressure (15 pounds), and cook for 20 minutes. Remove from heat and allow pressure to come down naturally. Open cooker and shake jar to break up rice clumps. Allow to cool to body temperature, open jar, and quickly sprinkle in starter. Put back the filter and screw on top immediately, then mix well by shaking. Lay jar on its side to spread out rice; incubate at 86° to 88° F. (30–31° C.) for 4 to 4½ days, or until rice is covered with black spores. Now pulverize rice in a clean (preferably sterilized), dry blender for 1 to 2 minutes until it looks like uniform dark-gray granules. The starter is now ready to use. Return unused portions to the mason jar and cover jar with filter, metal lid, and screw top (or store in a small unused polyethylene bag sealed in a jar with a desiccant). Store as described at the beginning of this chapter, or freeze. We use ½ teaspoon of this concentrated starter to inoculate 1 pound of dry soybeans that have been cooked; simply mix the starter with the beans. To use less, mix ⅛ teaspoon starter with 1 to 2 teaspoons cool, boiled water and mix with 1 pound of beans. Pack and incubate as usual.

Fig. 7.4: *Sporulated rice, pressure cooker, and Mason jar method of making tempeh starter*

**¼ cup white (polished) rice, preferably long-grain; do not use rice with talc (magnesium silicate) on its surface**

**2 tablespoons (1 fluid ounce) water**

**½ to ⅛ teaspoon commercial (freeze-dried) tempeh starter**

### Indonesian Mixed-Culture Tempeh Starters

Most of the tempeh starter in Indonesia is prepared weekly by tempeh craftsmen in their shops, with spores from one batch generally being used to inoculate the next. Because of the use of natural materials (such as tree leaves or river water) and the lack of a sterile environment, the resulting starters have traditionally consisted of mixed rather than pure cultures. In spite of this, however, and for reasons not yet well understood, contamination of the tempeh by unwanted micro-organisms almost never seems to be a problem.

Six methods of making tempeh starter are now used in Indonesia; the first five are traditional, having evolved over hundreds of years, while the sixth is a recent development. All yield mixed-culture starters. In order of popularity they are:

**1. Sandwiched-hibiscus-leaf method:** Cooked soybeans inoculated with *Rhizopus* molds (during the process of making tempeh) are sandwiched between hibiscus leaves and incubated until the molds sporulate. In some areas, teak leaves are used in place of hibiscus. The finished product is known as *laru, waru,* or *usar.* We would estimate that more than 80 percent of all Indonesian tempeh starter is prepared in this way.

**2. Contact-leaf method:** Hibiscus, banana, or other leaves are laid atop the inoculated beans during fermentation so that a mycelium grows on the leaves, which are then used to inoculate the next batch of tempeh.

**3. Sporulated-tempeh method:** Thin (surface) slices of tempeh are allowed to sporulate, and then used directly or sun dried and ground to a powder, which is used to inoculate subsequent batches.

**4. Unsporulated-fresh-tempeh method:** Fresh pieces of tempeh are simply crushed or broken and mixed in with cooked beans ready to be inoculated. Contrary to widespread reports in Western journal articles on tempeh, this method is rarely used.

**5. Ragi method:** Cooked soybeans are inoculated by crumbling small dry yeast cakes (called *ragi*) over them. Ragi is used mostly, however, for inoculating tapeh (see Glossary).

**6. Modern rice-substrate method:** *Rhizopus* molds are grown to sporulation on a substrate of cooked rice rather than the usual cooked soybeans. Both mixed-culture village methods and pure-culture laboratory methods have been developed.

Only the first and most common of these methods will be described here; the rest are given in our companion technical manual *Tempeh Production.*

Most Indonesian tempeh starter is grown on leaves. In Central and East Java, such leaf-grown tempeh inocula are called *laru,* or *waru,* while in West Java they are called *usar.* The first term, more widely used, is now considered standard Indonesian. Two general Indonesian terms for all types of starters (i.e., for bread, wine, tapeh, tempeh, etc.) are *ragi* and *bibit.* Ragi may also be used specifically to refer to "Indonesian yeast cakes," widely sold in the form of small beige discs and used primarily to inoculate *tapeh.* The term *ragi tempeh* (i.e., ragi for tempeh), which one hears frequently in Indonesia, is virtually always used as a synonym for *laru* or *waru,* rather than to refer to these yeast cakes used to inoculate tempeh.

While there are both advantages and disadvantages to the use of mixed cultures (see Appendix E in the professional edition), the methods themselves yield starters that can be prepared locally at little or no cost and that produce high-quality tempeh. Since volume or weight of the inoculum, however, cannot be easily measured, the amount used must be estimated, which requires considerable experience and skill, and can occasionally result in failures. Moreover, there may be damage from insects when starters are stored for longer than 1 to 2 weeks. To maintain maximum spore vitality, starters should be used within one week after preparation, since the warm and humid climate causes a rapid decline in their potency.

---

### Sandwiched-Hibiscus-Leaf Method (Laru, Waru, or Usar)

The most popular way of making tempeh starter in Indonesia involves the use of hibiscus leaves *(Hibiscus tiliaceus Linn),* whose common name in Java is *waru puteh.* The green leaves are 6 to 8 inches in diameter and come from a 15- to 30-foot-tall tree that grows throughout the country. Hibiscus is used because the underside

of each leaf is covered with downy hairs (known technically as *trichomes),* to which the mold mycelium and spores can adhere. (In some regions other leaves with hairs, especially teak — *Tectona grandis;* called *jati* in Indonesia — are also used.) This ingenious method is probably so popular because it produces the purest traditional starter. When the mold is grown on soybeans sandwiched between two leaves, the leaves shield the mold from outside alien microorganisms, first during sporulation and later during drying and storage. Most tempeh makers produce a batch of their own hibiscus-leaf starter once each week, using leaves from trees that grow near their shops. In some markets, however, ready-made hibiscus starter leaves (called *laru, waru,* or *usar)* are also sold, and some makers are willing to pay the extra money to save the time and work of having to prepare their own.

*Fig. 7.5: Picking leaves from a hibiscus tree for tempeh starter*

The methods for preparing hibiscus starter leaves vary from shop to shop. The following is that used by the fairly large Oeben shop in Bandung (see Chapter 8).

Line a large slatted wooden tray (approximately 36 by 16 by 1¼ inches) with a sheet of perforated plastic. Place a hibiscus leaf on the bottom of the tray with the underside of the leaf facing up. Sprinkle 30 to 40 cooked, inoculated soybean halves (or whole beans) over the surface of the leaf.

*Fig. 7.6: Arranging inoculated soybeans on hibiscus leaves*

Now unstack the trays and arrange them in a single layer on special starter incubation racks. Allow them to stand for about 24 hours more, or until the mold has sporulated. Transfer the leaf sandwiches to large wooden trays which are not lined or covered with plastic, arranging the sandwiches in a single layer over the bottom of each tray. Place the trays back in the racks and allow to stand for 3 to 6 more days, or until the leaves are well dried (which facilitates removal of the beans and spores) and the edges of each leaf are slightly upturned. (In many shops, the leaves are then briefly sun-dried.)

*Fig. 7.8: Hibiscus leaves for tempeh starter ready to use*

Then place a second leaf of about the same size congruently, underside down, atop the first leaf to form a sandwich of two leaves with inoculated soybeans between them. Proceed to make 50 or 60 such sandwiches until the tray is full, with 4 to 6 sandwich layers. Fold the sides and ends of the plastic over the top of the leaves, and place the tray on a double thickness of gunny sacking (jute bags) on a clean section of floor or on a sturdy bench or shelf. Prepare 4 to 6 more similar trays filled with hibiscus-leaf "sandwiches" and stack them atop the first tray, then cover the stack with a single gunny sack and allow to stand for 5 to 6 hours, while the heat of fermentation develops.

*Fig. 7.7: Covering hibiscus leaf sandwiches in trays*

To test the leaves for doneness, pull one pair gently apart. The two under surfaces should be bound together by and covered with a black or dark-gray mycelium. The individual soybean particles, which are also covered with this mycelium, should be firm and quite dry.

A large leaf used during a warm season will inoculate up to 20 pounds of dry soybeans that have been cooked; a small leaf used during a cool or cold season will inoculate only about 6 pounds.

VARIATIONS

While basically the same method is used in most shops, there are a number of interesting variations suited to differences of scale and available equipment. The following are used at one or more shops we visited: 20-inch-diameter trays made of woven raffia or split bamboo; 7 to 10 pairs of leaves are placed in a single layer on each tray.

In cold weather, cakes of finished tempeh are crumbled and spread over the surface of each leaf in place of the inoculated soybeans; thin boards are placed atop the layer of leaves on the tray and allowed to press for 2 days to prevent the leaf edges from curling; the trays and leaves are then placed on the rooftop in direct sunlight for 4 more days, being taken in each night after sunset. They are now ready to use.

Fig. 7.10: Drying inoculum leaves in sun on roof

To store leaves that will be used later, a piece of string is tied to the stems of each pair and they are hung indoors from the shop's rafters.

Fig. 7.11: Tying inoculum leaves under rafters to dry

In 1935 the Dutchman Burkill wrote that to make the inoculum "a portion of the older preparation [tempeh] is wrapped in a rather young teak leaf freely punctured with holes; this preparation is allowed to dry somewhat for two days, during which the fungus spreads to the leaf. Next, the cooked soybeans being ready, the teak leaf is emptied of its contents, cut fine, and sprinkled over the beans in order to convey the fungus." Finally, the beans are wrapped in banana-leaf packets, heaped together, and covered for 24 hours, after which they are uncovered, cooled, and sold.

Today, finished teak-leaf pairs are often dried in front of a fire (which, unfortunately, kills some of the spores), then crumbled over the cooked beans, the leaf being added to the beans together with the spores.

# 8

# The Indonesian Tempeh Shop

Having developed over a span of many centuries under fairly diverse conditions, Indonesia's tempeh shops can serve as a model for low-technology tempeh production in a tropical climate. Such a climate is ideal for making tempeh since it can be incubated at the natural temperature of the environment, the required microorganisms grow quickly and profusely, and organic wrapping materials such as banana leaves are available in abundance. Perhaps most important in the context of tempeh's helping to solve the world hunger problem, is that the great majority of people facing severe protein malnutrition live in tropical regions.

Most of Indonesia's 41,000 tempeh shops are small cottage industries that have an average of 3 workers per shop and use 11½ pounds of dry soybeans each day to produce about 21 pounds of fresh tempeh. Located in or adjacent to the craftsman's home, the shop does not require the use of machines or special equipment. Less than 1 percent of all shops employ 5 workers or more; these use an average of about 78 pounds of dry soybeans per day to produce 137 pounds of tempeh. The largest shops we know of (of which there are but a handful), employ from 10 to 20 workers, use 600 to 1,100 pounds of dry soybeans per day, and produce 1,000 to 2,000 pounds of fresh tempeh.

The overall feeling of the craftsmanship in the Indonesian tempeh shops that we have visited shares that carefree, often happy, and usually hectic feeling that is the carnival of everyday life. There is little of the mindfulness, the feeling of work as a spiritual practice, and the rich aesthetic dimension so evident among the traditional tofu and miso craftsmen in Japan. Even more conspicuous is the lack of attention to cleanliness and sanitation. And it has been a continual source of amazement to us that, whereas in America even the slightest lack of cleanliness seems to greatly affect the tempeh quality, in Indonesia the microorganisms are apparently so hardy and the climate so favorable, that even shops which would horrify U.S. health inspectors produce excellent, delicious and health-giving tempeh day after day.

Although the work is generally done with little sense of artistry and the process appears simple, making tempeh is definitely an art which requires a great deal of experience and practice to master, in order to obtain fine tempeh every time. Each step influences the quality of the final product, but the key to the art is mastery of the technique of making and using the starter (inoculum). This step is usually done by the master of the shop or his most experienced assistant. Intuition and sensitivity are required to know just how much inoculum to add under various conditions of temperature and humidity. Most of Indonesia (particularly Java) has a remarkably uniform climate, with an average temperature of about 78° F. (25° C.), a relative humidity of 79 percent, and roughly 200 days of rainfall per year. The temperature drops about 5° F. for each 1,000-foot rise in elevation above sea level, and some areas receive considerably less rain.

Tempeh is made in basically the same way throughout Java, except that in West Java polyethylene bags are used in place of banana leaves as tempeh containers and, in Central and East Java, where fuel is scarce, the beans are generally soaked for 1 to 14 hours before the first boil. A flowchart of the basic method is given in Fig. 8.1.:

Fig. 8.1. Flowchart for Basic Indonesian Soy Tempeh Method

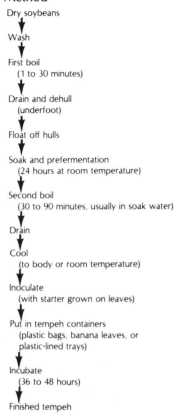

Dry soybeans
↓
Wash
↓
First boil
    (1 to 30 minutes)
↓
Drain and dehull
    (underfoot)
↓
Float off hulls
↓
Soak and prefermentation
    (24 hours at room temperature)
↓
Second boil
    (30 to 90 minutes, usually in soak water)
↓
Drain
↓
Cool
    (to body or room temperature)
↓
Inoculate
    (with starter grown on leaves)
↓
Put in tempeh containers
    (plastic bags, banana leaves, or
    plastic-lined trays)
↓
Incubate
    (36 to 48 hours)
↓
Finished tempeh

In much of the English-language literature describing tempeh production in Indonesia, there have been a number of serious mistakes which, having occurred once, are repeated by subsequent writers who have not studied the process firsthand. It is important to correct these at this time. In virtually all of the tempeh shops that we visited throughout Java and according to the many Indonesian tempeh researchers with whom we talked: (1) the soybeans are dehulled *after* they have been boiled once; (2) an important prefermentation takes place during the 24-hour soak before the second cook; (3) a number of microorganisms in addition to *Rhizopus* molds are considered essential to the preparation of fine tempeh; and (4) the tempeh is rarely inoculated with "some tempeh from a previous fermentation" but rather with a sporulated starter grown on soybeans sandwiched between various types of leaves (see Chapter 7).

In Indonesia, tempeh making is hard work, which starts early in the morning, usually no later than 4:30 A.M. and in some larger shops shortly after midnight. The entire family usually rises and works together. After an hour or two of work, the head of the family will generally take the fresh mature tempeh, which was started three days earlier, to the market, where he sells it before returning home. In the afternoon, he will often supervise inoculation of the next batch of soybeans to be incubated.

Most tempeh shops make a decent income by Indonesian standards, especially considering that the average per capita income for the population as a whole is $180 to $240 per year, or 50 to 65 cents per person per day. A typical shop making 20 pounds of fresh tempeh and retailing it for 27 cents (U.S.) per pound will have an income of $5.40. The cost of the beans (at 19 cents per pound) is $2.19, while the cost of firewood and banana leaves or polyethylene bags raises the total cost to about $2.48. This leaves a daily profit of $2.92 for a family of four or five, which is about the national average. Shops using roughly 65 pounds of dry soybeans per day will have a typical daily profit of $11.25 or 4 to 5 times the national average. We have observed that the families in even the smaller tempeh shops generally seem to be well nourished and happy, and to live comfortably by Indonesian standards.

In this chapter we will begin by describing briefly the methods used in a small-scale tempeh pilot plant now operated as a model in Indonesia; this will help us to develop a familiarity with the basic process and introduce a working model for similar pilot projects in other developing countries. We will then describe the method used to make tempeh at the relatively large-scale and well-known Oeben shop in Bandung. Next we elaborate on a number of popular variations on this basic method, and conclude with a description of the method for making Malang Tempeh, Indonesia's most distinguished variety.

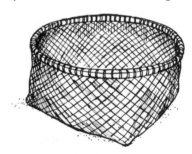

## Preparing Traditional Soy Tempeh *(Small Scale)*

MAKES 27 POUNDS

The scale of this method is about the same as that of a typical Indonesian cottage-industry tempeh shop. The process differs in several minor ways that make it more adaptable to model pilot projects: the beans are cooked in large kitchen kettles (rather than cutoff drum cans) over a gas burner (rather than a wood fire) and a ready-made pure-culture starter (inoculum) is used instead of the typical starter grown in tempeh shops on hibiscus leaves. This process was developed and is used at the prestigious Nutrition Research and Development Institute (GIZI) in Bogor. One man, working several hours each morning, makes the tempeh, which is then sold to a nearby hospital, where it is served to the patients. The floorplan of a traditional shop of this scale, where the beans are cooked in a caldron over a wood fire and the tempeh is incubated on overhead racks under the eaves, is shown in Fig. 8.2.

Fig. 8.2. A small Indonesian tempeh shop

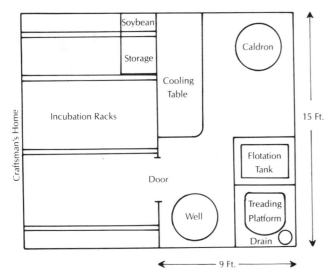

One-half the contents of one pot is poured into an 18-inch-diameter, 4-inch-deep woven bamboo colander, set over a woven basket to raise the colander off the floor; the beans are allowed to cool and drain for several minutes.

Lest the utensils in the following sequence look larger than we describe them, it must be pointed out that the craftsman was of small stature.

**15.4 pounds (7 kilograms) whole dry soybeans, rinsed several times and drained**
**Water**
**8 tablespoons tempeh starter**

The soybeans are divided among two 5-gallon pots and water is added to cover them by 1 to 2 inches. The beans are brought to a boil over two gas burners and simmered for 30 minutes. Any foam that rises to the surface is skimmed off with a colander.

A woven bamboo "treading basket," 16 inches in diameter and 14 inches deep, is placed on a well-washed, sloping tile (or cement) floor near a drain. The drained beans are poured into this basket and doused with water to cool them until they can be touched without discomfort. The craftsman then rinses his feet thoroughly, steps barefooted into the basket, and, supporting himself occasionally against the wall, treads the beans underfoot for 2½ to 3 minutes to dehull them.

THE INDONESIAN TEMPEH SHOP / 127

Using a metal plate or a small bucket, he scoops one-half the dehulled beans into the colander described above. He slowly immerses this in a 3-foot-diameter pot filled with water and floats off the hulls by lifting up the near corner of the colander and stirring or fanning the floating (or partially floating) hulls toward the far edge, where they slowly sink. He then raises the colander, stirs the beans gently, reimmerses the colander and floats off more hulls. This process is repeated four or five times until most of the hulls have been removed.

Meanwhile, the craftsman prepares the tempeh containers, using 22 polyethylene bags, each 11 inches long and 4½ inches wide (an 8-inch-long bag also works well for making smaller tempeh cakes). He stacks the bags congruently atop 4 layers of (terrycloth) toweling on a table, then uses a "bed of nails" (a piece of plywood a little larger than the bags with many nails driven through it ¾ inch apart) to make holes in the bags when the nails are pushed down through the stack. (Or the bags may be perforated with an icepick, using the same hole spacing.)

When the beans have cooled, he washes his hands and sprinkles the tempeh starter evenly over their surface.

He returns these dehulled beans to one of the cooking pots and floats off the hulls from the second half of the treaded beans. Finally he treads and dehulls the beans from the second cooking pot, then transfers them back into that pot. He adds water to cover the beans in each pot by 1 or 2 inches, then adds back several tablespoons of hulls (the natural *Lactobacillus* or *Pediococcus* bacteria on their surface aid the prefermentation, which acidifies the soak water). The beans are now allowed to soak for 24 hours.

The beans are then brought to a boil in their soak water (which may be slightly foamy on the surface) and simmered for 30 minutes. They are poured into the colander and drained for 5 minutes, then transferred to the cooling tray, a 4-foot-diameter, 5-inch-deep woven bamboo colander. They are then spread in an even layer with a wooden spoon and allowed to stand uncovered by an open window for 1½ to 2 hours, being mixed and/or fanned occasionally while they cool and dry.

Then he mixes the beans with both hands for about 2 minutes in order to distribute the starter uniformly.

Using one hand, he now fills each bag with 500 grams (1 pound, 1½ ounces) of inoculated beans, checking the weight on a scale.

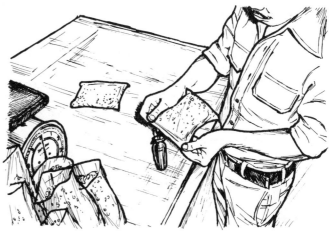

Now he arranges all the bags side by side on a slatted wooden rack (½-inch-wide slats separated by ½-inch spaces) which allows good air circulation, and allows them to stand, uncovered, overnight at room temperature (77° F. or 25° C.). He leaves the bags, which contain a fairly thick layer of beans, unpressed, so that more oxygen can reach the center to aid mold growth.

When all bags have been filled, he folds over the mouth of one bag parallel to the front edge and passes the folded edge slowly (for about 6 to 7 seconds) above the flame from a small alcohol burner or a candle in order to seal the mouth tightly. He proceeds to seal all the bags in this way.

The next morning he wipes the surface of each bag with a towel to remove any excess moisture that may have accumulated around the perforations, then presses the upper surface of each bag firmly by hand to compact the beans and make them into fairly flat cakes that are about 1.4 inches thick at the center and 1 inch thick around the edges. He then allows the beans to incubate for 24 more hours, until the next morning, when the tempeh is ready to be sold.

### Preparing Traditional Soy Tempeh *(Large Scale)*

MAKES 1,155 POUNDS

One of Indonesia's larger and better-known shops is the Oeben tempeh shop in Bandung, which uses 660 pounds of dry beans daily and employs 4 men. The shop is 50 feet long and 25 feet wide; it has a rough (slip-proof) tile floor that slopes toward several drains; the walls are made of cement blocks and the roof of corrugated aluminum sheeting. A floor plan is shown below:

Fig. 8.3. Floor plan of the large Oeben tempeh shop in Bandung, Java

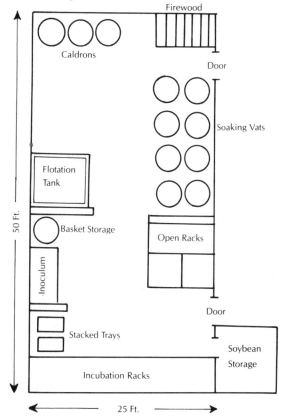

The process used at this shop has a number of unique and interesting characteristics: (1) the tempeh is made with virtually all the hulls mixed in with the dehulled beans; this rather uncommon practice saves time, money, and what little nutrients and dietary fiber are found in the hulls. The tempeh can thus be sold at a lower price, but the quality is also considered lower; (2) the beans are soaked in boiling water before they are dehulled, whereas in most shops they are boiled, dehulled, soaked and then reboiled; not boiling first keeps more bacteria alive to activate the prefermentation, as does soaking the beans with the hulls on; (3) the use of a relatively small amount of water in the caldron during the second cooking means that most of the beans are actual-

ly steamed rather than boiled. Steaming is said to produce more nutritious tempeh, having a less beany flavor; (4) tapioca is added as a growth promoter and moisture absorber; it may be essential in tempeh which contains the hulls; (5) a modern system of using large plastic-lined wooden trays facilitates large-scale production; and (6) because of the large production, the work must start at midnight in order to be finished on time.

The equipment is all inexpensive, locally made, and nonmechanized. The only energy input is from logs used to heat the fire for cooking the soybeans. A detailed description of the various pieces of equipment is given in *Tempeh Production: The Book of Tempeh, Volume II.* Since Indonesian tempeh generally requires three to four days to come to completion (from the beginning of soaking the beans until the end of fermentation) and since most shops make one large batch each day, the workers are actually working on *five* different batches of tempeh at the same time, each batch being at a different stage in its development. In this section, for the sake of simplicity, we will describe the basic process as if only one batch were being made. The ingredients used in making one batch at the Oeben shop are:

**660 pounds (300 kilograms) whole dry U.S. soybeans**
**Water**
**8.8 pounds (4 kilograms) tapioca**
**10 to 14 sets of inoculum-covered hibiscus leaves (laru)**

### THE FIRST DAY

At 3 A.M. the whole dry soybeans are measured out of their sacks into the bamboo colanders, where stones and debris are removed by hand. They are then transferred to a washing tub partially filled with water, and churned well. Any floating debris or chaff is skimmed off with a bamboo colander and discarded; the colander is then used to transfer equal quantities of beans into four wooden soaking vats, which as yet contain no water.

Three 55-gallon drum caldrons are partially filled with water, which is brought to a boil over a wood fire fueled by 12-inch-long pieces of firewood.

At 8 A.M. about 50 gallons of boiling water are transferred with a bucket from the caldrons into each of the four wooden soaking vats in order to just cover the beans, which are then allowed to soak until about 1 P.M. At that time enough cold water is added to each vat to fill it to within several inches of its rim. The beans are then allowed to soak overnight. Bacteria in the walls of the tub from previous batches aid the present prefermentation, and by the next morning a head of foam will usually have risen in each vat.

## THE SECOND DAY

At about 15 minutes after midnight the caldrons are filled about half full of water. A wood fire is started under each and the water brought to a boil. The foam that may have formed atop each soaking vat is skimmed off and discarded, revealing the soak water, which, as a result of the prefermentation, is milky yellow in color and slightly sour or acidic in flavor. Using a bamboo colander, the soaked beans are transferred into 11-inch-deep, 24-inch-diameter bamboo baskets until they fill each basket to about three-fourths of its capacity.

Several bucketfuls of water from a large concrete washing tank are poured over the beans in the baskets to rinse them. Two workers then wash their feet with water and each steps barefoot into a basket. Supporting themselves on the side of the washing tank, they tread the beans underfoot (as if walking in place) for 30 to 50 minutes, in order to dehull them. Every 3 to 5 minutes, without leaving the basket, they pour in 3 to 5 bucketfuls of water and rock the basket vigorously by quickly lifting up one side and then dropping it in order to shift and mix the contents.

When the treading of the beans is finished, the baskets are lifted onto one edge of the washing tank and the beans are rinsed with 3 to 4 bucketfuls of water. They are then hand mixed and plenty of water is run over them with a hose.

Using a bucket, the treaded beans (with or without the hulls mixed in) are transferred into the boiling water in the caldrons until they completely fill each caldron and are mounded at the top. The mounded surface is covered with moistened and well-wrung gunny sacks (jute bags). The beans are now partially boiled and partially steamed over high heat for 1½ to 2 hours, starting from the time they are transferred to the caldron.

At about 4 A.M. a draining rack is placed across the mouth of each soaking vat (which now contains the next day's batch of washed beans). A bamboo basket is placed on each rack and a bucket is used to transfer the cooked beans from the caldron into the basket, where the beans are mounded high. All four baskets are filled on the racks in this way.

Each mound of beans is covered with a gunny sack and a round wooden lid, then topped with a 30- to 40-pound pressing weight (one or more large clean stones); the beans are pressed for about 20 minutes to rid them of excess moisture.

While the beans are cooling, one worker takes 10 to 14 sets of inoculum-covered hibiscus leaves (see Chapter 7) and pulls apart the two leaves in each set. He picks off and discards the old dried beans, leaving mostly black mycelium (mixed with a little white) attached to the underside of each leaf.

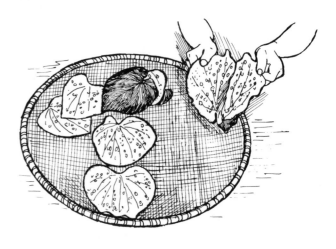

While the beans are draining, the floor of the cooling area is cleared and swept, and the shallow 65-inch-diameter woven bamboo cooling tray is placed at the center of the clean area. Two workers now lift up a basket of drained beans on its draining rack, carry it to the cooling tray, and pour the beans into the tray. A second basket is added in the same way, so that the soybeans are now mounded fairly high on the tray. An electric fan placed one foot from the tray is turned on and used to cool and dry the beans. The beans are mixed constantly with a wooden hoelike mixer for about 20 minutes, or until they are well dried.

When the beans have cooled to body temperature and are fairly well dried, the fan is turned off and 2.2 pounds (1 kilogram) of tapioca are sprinkled over the surface of the beans and mixed in thoroughly with the mixer. The fan is now turned on for about 7 minutes more and then turned off.

Two workers now each rub two hibiscus inoculum leaves against each other above the pile of beans in order to rub off some of the spores. They then rub the leaves face down over the surface of the beans for 3 to 4 minutes.

Now they mix the beans by hand for 6 to 7 minutes in order to distribute the spores evenly throughout. (During this mixing process — as throughout the entire process of making tempeh — there is no special washing of hands or other apparent efforts to ensure cleanliness.) Finally, a bamboo colander is used to scoop the inoculated beans into a bamboo basket (which has previously been dried over an empty caldron).

Either trays or polyethylene bags are now filled with the beans. Wooden tempeh trays (each 36 by 17 by 1.3 inches deep and having a slatted bottom) are lined with a piece of perforated plastic sheeting. The inoculated beans are ladled in so that the tray is just filled, then they are packed lightly and the surface is smoothed with a flat stick. Both sides and then both ends of the sheeting are folded over neatly to cover the beans' surface. A double layer of gunny sacking is placed on the floor for insulation in an out-of-the way place; the first filled tray is placed on the sacking.

Additional trays are filled in the same way and stacked atop the first tray.

When about 20 trays have been stacked in this way, an empty tray is inverted on the top tray and then covered with a gunny sack. Two to four such stacks are made until all of the beans have been used. It is now 9 A.M. The beans are allowed to incubate in the trays for about 8 hours at roughly 77° F. (25° C.), the temperature inside the shop, while they develop their own heat of fermentation.

At 5 P.M., the trays are unstacked and arranged in single layers in floor-to-ceiling incubation racks. Here they are left to incubate for 14 hours.

## THE THIRD DAY

At 7 A.M. two workers remove the first tempeh tray from the incubation racks, place an empty tray of the same size bottom down atop the tempeh-filled tray, and then quickly invert the two trays, leaving the tempeh resting upside down on the bottom of the inverted empty tray, which is placed on any open surface inside the shop. All the trays are inverted in this way in various places and allowed to stand for about 17 hours.

## THE FOURTH DAY

Shortly after midnight, after the tempeh has incubated for a total of 39 hours, it is finished and ready to be cut and sold. Each of the trays is again inverted on the bottom of an empty tray, the plastic sheeting unwrapped, and the trays placed outside on the concrete-surfaced, well-swept courtyard. Each pallet of tempeh is now about 1¼ inches (3.2 centimeters) thick. Using a wooden cutting guide and a knife, the tempeh from the large trays is cut vertically lengthwise into halves and then crosswise into sevenths to make a total of fourteen cakes from each pallet.

Each cake weighs about 1.3 pounds (600 grams), wholesales for 18 cents (U.S.) and retails for 22 to 24 cents. Some pallets are sold in their uncut form. For others, after the individual cakes are cut, the plastic sheeting is folded back over them (to provide optimum sanitary protection) and they are sent out on the tray to market.

At about 1 in the morning, as soon as the tempeh has been cut, men who will sell the tempeh in local marketplaces begin to arrive with their *bechaks* (3-wheeled pedicycles), which they load with cut or uncut pallets.

After work, after the previous day's plastic sheeting has been returned by the market sellers, the sheeting together with all tools are washed in cold water. The sheets are then dried in the sun on a clothesline. After a sheet has been used for one week it is discarded.

## Variations in Dehulling and Floating off the Hulls

**Variation 1. Floating off the hulls in a washing tank:** After the beans have been dehulled, the mixed beans and hulls may be placed in a shallow bamboo colander, which is immersed in a water-filled drum.

The hulls are then floated off in somewhat the same way as described on page 128.

**Variation 2. Dehulling the beans and floating off the hulls in a stream:** The precooked beans are treaded underfoot on a wooden or concrete platform above a riverbank. The basket is then taken down to the river and gently immersed until it just fills with water. The craftsman stirs the beans with one hand to help the hulls float to the surface, then lowers the downstream corner of the basket about 1 inch below the water surface while lifting the upstream corner so that the hulls float out of the basket and are carried away downstream by the current.

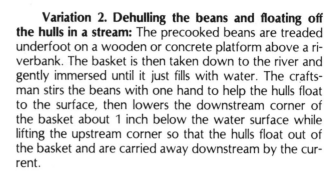

**Variation 3. Dehulling by hand and pouring off the hulls from a drum can:** About 10 gallons of precooked soybeans are put in a watertight cutoff drum can. Enough water is added to cover the beans and then the craftsman vigorously rubs them between his hands to dehull them.

After repeated rubbings, additional water is added to the container, the beans are stirred, and then the water is poured off, carrying the hulls with it. The process of stirring and pouring off is repeated several times.

A low-technology dehuller-separator has been developed at the Bandung Institute of Technology. The beans are placed in a hopper and are then dehulled between a hand-turned scored wooden roller and a fixed plate, whence they fall into a large compartment into which water is piped from below. The beans sink onto a screen at the bottom of the compartment while the hulls float over a spillway with the overflowing water and are caught on a separate screen.

### Modern Dehullers and Dehuller-Separators

The key piece of equipment needed to transform a small-scale shop into a larger operation is a dehuller or, even better, a dehuller-separator, which separates and removes the hulls (seed coats) from the beans (cotyledons). The motor-driven wet dehuller shown below is used at the Tempeh Murni shop in Yogyakarta and was built by the owner for about $125 (U.S.). It handles 165 pounds of dry soybeans (that have been soaked) per hour and contains a ½-horsepower electric motor that drives one of a pair of millstones at low rpm. The precooked beans fall from the hopper down between the stones into a rectangular metal catch box. The hulls are separated from the cotyledons, using the method described in Variation 1, above.

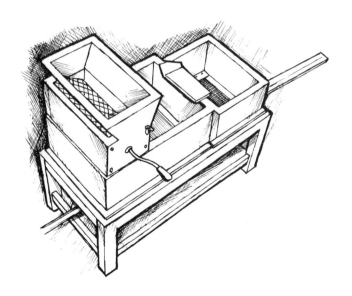

Dry dehullers, in which the beans are cracked between wooden rollers, have also been developed, but they are not as widely used since the grits they yield give tempeh with texture that is not popular in Indonesia; and while the grits are soaking, there is a relatively large loss of solid nutrients and protein.

### Variations in Tempeh Containers and Incubation

**Variation 1. Incubating and selling the tempeh in plastic bags:** At present, an estimated 90 percent of the tempeh made in West Java is sold in polyethylene bags; the figure drops to about 30 percent in Bali and 10 percent in Central and East Java. The bags range in width from 4 to 14 inches and in length from 6 to 17 inches. The basic method of use was described in the section on the small-scale shop earlier in this chapter. In most commercial shops, however, after the bags have been filled and sealed, they are stacked and tapped flat to a thickness of ½ to 1 inch with a wooden tool shaped like a mason's trowel.

In some shops, they are then placed in a simple no-energy-input incubator, which is covered with gunny sacking to keep in the heat that develops during fermentation. Here they are kept for 20 to 24 hours.

Finally, they are arranged on long incubation trays made of woven bamboo and allowed to finish the incubation period in open shelves at room temperature.

The finished tempeh may then be packed into a number of 5-gallon cans attached to a rack on the back of a bicycle. In one of his four cans the tempeh maker shown below has put a bucket containing fresh tofu immersed in water. He purchased the tofu from a nearby tofu shop and will sell it at the market together with his tempeh since both are popular low-cost soy-protein foods.

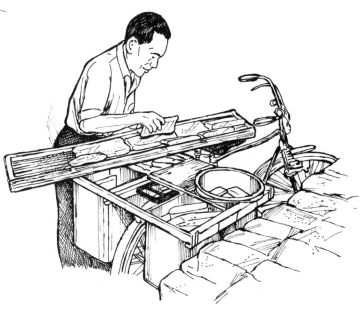

In some shops, a boy from the family will fill two baskets with the fresh tempeh, attach the baskets to the ends of a shoulder pole, and sell his wares along the streets or in a market.

each other as shown below, then ¾ cup of inoculated beans are placed at the center of the leaves. First both sides and then both ends are folded over to form a compact packet. In some shops the packets are tied shut with a strand of rice straw or split bamboo.

**Variation 2. Incubating the tempeh in banana-leaf wrappers:** Since ancient times, especially in Central and East Java, tempeh has been incubated wrapped in banana leaves. Various sizes and shapes are produced as shown below. The most popular weighs 15 to 25 grams (0.5 to 0.9 ounces), while the largest weighs 95 to 100 grams.

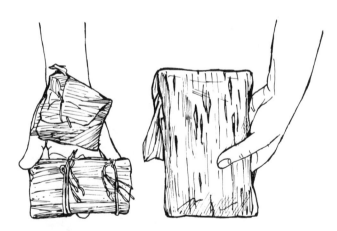

To make the larger size, fresh whole banana leaves (gathered locally or purchased at the market) are arranged in a stack 1 to 2 inches thick and perforated with an icepick at intervals of ¾ to 1 inch. They are then cut crosswise into sections about 12 inches long. Five to six such rectangles are arranged on a low table overlapping

Finally the packets are arranged slightly overlapping one another on slatted shelves and allowed to incubate at air temperature for 36 to 48 hours.

Other incubation wrappers include the leaves of hibiscus *(Hibiscus tiliaceus),* teak *(Tectona grandis; jati),* and other large-leafed plants; sections of hollow bamboo; banana-stalk sheaves formed into hollow cylinders; and shallow woven bamboo trays lined with banana leaves.

## Preparing Malang Tempeh
### (Firm & Thick Soy Tempeh)
MAKES 90 TO 115 POUNDS

The aristocrat of Indonesian tempehs, this variety is firm, thick, and white, being made and sold in large pieces, as described in Appendix C of the professional edition. The method of preparation is somewhat similar to that of regular soy tempeh up to the stage of inoculation and incubation, except that the soaking period and prefermentation between the two cookings is often omitted or is relatively short, and the beans are cooled and dried for a relatively long time. Thereafter, the process is unique and well suited for large-scale production. Thus we will discuss the first half of the process only briefly, emphasizing its key points, then go into a detailed illustrated discussion of the latter half. The fact that banana leaves play an important role in the incubation technique does not mean, necessarily, that this type of tempeh could not be made in areas such as North America, where such leaves are not generally available. The combined use of perforated plastic sheeting as a liner and gunny sacking for insulation would probably make a good substitute.

A flowchart for the Malang Tempeh process looks like this:

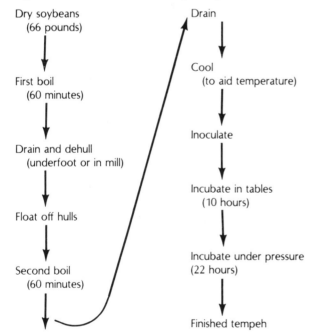

Fig. 8.4.   Flowchart for Malang Tempeh

Dry soybeans
(66 pounds)

First boil
(60 minutes)

Drain and dehull
(underfoot or in mill)

Float off hulls

Second boil
(60 minutes)

Drain

Cool
(to aid temperature)

Inoculate

Incubate in tables
(10 hours)

Incubate under pressure
(22 hours)

Finished tempeh

There are some important possible variations in the process; at the Manan B. Samun shop in Bogor (see Appendix F, professional edition) we observed the following: the beans were washed before the first boil; the first boil took only 20 minutes; after the hulls were floated off, the beans were given a short soak-and-prefermentation of 4 to 6 hours in fairly hot water; the second boil in the soak water took only 40 minutes. Since there was no communal dehulling pond nearby, the beans were treaded in baskets on the sloping brick floor of the dehulling room (located adjacent to the shop) and the hulls floated off in a water-filled concrete washing tank, 4½ feet square and 18 inches deep, located in one corner of the same room. The quality of the tempeh was excellent.

In the village of Sanan north of Malang there is a group of more than 25 shops known as the Malang Tempeh Cooperative. Some of the producers use stone wheel dehullers with a single 12-inch-diameter movable stone, while others dehull their beans by treading them underwater in a shallow communal pool. The relatively cool and very pure air in Malang is considered to be a key factor in the production of the fine tempeh.

In the method described here, the tempeh is made by a master with the help of two or three assistants, in this case his wife and two daughters. One batch is prepared daily and harvested two days later.

**66 pounds (30 kilograms) American-grown soybeans**
**2 hibiscus leaves with sporulated *Rhizopus* mold spores**
**attached to them (see Chapter 7)**

THE FIRST DAY

At 4 A.M. the master measures out the soybeans and cooks them for 60 minutes in excess water indoors over a wood fire in a metal drum-can caldron 14½ inches in diameter and 21¼ inches deep. Using a perforated cone-shaped scoop attached to a wooden handle, he drains and transfers the beans from the caldron into a number of woven bamboo baskets, which are carried to the nearby communal dehulling pool. At about 6:30 the

master's wife tucks up her dress and wades into the 10- to 20-foot-diameter natural (or brick-lined) pond, which is about 15 inches deep. Well water enters the pond at one side and flows out a channel drain at the other. She washes her feet in the 77° F. (25° C.) water, places the bean-filled basket on one of the many small raised platforms 4 to 6 inches below the water surface, steps into the basket, and treads the beans underfoot for about 5 minutes, or until they are dehulled.

She then pours the dehulled beans into a bamboo flotation tray 3 feet in diameter and 1½ inches deep. Still standing in the pool, she proceeds to float off the hulls, then carries the dehulled beans in baskets back to the shop. (At the end of the day the communal pond is drained and the hulls, caught in a bamboo screen at the drain, are saved for use as fodder.) Combining the dehulled beans in the caldron with excess fresh water, she gives them a second cooking for 60 minutes. They are then drained and transferred to a 5-foot-diameter, 6-inch-deep cooling tray, which is set on a table next to a small electric fan. The beans are mixed from time to time to aid cooling and evaporation of excess moisture. The fan is run for 2 to 6 hours, depending on the weather, until the beans are cooled to air temperature (77° F., 25° C.) and well dried.

Now he mixes all the beans together for 1 minute more, tossing them with both hands lightly into the air over the tray.

The inoculated beans will be incubated in a large traylike incubation table lined with perforated banana leaves. The table is 10 feet 10¾ inches long, 24¾ inches wide, and has sides 2¼ inches deep. The bottom is made of 1-inch-wide bamboo slats that run lengthwise and are spaced 1 inch apart. The shop contains two such tables, each of which stands about 38 inches tall; they are filled on alternate days. In some shops, 20-inch-diameter trays (called *tampahs*) are used in place of the incubation tables.

In the evening, usually about 7:30, the master inoculates the beans. He takes two hibiscus leaves with sporulated beans attached to them and holds them against the palms of his two hands, somewhat as one might hold two potholders. Then, in a motion resembling that of washing hands, he vigorously rubs the freshly cooled beans between the two leaves for about 1 minute.

To perforate the banana leaves (a variety called *batu,* which are picked fresh that day near the shop or purchased in the market), the master arranges 15 to 20 at a time in a neat stack and folds the stack in half end to end. Sitting on a workbench and placing the bundle before him, he uses an icepick to perforate all the leaves at ¾-inch intervals.

A second identical layer is placed atop the first, and then a third is placed on the second, this time with the smooth side facing down.

The inoculated beans are then transferred from the cooling tray to baskets and poured onto the leaves in the incubation table.

He now proceeds to line the table with one layer of perforated leaves, having the smooth, glossy bright-green upper surface of each leaf facing upward. The leaves are arranged lengthwise on the table with each leaf overlapping the one beside it by about 3 inches.

Using one hand, the master smooths the beans over the entire area to form a uniform layer 1.4 inches (3.5 centimeters) deep. He repeatedly gauges the depth of the layer by holding the palm of his hand flat against the surface of the beans, pressing his middle finger down until it touches the leaf-lined bottom.

The beans are now covered with a layer of perforated banana leaves having the smooth sides up. The edges of the bottom layers of banana leaves sticking up around the sides of the table are folded over and the beans are covered with two more layers of leaves, each smooth side down. The beans are allowed to incubate at room temperature (77° F., 25° C.) overnight for about 10 hours.

## THE SECOND DAY

At about 6 A.M., after cooking the day's soybeans, the master covers the surface of the banana leaves on top of the tempeh with a single layer of about 70 bricks, spaced ¾ to 1 inch apart on all sides to allow some air to continue to permeate the leaves. The incubation then continues all day and night for about 22 hours.

In the evening, the next batch of inoculated beans is incubated in the second table.

## THE THIRD DAY

At about 4 A.M. the master removes the bricks and upper layer of banana leaves. Using a measuring board about 5 feet long, 2¼ inches wide, and ⅜ inch thick, and a sharp-pointed knife, he cuts the tempeh lengthwise into strips about 5¼ inches wide. He then uses a similar board 24 inches long to cut it crosswise into 14-inch-wide strips.

While cutting he always holds the knife at a steep angle (about 45 degrees to the surface of the tempeh); cutting in this way makes the pieces look larger than they actually are, aids permeation of flavors while cooking, and allows many large slices to be cut at home from a single piece (see Preparatory Techniques for Tempeh).

Each day's production of tempeh (1 tableful) yields about 48 cakes, each 7 by 13 by 1 inch and weighing 28.6 ounces (810 grams). The cakes at the edges of the table are as thick as 1¼ inches; the mycelium is densest and whitest within 4 inches of the sides of the table and somewhat sparser near the table's center. Sold whole, or cut into halves or quarters, a typical piece retailed in the Bogor market in 1977 for 48 cents (U.S.) or the equivalent of 26.9 cents per pound.

The master packs the cut tempeh pieces carefully into a slatted wooden box or a cardboard box, preferably with holes in the sides; the slats or holes provide good air circulation, which prevents overheating.

The master and his wife carry the tempeh-filled boxes to a nearby road. The master hails a microbus, loads on the tempeh, and is off to market. If all goes well, he will have sold it all by noon and be back to help with the afternoon work. The wife stays at home to attend to the dehulling and second cooking of the next batch of soybeans.

Southeast Asia

Java, Madura, and Bali

# Appendix A

## A Brief History of Tempeh East and West

### History of Tempeh in Indonesia

**Early History (pre 1875):** Since ancient times the Malay language has been the lingua franca of the archipelago that includes today's Malaysia and Indonesia. The people of Java have had a written language since antiquity, with existing stone inscriptions dating from the seventh century A.D. This early literature concerned primarily religion, philosophy, and culture, with very little information about food.

The world's earliest known reference to tempeh appeared in the *Serat Centini,* which was probably written around A.D. 1815 on the orders of Sunan Sugih, then Crown Prince and later Pakubuwana V of Surakarta, in today's eastern Central Java. The main author was probably Rangga Sutrasna. This classic work of Modern Javanese literature contains one line mentioning "onions and uncooked *témpé.*"

Although the *Serat Centini* was written in about 1815, it is quite possibly based on much older sources; the story is *set* in the reign of Sultan Agung (1613-45), and the descriptions purport to be of that time. Thus tempeh may well have existed in the early 1600s. However the actual document in which this reference appears (Codex Orientalis 1814 of the Leiden University Library) bears the date 1846, making it conceivable (but highly unlikely) that reference to tempeh was added just prior to publication.

The *Serat Centini,* written in verse, tells of the adventures of "students" wandering in the Javanese countryside in search of truth. In the course of the story, detailed information is given on many subjects including Javanese culture and life. The passage mentioning tempeh occurs in a description of Wanamarta, a prosperous place, in the context of a reception given to Jayengwesti, and involving all sorts of foods. These, including "onions and uncooked *témpé,*" are simply listed without further information.

Conservative estimates that tempeh originated at least several centuries ago are also supported by evidence based on the food's present widespread geographical distribution, popularity, and large number of varieties. Tempeh is known in even the most remote rural areas throughout most of Java, is an integral part of the cuisine served in a wide variety of popular dishes (90 named Indonesian recipes are given in this book), and by the mid-1970s it was being made from at least 17 indigenous seeds and presscakes by more than 41,000 shops, using simple, traditional methods.

But where did tempeh come from? The earliest known written record of soybeans in Indonesia was by the Dutch botanist Rumphius (1747), who reported that soybeans were being used in Java for food and as green manure. Yet soybeans may well have been introduced to Indonesia at the time that regular trade started with south China in about 1000 A.D. Tempeh may have developed from an earlier fermentation used for making the now famous coconut presscake tempeh *(tempeh bongkrek).* Or tempeh may have been introduced by the Chinese who were making a similar product, the soybean koji for their soy sauce, produced by inoculating cooked dehulled soybeans with wild molds such as *Aspergillus oryzae;* the use of *Rhizopus* in Indonesia may have been due to its better adaptation to the Indonesian climate. One piece of evidence supporting the Chinese Introduction Theory is that in 1931, in Peking, William Morse of the USDA observed, described, and photographed a fermented soyfood closely resembling tempeh and called *tou chiah ping* ("soybean fried cake"; Dorsett and Morse 1928–31). It would be very interesting to know more about *tou chiah ping*: Is it fermented with *Rhizopus?* What is known of its history? Unfortunately there are no known references to it after 1931.

The rise of tempeh's popularity in West Java (where the culture is Sundanese), and its spread to other Indonesian islands and other countries of the world, probably began in the 20th century. We hope that Indonesian scholars will soon begin a serious search of their literature to help us construct a more reliable picture of tempeh's early history.

**Early European References (1875–1939):** Since Indonesia (formerly the Dutch or Netherlands East Indies) had been a Dutch colony since the late 1600s, it was only natural that the first Westerners to study tempeh came from Holland. The earliest known reference to tempeh (actually tempe) by a European appeared in 1875 in a Javanese-Dutch dictionary, the *Javaansch-Nederduitsch Handwoordenboek* by J. F. C. Gericke and T. Roorda. The term was defined as "Fermented soybeans or presscake *(bunkil)* baked or fried in flat pressed cakes. It is well-liked as a side dish with rice."

In 1895 the Dutch microbiologist and chemist H. C. Prinsen Geerligs made the first attempt to identify the tempeh mold in his classic article titled "Eenige Chineesche voedingsmiddelen uit sojaboonen bereid" (Some Chinese Foods Made with Soybeans). After describing Indonesian soy sauce, and miso *(taucho)*, he noted: "In a similar way, in Java, other molds are used to make leguminous seeds into more digestible foods. Thus the presscake, which remains after making peanut oil and would be indigestible without further preparation, is subjected to the action of molds. In central and eastern Java *Chlamydomucor Oryzae* [now known as *Amylomyces rouxii*] is used, whereas in western Java an orange mold of the family *Oospore [Neurospora]* is used. In the

former case, the food is called 'bongkrek,' and in the latter 'ontjom.' If soybeans are molded with *Chlamydomucor* the spice is called 'tempets' [sic]. In the preparation the seeds are boiled, spread, mixed with a little molded cake from a former batch, and left alone for a while, until the mass is bound into a solid cake." A year later, when this article was published in German, he corrected two mistakes he had made in the 1895 Dutch version. He changed the name of the mold from *Chlamydomucor Oryzae* to *Rhizopus Oryzae* and he changed the name of the product from "tempets" to "tempeh." Prinsen Geerligs' two articles ushered in the era of scientific research on tempeh by European microbiologists and food scientists.

In 1900 the Dutchman Dr. P. A. Boorsma, who lived in Java and did original laboratory tests, published an excellent 13-page article on soybeans, including a detailed 4-page description of the traditional process for making *Tempe kedeleh* (kedeleh = soybeans), and of the chemical and compositional changes taking place during tempeh fermentation.

In 1902 the Dutch microbiologist Vorderman gave the first descriptions of other varieties of Indonesian tempeh and onchom, including those made with coconuts and coconut-soy mixtures. He noted that coconut tempehs had led to several cases of fatal food poisoning.

In 1913, K. Heyne published a lengthy review of earlier literature on tempeh. The Dutchman Jansen (1923) showed that vitamin $B_1$ in tempeh was reduced during fermentation. Jansen and Donath (1924) showed that tempeh protein is of good quality and makes a good supplement to the protein in rice. The content of vitamins $B_1$ and $B_2$ in tempeh was further investigated by A. G. van Veen (1932, 1935); he found it to be a good source of both.

The first English-language information about tempeh appeared in 1931 in J. J. Ochse's *Vegetables of Dutch East Indies,* an excellent 1005-page tome published in Buitenzorg (today's Bogor), Java. Ochse, a Dutchman, described the tempeh-making process in detail, saying that the mold used was *Rhizopus oryzae,* and that it was obtained from a former batch of tempeh. The next English-language reference appeared in 1935 in I. H. Burkill's *A Dictionary of the Economic Products of the Malay Peninsula,* a two-volume, 2,400-page work published in England. It contained six pages of information about tempeh and other soyfoods, including a description of the tempeh-making process. Burkill was a British authority on the flora of southern and southeastern Asia.

**Tempeh During World War II and the Postwar Era (1940–1959):** During World War II tempeh served as an important food in Indonesia and surrounding countries, both for the native population and for foreigners in Japanese prisoner-of-war (POW) camps there.

The first English-language article specifically about tempeh was written in 1946 by Gerold Stahel, director of the Agricultural Experiment Station in Paramaribo, Surinam (a Dutch colony). Stahel described how, during

World War II, the United States shipped soybeans to New Guinea in order to feed the Europeans and Indonesians living there. During the Japanese occupation of New Guinea, tempeh production had stopped and the local New Guinea starter cultures had, therefore, all been lost. Stahel, asked to furnish new cultures from Surinam, sent both fresh tempeh cakes and pure-culture starters to the Netherlands Indies Civil Administration (NICA) in New Guinea. Soon NICA kitchens all over the territory started using the U.S. soybeans to make tempeh for the people. Also in his 1947 article Stahel described in detail the way Javanese women in Surinam made and sold tempeh.

Important studies on tempeh in POW camps (a number of them written by POWs) were published by Roelofsen (1946,1964), van Veen (1946), De Bruyn, van Dulst, and van Veen (1947), Smith and Woodruff (1951), and Grant (1952). Most concluded that many POWs owed their very survival to tempeh.

The first study in English on the chemical and microbiological changes occurring during tempeh fermentation was published in 1950 by the Dutch microbiologists van Veen and Schaefer. This classic paper was based partly on van Veen's experiences in a POW camp.

Other than the *Serat Centini* (1815, 1846), the earliest known reference to tempeh in Indonesian or by an Indonesian appeared in 1956, when Soetan mentioned it briefly in a booklet entitled *Kedelai* (Soybeans).

It is curious to note that, despite the fact that tempeh has long been a very important and widely used Indonesian food, all of the scientific studies on tempeh from 1895 to 1960 (and virtually all of the references to it in any language) were done by Europeans living in Indonesia. There are several reasons for this: First, while Indonesia was a Dutch colony there were very few Indonesian food scientists or microbiologists, and these were not encouraged to study indigenous foods. Second, during Dutch colonial rule Western foods were prized and an indigenous food such as tempeh acquired the image of an inferior, lower-class, or even poor-people's food, even though it was consumed by people of all classes. Indonesian scientists did not feel it was worthy of their attention or research. Unfortunately, this attitude persisted even after independence. Sukarno, President of the Indonesian Republic from 1945–1967, admonished his fellow citizens on numerous occasions, saying "Don't be a tempeh nation," or "Don't be a tempeh scientist," implying that tempeh was somehow second class or inferior. Only by the mid-1960s did that image begin to change. And third, there was little interest in tempeh outside of Indonesia to stimulate interest inside.

**New Interest in Tempeh (1960–82):** A new wave of worldwide interest in tempeh began in the early 1960s, sparked largely by the initiation of tempeh research on the part of two groups of American microbiologists and food scientists: one at Cornell University's New York State Agricultural Experiment Station in Geneva, New York, and the other at the USDA Northern Regional Research Center at Peoria, Illinois. Each group had an Indonesian as a catalyst and co-worker for its tempeh research. The Cornell group worked with Ms. Yap Bwee Hwa, the first Indonesian to do scientific research on tempeh and to write a post-graduate thesis on the subject. She won a Fulbright scholarship to America and began work at Cornell in 1958, where she was the first to ever make tempeh in America. From the pulverized sample of tempeh that Yap brought with her from Indonesia, the group isolated the culture of *Rhizopus oligosporus,* which Dr. Hesseltine later identified and gave the number NRRL 2710. This is still the most widely used tempeh culture strain in the U.S.A.

The USDA group worked with Ko Swan Djien, starting in 1960. The second Indonesian to publish scientific research on tempeh, Ko was author or co-author of at least six important articles about tempeh. His first two articles (1961, 1964), both written in English, were packed with original information. Ko played a key role in introducing this food to the West, and in giving it a better image in Indonesia. In his 1964 article Ko signaled what he hoped would be the beginning of a new image for tempeh in Indonesia: "But there is no doubt that the time will come when Indonesians will be proud of their tempe, in the same way as the Japanese are proud of their sake, the French people of their wine, Italians of their macaroni, Indians of their curry, Russians of their caviar, the Dutch of their cheese, etc."

During the 1960s, at the microbiology laboratory in Bandung, Ko worked to stimulate new research on and interest in tempeh. When Indonesian newspaper reporters finally discovered that he had studied tempeh at a university and in the United States, they were simply astonished. Articles with bold headlines such as "Tempeh Steps to a Higher Throne" appeared in several widely read Indonesian newspapers in September 1965. This marked the beginning of a change in attitude toward tempeh in Indonesia.

Yap and Ko had pioneered the way for Indonesians to do research on tempeh in the United States. Many others (including Iljas, Sudarmadji, and Rivai) followed in their footsteps and published important works, starting in the late 1960s.

Interest in and publications about tempeh in Indonesia increased rapidly after the late 1960s. In 1967 the Indonesian Department of Agriculture published *Mustika Rasa* (Gems of Taste), a huge (1,123-page) cookbook of the best recipes from throughout the Indonesian archipelago. Referred to as the "Bible" of local cooks, it contained 35 Indonesian tempeh recipes and seven onchom recipes. Sastroamijoyo (1971) was the first Indonesian to suggest that tempeh offered an answer to the world food crisis. Hermana was senior author of six important articles between 1970 and 1974, and Indrawati Gandjar wrote the first two of her many publications on tempeh in 1972. Thio, Winarno, and numerous others published on tempeh in Indonesia during the 1970s.

The ASEAN Protein Project served as a major stimulus for additional research on tempeh by Indonesians, and numerous papers were published in its periodical progress reports (Saono et al. 1974, 1976, 1977; Suhadi 1979; Jutono 1979; Hartadi 1980).

Starting in the late 1960s and early 1970s a number of changes began to take place in the process for making tempeh in Indonesia. Polyethylene bags (and, to a more limited extent, wooden trays lined with plastic sheeting), first applied to tempeh production in the U.S.A. by Martinelli and Hesseltine in 1964, began to replace banana leaves as the container in which the tempeh was incubated and sold. Semi-pure culture rice-based tempeh starters began to replace traditional soybean-based starters. Imported soybeans rapidly replaced their smaller domestic counterparts.

The first detailed and comprehensive survey of the tempeh industry in Indonesia was published by Winarno and co-workers in 1976. It reported that, at that time, tempeh was the nation's most popular soyfood, making use of 64 percent of the country's total soybean production and imports. There were 41,201 tempeh manufacturers, which made fresh tempeh daily. They employed a total of 128,000 workers, who produced each year 153,895 tonnes (metric tons) of tempeh having a retail value of U.S. $85.5 million. Most companies were small, run out of the home. The largest companies used no more than 100 kg of soybeans a day to make 175 kg (385 lb.) of tempeh. (This would be 1,050 kg or 2,310 lb. of tempeh per 6-day week.) Tempeh was an important source of high-quality, low-cost protein and vitamins in the diet of all Indonesian socio-economic groups, and especially in the diet of low-income families. Yet its importance should not be exaggerated. Per capita consumption for all Indonesians in 1976 was about 16 gm a day or 5.8 kg (12.8 lb.) a year.

On 11 March 1979 a key event took place in Indonesia with the organization of KOPTI, the Cooperative of Tempeh and Tofu Producers of Indonesia, with Achmad Rouzni Noor as director in Jakarta. By 1983 KOPTI had over 28,000 tempeh and tofu companies as members in Java; 72 percent of these ran home industries.

In part because of KOPTI, tempeh production was on the upswing in Indonesia by the early 1980s and the industry was modernizing. In 1984 Ko Swan Djien was able to write: "From my recent visit to Indonesia I get the satisfactory feeling that our efforts to have fermented foods valued in their right proportion are not in vain. Tempe is no longer considered an inferior food. Nowadays Indonesians are as proud of THEIR tempe as Japanese are of their sake, and French of their wine. . . !" (personal communication).

## History of Tempeh in Europe and Australasia

**History of Tempeh in Europe:** As noted previously, all of the references to and articles about tempeh written between 1875 and the early 1950s were written by Europeans, most of them Dutchmen. Yet, perhaps because Dutch was not a widely read or spoken language and tempeh was not known in countries more famous for soyfoods such as Japan and China, tempeh was rarely mentioned in the numerous articles about soyfoods published in French, German, and English prior to the 1950s. Relatively little was published about tempeh in Europe between 1940 and 1959, and most articles focused on its role in prisoner-of-war camps in Southeast Asia.

All of the first tempeh companies in Europe were started in the Netherlands by immigrants from Indonesia. The earliest of these, called ENTI, was founded in April 1946 by a Dutch couple whose last name was Wedding. While living in Indonesia, they had learned to make tempeh. Bringing their starter culture and tempeh recipe to the Netherlands, they began to make Europe's earliest known tempeh on a home scale for friends and relatives. Gradually ENTI grew and became a commercial operation, making 2,000 lb. of tempeh a day by the early 1970s. In about 1974 the Weddings sold their company (located in Zevenhuizen) to Mrs. L. J. Duson, who ran it until January 1984, when she closed it. Firma E. S. Lembekker, founded in January 1959, then became Europe's oldest existing tempeh company.

Interest in tempeh in Europe began to increase during the 1960s. At least eight scientific articles were published by researchers between 1964 and 1984. Among these researchers, Thio Goan Loo from Indonesia was especially active in teaching people in Third World countries about tempeh. He wrote about tempeh for use in Zambia (Africa) and spent three months in 1979 teaching tempeh production and recipes in Sri Lanka. The earliest known popular article on tempeh was an excellent 7-page feature story with nine photographs published in 1982 in *Le Compas* in French. In 1982 *Soja Total*, a translation of the *Farm Vegetarian Cookbook* (Hagler 1978), containing 13 pages of information on tempeh, was published in Germany. *Das Tempeh Buch*, an updated and expanded translation of this *Book of Tempeh*, was scheduled for publication in Germany in early 1985.

Europe's largest tempeh company, Tempe Production Inc. (called Handelsonderneming van Dappern until 1983) was founded in 1969 in Rotterdam by Robert van Dappern. In about 1972 the thriving family business moved to Kerkrade, in southern Holland, and by early 1984 it was producing 6,000 kg (13,200 lb.) of tempeh a week, making it the world's second largest tempeh manufacturer, after Marusan-Ai in Japan.

Prior to early 1981 all of Europe's tempeh companies were located in the Netherlands and run by older Dutchmen catering largely to an Indonesian clientele. Europe's first generation of "New Age" tempeh shops was started from 1981 by young people interested in natural foods and/or macrobiotics. Europe's earliest known New Age tempeh company was Paul's Tofu & Tempeh, which was in operation by January 1981 in

London. JAKSO, the first New Age shop in the Netherlands, started in July 1981. By January 1982 there were seven tempeh shops operating in Europe; by January 1984 there were 18. Of these, seven were in the Netherlands, three in Austria, two each in England and West Germany, and one each in Belgium, France, Italy, and Sweden. Total tempeh production in the Netherlands was about 4,500 kg a week in 1982, rising to 12,000 kg a week in 1984.

By 1980 another center of interest in tempeh had developed at the Department of Botany and Microbiology, University College of Wales, Aberystwyth, Wales, U.K. There Dr. J. Hedger and Mr. T. Basuki (from Indonesia) had produced a 4-page leaflet on "Tempe—An Indonesian Fermented Soybean Food," and had written a script for a BBC program "Tomorrow's World," on tempeh, which was broadcast in the summer of 1979. In 1982 Hedger wrote a brief article on tempeh production.

**History of Tempeh in Australia:** Australian interest in tempeh began in about 1977, when McComb published an excellent B.S. thesis on the use of sweet narrow-leafed lupins to make tempeh. The earliest known Australian tempeh companies were started in about 1980, and by March 1981 there were three small ones, all run by young "New Age" people, interested in natural foods, meatless diets, and alternative lifestyles. The first two to start were Dharma, part of Earth Foods in Waverley, run by Swami Veetdharma, and a small shop at Bodhi Farm in Channon, New South Wales, run by John Seed. Cyril and Elly Cain founded Beancoast Soyfoods in Eumundi, Queensland, and started making tempeh in July 1982. Because of Australia's proximity to Indonesia, both countries could learn much from each other about traditional and modern tempeh making.

## History of Tempeh in the United States and Canada

**Early Years in America (1954–1969):** Interest in tempeh in the United States began at a surprisingly late date. As noted previously, early English-language articles on tempeh had been written by Ochse (1931) and Burkill (1935), both published outside the U.S. The earliest known reference to tempeh in a U.S. publication appeared in 1946, when an article by Gerold Stahel, writing from Surinam in South America about tempeh in Surinam and in New Guinea, was published in the *Journal of the New York Botanical Garden.* A summary appeared in November of that year in *Soybean Digest.* These articles appeared just 50 years after the first reference to tempeh in Europe by Prinsen Geerligs.

In 1955 Autret and van Veen (both working for the Nutrition Division of the Food and Agriculture Organization of the United Nations, outside the U.S.A. but writing in the *American Journal of Clinical Nutrition*) were the first to suggest tempeh as a protein-rich, nutritious, and low-cost food for infants and children in Third World countries.

Research on tempeh in the U.S. was started in 1954 by Dr. Paul György, a pediatrician and researcher at the Philadelphia General Hospital, and Professor of Pediatrics at the University of Pennsylvania. György had been to Indonesia many times, knew tempeh well, and (like Autret and van Veen) thought that it offered a way of improving the diets of infants and children in Third World countries. György received his first tempeh from Indonesia and Southern Rhodesia in 1954 and 1955. Ms. Kiku Murata of Japan worked with György in the U.S. investigating tempeh during 1959 and 1960. Following largely futile attempts to make tempeh in his own laboratory and lacking adequate facilities for making larger quantities of fermented foods, György worked out a cooperative arrangement in 1959 to have the tempeh made at Cornell University's New York State Agricultural Experiment Station. The first publication from György's work appeared in 1961, when he wrote "The Nutritive Value of Tempeh."

As noted earlier in the section on Indonesia, a great expansion of interest in tempeh began in the early 1960s, largely because of the pioneering, in-depth research at two centers: Cornell University's New York State Agricultural Experiment Station at Geneva, New York, under the leadership of Dr. Keith H. Steinkraus; and the USDA Northern Regional Research Center at Peoria, Illinois, under the leadership of Dr. Clifford W. Hesseltine. Each center became actively interested in tempeh because of the arrival of an Indonesian researcher: Ms. YAP Bwee Hwa at Cornell and KO Swan Djien at Peoria. They did important, original investigations on pure culture fermentations, microbiological and biochemical changes during tempeh fermentation, tempeh's nutritional value, and industrial production of tempeh. This research awakened a new interest in tempeh among microbiologists and food scientists worldwide. Moreover, with this research, the world center of interest in and research on tempeh shifted from Indonesia and the Netherlands to the U.S.A.

In the summer of 1958 Miss Yap started her research on tempeh in New York. In 1959 Steinkraus became the first American ever to study tempeh in its homeland, and his group began making tempeh for Dr. György in Pennsylvania. The first article on tempeh by Americans was written in 1960 when Steinkraus, Yap, van Buren, Provvidenti, and Hand published their now classic "Studies on Tempeh—An Indonesian Fermented Food." That same year Miss Yap, as part of her graduate degree in nutrition, submitted her M.S. thesis titled "Nutritional and Chemical Studies on Tempeh, an Indonesian Soybean Product."

During the 1960s the Cornell University Group, consisting of interdisciplinary scientists from both the Agricultural Experiment Station and Cornell University, published at least 13 original scientific articles on all aspects of tempeh. Steinkraus was the senior author of six of these papers. Particularly important for the coming new generation of U.S. tempeh manufacturers were his "Pilot Plant Studies on Tempeh" (1962), "Research on Tempeh Technology in the United States" (1964), and "A Pilot Plant Process for the Production of Dehydrated Tempeh" (1965). These represented the first attempts to develop a process for making tempeh in an industrialized country with a temperate climate. Van Veen, who had done pioneering research on tempeh in Indonesia as early as 1932 and had arrived at Cornell in 1962 as a professor of International Nutrition, was senior author of seven papers related to tempeh between 1962 and 1970.

In 1960 a second U.S. tempeh research program was started under the direction of Dr. Clifford W. Hesseltine at the USDA Northern Regional Research Center (NRRC) at Peoria, Illinois. Initial interest was sparked by the arrival of KO Swan Djien of the Bandung Institute of Technology's Laboratory of Microbiology. The first publication appeared in 1961 with Ko and Hesseltine's "Indonesian Fermented Foods." During the 1960s the USDA Peoria group published 17 original scientific papers (including two public service patents) about tempeh, plus four derivative articles; Hesseltine was senior author of 12 of these, Wang of four, Ko and A.K. Smith of two each, and Martinelli and Sorenson of one each.

In 1963 Hesseltine and co-workers published their first major tempeh study "Investigations of Tempeh, an Indonesian Food." In 1964 Dr. Martinelli (a Brazilian scientist studying tempeh at the NRRC) and Hesseltine developed a new method for incubating tempeh in perforated plastic bags. It soon became widely used by commercial tempeh producers in both Indonesia and North America, a nice example of cultural cross-fertilization. In 1966 and 1967 Hesseltine and Wang published the world's first studies showing that delicious tempeh containing higher quality protein could be prepared using soy-and-grain mixtures (including wheat and rice) or cereal grains alone.

A key component of the tempeh research at the NRRC concerned identification of the main microorganisms in the fermentation. In 1962, after observing 50 tempeh strains from various tempeh sources, Hesseltine identified *Rhizopus oligosporus* as the chief tempeh mold. Ko (1965) reported collecting 81 samples of tempeh from various places in Java and Sumatra. Isolation of 116 pure cultures revealed that *Rhizopus oligosporus* was always present in good quality tempeh, thereby establishing without a doubt that it was the typical dominant species used.

It is not known for sure when the first commercial tempeh was made in the U.S. After the long and bloody war that drove the Dutch out of Indonesia and led to Indonesian independence in 1949, many people from Indonesia emigrated to America. In 1950 an estimated 500 of these settlers arrived in California. The first of these known to have started a tempeh shop was Mary Otten, who in 1961 began making tempeh in her basement on Stannage Avenue in Albany, California. She sold it to her friends and served it at parties that she catered. In 1967 she started Java Restaurant and served many tempeh dishes. Then in 1974 she and her daugh-

ter, Irene, started Otten's Indonesian Foods, which made and sold a line of tempeh products.

Other early U.S. tempeh shops (all located in Los Angeles) were Runnels Foods (started in 1962), Toko Baru (1969), and Bali Foods (1975). Thus America's first generation of tempeh shops were all located in California and all run by Indonesian-Americans.

**The Americanization of Tempeh (1970 to 1980s):** The 1960s, a decade of creative scientific research on tempeh, laid the foundation for the 1970s, when tempeh began to enter the American diet. The main forces spurring increased production and consumption of tempeh after 1970 were the three closely related movements working to popularize natural foods, meatless and vegetarian diets, and soyfoods. From the late 1970s on there was a rapid growth of interest among many Americans in health, nutrition, and fitness, in low-cost protein sources, meatless diets, and world hunger, in ecology, and simpler, more satisfying lifestyles. Specific factors popularizing tempeh were the various promotional efforts, books, media coverage, and increased availability of good fresh tempeh. By the early 1980s the growing mainstream concern with cholesterol and saturated fats had also became a significant factor.

During the 1960s the Cornell University group under Dr. Steinkraus and the USDA Peoria group under Dr. Hesseltine and Dr. Wang had completed most of their basic research on tempeh. But during the 1970s the Cornell group made important discoveries relating to vitamin B-12 in tempeh and the USDA Peoria group developed improved, larger-scale methods for making rice-based tempeh starter cultures.

But much more important than the research work of these two groups during the 1970s and early 1980s was their "extension" work. Members of both groups summarized the results of their research on tempeh in at least 35 articles, both scientific and popular. They also gave many speeches. This work brought tempeh to the attention of many more scientists and consumers. Between 1976 and 1981 the Peoria group sent out some 35,000 tempeh starter cultures and instructions for making tempeh, free of charge to people and organizations requesting them. Steinkraus organized a Symposium on Indigenous Fermented Foods, held in Bangkok, Thailand, in November 1977 in conjunction with the fifth United Nations-sponsored conference on the Global Impacts of Applied Microbiology (GIAM V). It was attended by over 450 scientists from around the world and 17 papers were presented on tempeh, more than on any other single food. In 1983 Steinkraus edited the monumental *Handbook of Indigenous Fermented Foods*, containing 94 pages of information about tempeh, much of it from the 1977 Symposium. Hesseltine, Wang, and Steinkraus also did a great deal to help America's first generation of Caucasian tempeh manufacturers start their businesses and deal with their production problems. For their two decades of pioneering research, more than 65 publications on tempeh, and highly effective extension work, the U.S. tempeh industry owes the Peoria and Cornell groups an immense debt of gratitude.

During the early and mid-1970s, in addition to the groups at Cornell and Peoria, there were four other main groups that played leading roles in introducing tempeh to America: The Farm in Tennessee, The Soyfoods Center in California, Rodale Press in Pennsylvania, and the food- and counter-culture media.

A great deal of the credit for introducing tempeh to the American public goes to The Farm, a large spiritual and farming community of "long-hairs" living on 1,700 acres in Summertown, Tennessee. In late 1971 Alexander Lyon, a member of The Farm with a Ph.D. in biochemistry, learned about tempeh while doing library research on soy-based weaning foods. In 1972 he helped The Farm to set up a small "soy dairy" and to make occasional small batches of tempeh. This was America's first Caucasian-run tempeh shop, although it was not a commercial shop. Tempeh was an immediate hit in The Farm's vegan or total vegetarian diet—a diet containing no dairy or other animal foods. By late 1974 Cynthia Bates was making 20–30 pound batches of okara tempeh, using the soy pulp (okara) left over after making soymilk. By January 1975 The Farm was making 80-200 pounds of tempeh a week. In 1975, in order to share their discovery with people across America and around the world, the community (now having 1,100 members) featured a section on tempeh (written by Bates) in their popular *Farm Vegetarian Cookbook,* including the first tempeh recipes to be published in any European language.

In 1975 Bates set up a little laboratory and began making powdered, pure-culture tempeh starter for use on The Farm. By 1976 it was being sent out or sold to interested people. Starting in late 1975 a number of pamphlets were published describing how to use the starter to make and cook with tempeh. One contained the first instructions in any European language for making tempeh at home; another the world's first Tempeh Burger recipe. A number of America's early tempeh shops (such as The Tempeh Works in Massachusetts or Surata Soyfoods in Oregon) were started by people who learned the process on The Farm. America's first soy deli, set up in August 1976 at the Farm Food Company's storefront restaurant in San Rafael, featured tempeh in Tempeh Burgers, Deep-fried Tempeh Cutlets, and Tempeh with Creamy Tofu Topping, the first tempeh dishes sold in an American-style restaurant.

In the many media articles on tempeh from 1977 on The Farm was listed as the best source of tempeh starter and split dehulled soybeans. This created a booming little business on The Farm. Numerous articles on tempeh written by Bates and others were published in national magazines, starting in 1977. In 1977 Farm Foods was founded. It actively and creatively promoted and sold The Farm's line of tempeh products. In 1978 Hagler edited a revised edition of the *Farm Vegetarian Cookbook,* containing 12 pages on tempeh, including many recipes. Then in March 1984 The Farm published

*Tempeh Cookery,* America's fourth popular book about tempeh and the first with full-page color photos (Pride 1984). By mid-1984 the Tempeh Lab at the farm was supplying more than half of all tempeh starter used in the U.S.A.

William Shurtleff and Akiko Aoyagi of The Soyfoods Center in California were also active in helping to introduce tempeh to America. They first became interested in tempeh in March 1975 after reading the *Farm Vegetarian Cookbook.* In July 1979 Harper & Row published their *Book of Tempeh,* the first book in the world devoted entirely to tempeh. It contained the first sizable collection of American-style and Indonesian tempeh recipes (130 in all), the first illustrated descriptions of making tempeh, tempeh starter, and onchom on various scales in Indonesian tempeh shops, the first history of tempeh, and the first recommendations for commercial names for the more than 30 types of tempeh that could easily be made in the West. It also contained the largest bibliography on tempeh to date (including many new Indonesian references), an annotated listing of 61 people and organizations around the world connected with tempeh, and the first list of tempeh companies in the West. By early 1984 over 21,000 copies of the book had been sold.

Between 1976 and 1982 Shurtleff and Aoyagi wrote eight articles on tempeh for popular and trade magazines. In March 1980 The Soyfoods Center published *Tempeh Production,* the first book describing how to start and run a commercial tempeh plant in industrialized or Third World countries. Starting in 1982 Shurtleff did extensive annual surveys of the tempeh industry and market in the U.S.A., which were published yearly by The Soyfoods Center in *Soyfoods Industry and Market: Directory and Databook.*

Another early pioneer of tempeh in America was Rodale Press in Emmaus, Pennsylvania, best known as the publisher of *Organic Gardening* and *Prevention* magazines. In the spring of 1975 Rodale's R&D department began to study tempeh, and this work led to four major articles in their two magazines during 1976 and 1977. These were the first major popular articles on tempeh published in America and their impact was immense. In addition Rodale Press published books with extensive information on tempeh: *Home Soyfood Equipment* (Wolf 1981) and *Tofu, Tempeh, & Other Soy Delights* (Cusumano 1984).

Starting in 1971, the American media first began to take an interest in tempeh, when *Food Processing* magazine predicted a bright future for tempeh. Then between 1974 and 1979 twelve major articles on tempeh were published in magazines such as *Mother Earth News* (May 1976, September 1977), *Vegetarian Times* (November 1977), and *East West Journal* (July 1978) . . . a veritable media blitz for a little-known but increasingly popular fermented soyfood.

The first commercial Caucasian American tempeh shop was started in the winter of 1975 by Mr. Gale Randall in Unadilla, Nebraska. Robert Rodale's article in *Prevention* in June 1977 brought him and his shop instant national prominence.

The macrobiotic movement in America (including the *East West Journal*) took a strong interest in tempeh, starting in the late 1970s. Aveline Kushi's laudatory "My Favorite Tempeh Recipes" was published in August 1981. Aveline used tempeh extensively in diets for cancer patients. Those practicing a macrobiotic diet increasingly used tempeh as a basic daily protein source, and a number of them started tempeh companies.

Whereas the first generation of American tempeh makers had been Indonesian-Americans, the second generation were mostly long-haired, bearded hippies or other members of the counter-culture, who were interested in natural foods, meatless diets, alternative lifestyles, and right livelihood as part of a spiritual life. In the midst of the world's largest meat-producing country, they began to make tempeh as an alternative source of high-quality, low-cost protein. Many considered tempeh to be the finest "meatless meat" available.

By early 1979 there were 13 tempeh shops in operation in the U.S. and one in Canada. But all of these were either run out of a home, restaurant, or food retail store, or they were run in connection with a tofu shop.

America's first bona fide commercial company, having its own building and specializing in tempeh, was The Tempeh Works, started in 1979 in Greenfield, Massachusetts, by Michael Cohen (who had formerly lived on The Farm in Tennessee). The company began commercial production in September 1979 and quickly grew to be the biggest tempeh maker in America. By September 1980 they were making 3,000 pounds a week, rising to a peak of 6,800 pounds a week in September 1981—when competition began to grow fierce. Other early companies making only tempeh in a commercial building were Pacific Tempeh in California (from August 1980), Higher Ground Cultured Foods in Wisconsin (August 1980), Soyfoods Unlimited in California (February 1981), and Appropriate Foods in New York (March 1981).

U.S. tempeh researchers developed a number of important innovations in processing equipment and techniques that helped greatly in adapting tempeh to America, and to temperate climate production in general. These included perforated polyethylene bags for use as tempeh incubation containers (developed by Martinelli and Hesseltine in 1964), mass production of pure-culture tempeh starter on rice using simple equipment and techniques (Hesseltine and Wang 1975–76), a low-tech soybean dehuller and hull remover (Nelson and Ferrier 1976), a centrifuge to dewater cooked soybeans prior to inoculation (The Farm 1977), and vacuum packaging of tempeh (Pacific Tempeh 1980).

New types of commercial tempeh and tempeh products soon began to proliferate. These included America's first commercial multi-ingredient soy-and-grain tempeh, Soy & Rice Tempeh (White Wave 1979), commercial tempeh burgers (Island Spring 1980), and Tempehroni and Five Grain Tempeh (the first multi-grain tempeh, made with soybeans, rice, millet, sunflower seeds, and sesame seeds; Turtle Island 1981). By February 1983 there were three brands of okara tempeh on the market; most contained some rice and/or soybeans too. Other distributed tempeh entrees included Chili Con Tempeh, Sweet & Sour Tempeh, Tempeh Enchiladas, Tempeh Pizza, and Tempeh Lasagna, Tempeh Mock-Chicken Salad, and Tempeh Mock-Tuna Salad. Of all these second generation tempeh products, tempeh burgers (first developed by The Farm in 1975) were by far the most popular. By early 1984 at least eight brands were available nationwide.

In addition, during the late 1970s and early 1980s, many new tempeh dishes began to appear at soyfoods delis, cafes, and restaurants. The most popular ones, in descending order of popularity, were tempeh burgers, tempeh salads (mock chicken or tuna), tempeh sandwiches (tempeh, lettuce & tomato, etc.), sloppy joe tempeh, tempeh cacciatore, tempeh cutlets, and tempeh stroganoff (Shurtleff and Aoyagi 1983).

As the tempeh industry expanded, the national media showed a new wave of interest. Three major media articles on tempeh were published during 1979, five in 1980, four in 1981, seven in 1982 (including a February *Soyfoods* magazine cover story titled "The Coming Tempeh Boom"), and three in 1983 (including feature articles in the *Los Angeles Times, Washington Post,* and *Sunset Magazine*). Complete references for each of these are given in *Soyfoods Industry and Market* (Shurtleff and Aoyagi 1984). Moreover, between 1977 and 1980 there were 18 new scientific journal articles on tempeh. And between 1979 and 1984, seven books on tempeh (or with tempeh mentioned in the title) were published.

During its early years, the U.S. tempeh industry grew rapidly. The period from early 1979 to early 1983 was one of rapid growth in the number of producers. There were 13 commercial tempeh manufacturers in January 1979, 17 in 1980, 31 in 1981, 43 in 1982, 56 in 1983, dropping back to 53 in January 1984. The four largest tempeh makers in North America, in descending order of weekly production in January 1984, were Quong Hop/Pacific Tempeh in California (7,000 lb./week), White Wave in Colorado (5,850), Soyfoods Unlimited in California (5,800), and The Tempeh Works in Massachusetts (5,500).

During 1983 the U.S. tempeh industry made an estimated two million pounds (900 tonnes) of tempeh, using 514 tonnes of soybeans and employing about 200 workers. It retailed for $5 million; the average retail price was $2.50 per pound. Tempeh production grew by 36 percent during 1982 and by 29 percent during 1983, making it the fastest growing soyfoods market in the U.S. The top four manfacturers produced 63 percent of the industry's tempeh. An estimated 95 percent of U.S. tempeh was consumed by Caucasian Americans, since there were only 5,000 to 6,000 Indonesian-Americans in the entire country. Second-generation products were seen as the wave of the future.

Tempeh, a new and healthful, low-cost food that gave Americans "all the sizzle without the steak," would appear to have a bright future in the land of the meat-centered diet.

**History of Tempeh in Canada:** Inspired by Robert Rodale's article on tempeh published in *Prevention* magazine in July 1977, Robert Walker founded Canada's first tempeh company in his home in Port Perry, Ontario. By early 1983 there were five tempeh companies in Canada, as there were in 1984. All were quite small, making less than 200 pounds of tempeh a week.

## History of Tempeh in Japan

**1905 to 1945:** In 1905 Kendo Saito, a professor in the Plant Physiology Laboratory of the Botanical Institute at Tokyo Imperial University, first described and illustrated what is today considered to be the main tempeh microorganism, *Rhizopus oligosporus*. He did not, however, mention tempeh.

The first Japanese to mention and to study tempeh was the great microbiologist, Dr. Ryoji Nakazawa, who was working at the Taiwan High Commissioner's Office Central Research Laboratory. In 1912 Nakazawa asked a person from Southeast Asia to bring him samples of tempeh and onchom (ontjom; made from peanut press-cake). He analyzed their microorganisms. In 1926 Nakazawa took a research trip to Java and Sumatra and carefully collected 33 samples of soy tempeh and onchom from various markets and small manufacturers. He and Takeda analyzed their microorganisms and in 1928 published "On the Filamentous Fungi Used to Make Ontjom and Tempeh in the South Pacific" in the Laboratory's Annual Report. This was the earliest known published reference to or study of tempeh by a Japanese.

One of Dr. Nakazawa's youngest but eventually best-known students was Dr. Masahiro Nakano, who worked at the Taiwan laboratory with Dr. Nakazawa from 1934 until the early 1940s and learned from him about tempeh. During World War II Nakano was in Indonesia for three years; there he did informal studies on tempeh. In about 1944 Dr. Nakano (then age 37) went to the National Food Research Institute (NFRI) and created a Department of Applied Microbiology.

**1946 to 1979:** After the war the Department of Applied Microbiology at NFRI, and especially Dr. Nakano and his student Teruo Ohta, introduced tempeh to Japan. They were the first to make and serve tempeh in Japan and they wrote numerous articles about this food, which was unknown to the Japanese. In the mid-1950s Nakano and Ohta (Ohta's specialty was natto fermentation) suggested to natto makers and the Japan Natto Association that they try to introduce tempeh as a commercial product. In the laboratory and in natto plants, big and small, the researchers showed natto makers how to make tempeh, but the idea did not take root.

In 1960 Mr. Hayashi, director of the new Japanese-American Soybean Institute (*Nichibei Daizu Chosa-kai;* forerunner of the American Soybean Association) in Tokyo, grew interested in tempeh. That year his institute published (in Japanese) a 4-page pamphlet describing Hesseltine's method for making tempeh. He convinced Mr. Haruo Kato of Marukin Shokuhin (Marukin Foods Industry Co. Ltd.), a large natto manufacturer in southern Japan, to make the first commercial tempeh trials in Japan. Samples were sent to JASI, where it was served fried and drew nice compliments. JASI planned to introduce tempeh to tempura restaurants to be served as tempeh tempura, but the project never got started, for Marukin reported that tempeh interfered with the natto fermentation and vice versa.

Another center of tempeh research in Japan developed during the early 1960s at the Food and Nutrition Laboratory, in the Faculty of Science of Living, at Osaka City University, in Osaka. Early research there was done by Dr. Kiku Murata, Dr. Hideo Ikehata, and co-workers. Between 1964 and 1980 Murata was senior author of eight publications on tempeh and co-author of five others.

**The 1980s:** Despite all the research and promotional work from the early 1960s until the early 1980s, tempeh had not yet been produced commercially in Japan. But starting in about 1983 new interest arose for various reasons: (1) the growing popularity of tempeh in the United States and Europe; (2) the activities of KOPTI, the new Indonesian trade association for tempeh and tofu makers; (3) the growing internationalization of the Japanese diet; and (4) the increasingly positive image of soyfoods as a source of high quality nutrition. Thus as Japan's first major postwar soyfoods revolution (soymilk) was getting into high gear, another new one was starting.

Five organizations deserve the lion's share of the credit for commercializing tempeh in Japan, starting in 1983: Torigoe Flour Milling Co., the National Food Research Institute (NFRI), The Japan Natto Association, Marukin Foods, and Marusan-Ai. Most of these were large, established Japanese food companies, and several had been studying tempeh for more than a decade. They seized the opportunity boldly in launching tempeh.

*Torigoe:* The first company in Japan to start commercial tempeh production was Torigoe Flour Milling Co. (*Torigoe Seifun*), the nation's fifth largest flour-milling company. In 1975 Torigoe, looking to diversify, began an 8-year research project on tempeh with Kazuhiro Takamine in charge. In June 1983, under Takamine's supervision, Torigoe started making tempeh at their Fukuoka flour mill in a pilot plant that cost $50,000 and had a capacity of 15 tonnes or 33,000 pounds of tempeh a month. They made the key decision not to sell plain tempeh, but rather to make two semi-prepared products, both called Gold Tempeh. These were seasoned and breaded tempeh "fingers" and patties, sold frozen together with little packets of tartar sauce. The consumer would then deep-fry them to serve crisp and crunchy with the sauce. The tasty and nutritious morsels were introduced like fish sticks or cutlets to a school lunch program in Kyushu, where they became very popular. They were also test-marketed at several natural food stores, again with good results.

In mid-July 1983 Torigoe started commercial-scale production, and by early 1984 was producing 10–12 tonnes of tempeh a month, or about 5,770 pounds a week, making it suddenly one of the world's largest tempeh manufacturers—yet the product was only in the test-market stage!

*Marusan-Ai:* Founded in 1952, Okazaki Marusan was primarily a miso maker during its first two decades. In February 1973 Mr. Naoki Kawai, manager of Marusan's development section, discovered tempeh in Indonesia, brought back tempeh starter from Bogor University, and began to study tempeh production and cookery in his laboratory. In 1974 the company launched soymilk, then in January 1983 the company changed its name to Marusan-Ai (Ai means "love"). In August 1983, with soymilk sales booming (Marusan-Ai was the second largest soymilk manufacturer in Japan), the company started pilot plant production of tempeh and built a tempeh plant with a capacity of two tonnes a day inside their main Okazaki miso plant.

In January 1984 they introduced their tempeh to dietitians and in February 1984 they began to distribute pasteurized, refrigerated bulk tempeh (500 gm and 2.5 kg) for use in local school lunch programs and hospitals, and by caterers. They published a 27-page booklet titled "A New Soy Protein Food for Tomorrow: Marusan's Healthy Tempeh," which included a description with photographs of their tempeh production process and eight Japanese-style recipes. In March they began retailing Sukoyaka Tenpe ("healthy-robust tempeh"). By May 1984 Marusan-Ai was producing 30 tonnes (66,000 lb.) of tempeh a month or 15,148 lb./week, making them the largest tempeh manufacturer in the world. Most of their tempeh was sold to local schools and hospitals, pasteurized, then refrigerated. Much of it was used as a meat replacer, as in Sweet-Sour Pork or Tempeh-Hijiki Saute. Marusan-Ai is planning a major expansion in tempeh production; the product will be distributed refrigerated through their extensive soymilk routes.

*National Food Research Institute (NFRI):* Work on tempeh at the NFRI picked up during the 1980s, with Ohta, Okada, and Katoh comprising the tempeh research team. The NFRI and Bogor University had a joint research program in microbiology, and top scientists were exchanged to study and teach tempeh. But perhaps most important the NFRI worked closely with the Japan Natto Association to interest that group in tempeh and to help individual members start production.

*Japan Natto Association:* The idea of Japanese natto makers making commercial tempeh, having smoldered for over 25 years, suddenly caught fire in about 1983. From 1958 to 1982 natto production had grown at the admirable compound rate of 6.2 pecent year. Yet per capita consumption in 1982 was still only about 1.4 kg per capita per year. So natto makers were looking for a new product, ideally a fermented soyfood, to help expand sales. The NFRI tempeh team suggested tempeh and provided extensive technical advice. Officers of the Japan Natto Association, Ose and Kanasugi, became very interested, and in May 1983 the Natto Association decided to go ahead with tempeh research and popularization, including recipe development. Many newspaper stories appeared praising tempeh, but surprisingly they referred to it repeatedly as "tempeh natto," a term that both the Natto Association and NFRI apparently wanted to popularize to aid sales of both tempeh and natto.

In June 1983 the Natto Association sent a team of three men to Indonesia to study tempeh, and later that year Kanasugi started a restaurant in Omiya city featuring many tempeh recipes. He also developed and sold rice crackers and confections containing tempeh. In 1983 and 1984 two large natto companies began making tempeh—both prompted by the Japan Natto Association's promotional work. They were Marukin Foods and Takashin.

*Marukin Foods:* The first natto company to start large-scale production of tempeh was Marukin Shokuhin (Marukin Foods Industry Co. Ltd.). Located in Kumamoto, Kyushu (Japan's southernmost main island), they were (by 1984) one of Japan's Big Five natto manufacturers. In about 1964, following the recommendations of Mr. Hayashi of the Japanese-American Soybean Institute, Marukin began to investigate tempeh and to produce small batches, but the product was never commercialized. Then in July 1982 Marukin and eleven other soyfoods manufacturers from Kyushu joined to establish the Kyushu Soyfoods Industry Association (*Kyushu Daizu Shokuhin Kyogyo Kumiai*), which built a large and modern plant. In April 1983 this new natto factory, the largest in Japan, started production, employing 85 workers. Marukin decided to use its former natto factory to make tempeh and in November 1983 they launched SunSeed brand tempeh. By May 1984 they were selling about 4,620 lb. (2,100 kg) a week or 9,150 kg a month. They were also developing secondary tempeh products, including snack foods, paste-type foods, and fried foods, and promoting their tempeh widely.

*Takashin Foods:* Takashin Shokuhin is a new company, a spinoff from a large natto manufactuter called Takato. In 1983 they built a new tempeh plant in Tokyo and began to sell their tempeh as "Tempeh Natto" in small plastic trays.

From 1983 on tempeh received extensive media coverage in Japan. In July 1984 the Japan Tempeh Research Society (*Tenpe Kenkyu-kai*) was founded in Tokyo to provide a forum for ongoing investigation, discussion, and popularization of tempeh.

The future of tempeh in Japan looks promising. As of mid-1984 a number of Japan's largest food companies were seriously considering tempeh production. And some industry watchers were predicting that tempeh might duplicate the remarkable success story of soymilk. When Japan's sophisticated fermented soyfoods technologies start to be applied to the tempeh-making process, exciting new developments are sure to result. Likewise, when fine Japanese chefs set out to create new tempeh recipes—from Teriyaki Tempeh to Tempeh Tempura—the world is sure to take notice. Indeed, Japan could soon lead the world in tempeh research and development—and in large-scale tempeh production.

As of 1984 three Japanese tempeh companies were among the world's eight largest, as shown below.

## History of Tempeh in China, and South and Southeast Asia

**History of Tempeh in China and Taiwan:** As mentioned previously in the Indonesian section, a tempeh-like product was observed in Beijing, China, in 1931 by William Morse. In the log of the Dorsett-Morse Expedition (Vol. 10, p. 6273) he gave a good photograph and the following description: "Soybean Cake. Peiping, China. A life-sized picture. Chinese name *Tou chiah ping* (soybean fried cake). Small cakes made from boiled soybeans. The beans are pressed into small round cakes which are allowed to develop a mold—taking about seven days. These cakes are broken into small pieces and fried in sesame oil." There are no known subsequent references to this Chinese product in any language. In 1983 Shurtleff asked at least 10 people in China connected with soybeans if they had ever heard of *tou chiah ping* and showed them the characters. None had. Tempeh could be produced easily by Chinese companies making soybean jiang or soy sauce and it would fit nicely into Chinese dietary patterns. But given the proverbial conservatism of most Chinese toward new foods, it might have difficulty in catching on.

In the early 1980s Dr. Steve Chen, Director of the American Soybean Association in Taiwan, wrote two books, *Soy Foods* (1980) and *Handbook of Soy Oil and Soy Foods* (1981), both in Chinese. In each book he discussed tempeh, writing its name with two Chinese characters (meaning heaven + sea shell), which he chose to sound like "tempeh." (In *pinyin* these are written and pronounced *tian-bei*.)

**History of Tempeh in Southeast Asia:** Tempeh migrated with Indonesian travelers to Malaysia and Singapore in early times, although no definite dates of introduction are known. The earliest known journal article about tempeh from Southeast Asia was written in 1968 by Diokno-Palo and Palo in the Philippines. It stated that there was no record of tempeh having ever been made or sold in the Philippines, but the authors recommended its introduction. In 1976 Bhumiratana described a pilot process for making tempeh developed at Kasetsart University in Thailand (Goodman 1976). In 1977 Noparatnaraporn from Kasetsart presented a paper on "Factors Affecting Fermentation and Vitamin B-12 in Tempeh." Also in 1977 Yeoh and Mercian from the Malaysian Agricultural Research and Development Institute (MARDI) presented the first known report on "Malaysian Tempeh." Both have been summarized by Steinkraus (1983). The ASEAN Protein Project stimulated research on tempeh. Noranizam (1979) in Malaysia studied *Rhizopus* isolated from tempeh.

**History of Tempeh in India:** In about 1936 (according to van Veen 1962) a group of missionaries from Travancore, a poor region from southern India, wanted to make and introduce soy tempeh. Van Veen gave them a 3-week course in production methods. When they returned to Travancore they made good quality tempeh, but the Indian people there had no interest in this unknown mold-fermented product, and the experiment failed.

From the 1950s on occasional papers on tempeh were published in India, by Nakano (1959), Prawiranegara (1960), Bai et al. (1975, 1979), Bhavani Shankar (1978), David and Verma (1981), Vaidehi (1981), and Vaidehi and Vijayakumari (1981). M. P. Vaidehi in the Department of Rural Home Science, University of Agricultraul Sciences at Bangalore, did the most to introduce and promote tempeh to Indian villages. She made her own tempeh, then did a study serving tempeh curry and tempeh chips to 100 villagers and 100 urban consumers. The two products were well received (Vaidehi 1984, personal communication). Yet as of 1984 tempeh had not been commercialized in India.

## World's Largest Tempeh Makers

| Company Name | Country | Year Started | Avg. Weekly Production lb./week | kg/week |
|---|---|---|---|---|
| Marusan-Ai | Japan | 1983 | 15,148 | 6,885 |
| Tempe Production Inc. | Netherlands | 1969 | 13,200 | 6,000 |
| Quong Hop/Pacific Tempeh | U.S.A./CA | 1980 | 7,000 | 3,182 |
| White Wave | U.S.A./CO | 1979 | 5,850 | 2,659 |
| Soyfoods Unlimited | U.S.A./CA | 1981 | 5,800 | 2,636 |
| Torigoe Flour Milling | Japan | 1983 | 5,770 | 2,623 |
| The Tempeh Works | U.S.A./MA | 1979 | 5,500 | 2,500 |
| Marukin Foods | Japan | 1983 | 4,620 | 2,100 |

**History of Tempeh in Sri Lanka:** From April to June 1979 Mr. Thio Goan Loo, a Chinese-Indonesian stationed at the Royal Tropical Institute in Amsterdam, taught many people in Sri Lanka, especially those associated with the Soyafoods Research Centre at Gannoruwa, how to make and serve tempeh. In January 1981 *Soyanews* introduced tempeh to its many Sri Lankan readers, with a description of how to make tempeh at home plus several recipes. Sun-dried tempeh came to be called "soya karawala," karawala being a popular type of dried fish, which tempeh apparently was found to resemble in texture and flavor. As *Soyanews* continued to praise the virtues of tempeh (May, June, July 1981; October 1983; April, June 1984) and the Soyafoods Research Centre at Gannoruwa, Peradeniya, continued to teach people how to make and use it, the new food caught on surprisingly rapidly, although by mid-1984 it was still not being produced commercially. This could be the beginning of the first successful introduction of tempeh to a Third World country having no Indonesian influence.

## History of Tempeh in Latin America and Africa

**History of Tempeh in Latin America:** In Latin America, tempeh was first mentioned by Stahel in Surinam in 1946. He wrote: "Here in Surinam, as in the East Indies, most of the soybeans are consumed in this form." In 1964 Martinelli, a Brazilian scientist working in the U.S. with Hesseltine at the USDA/NRRC in Peoria, Illinois, developed the extremely important technique for making tempeh in perforated polyethylene bags. In December 1980 *Natura* magazine in Mexico published a cover story on tempeh by Heltova (largely pirated from *The Book of Tempeh*). In December 1981 *Quadernos de Natura* in Mexico published an 86-page booklet on tofu, miso, and tempeh by Shurtleff and Aoyagi. Yet as of mid-1984 there were no tempeh shops in Latin America (except in Surinam) and no research was underway.

**History of Tempeh in Africa:** Shortly after World War II, tempeh was introduced in Southern Rhodesia. Van Veen recalled that one of his co-workers had gone there and noticed that a lot of locally grown soybeans were being exported rather than being eaten by the people. Working with the local food technology institute, he and the staff made tempeh for hospitals. The United Nations' FAO gave technical advice and the project looked promising, but it eventually failed since the product was not accepted by the general population, which had no experience with mold-fermented foods (Autret and van Veen 1955; van Veen 1962). György (1962) reported that in 1955 he had received tempeh from the Ministry of Health in Salisbury.

In 1971 Mr. Thio Goan Loo, a Chinese-Indonesian, helped to introduce tempeh (along with other low-tech soyfoods) to Zambia (Thio 1972), but it is not known what became of his efforts. In 1977 Bahi El-Din and co-workers, having done research on the basic tempeh fermentation process at Wageningen, Netherlands, published an article about it in the Sudan. Also in 1977 Djurtoft and Jensen investigated the production of tempeh from beans widely grown in Africa—broad beans and cowpeas, used as is or mixed with wheat or barley. Deep-fried tempeh slices were well received by a taste panel of 20 Africans. In 1980 Raintree in Ibadan, Nigeria, published a short article on how to make tempeh from the seeds of the fast-growing leucaena or ipil-ipil tree. The only place this tempeh was made traditionally was in the dry, inhospitable hill country of Gunung Kidul in south central Java. As of mid-1984 we know of no tempeh production or research in Africa.

## Outlook and Prospects for Tempeh

One indication of the widespread and growing interest in tempeh appeared at the United Nations-sponsored international Symposium on Indigenous Fermented Foods (SIFF), held in conjunction with the fifth international conference on the Global Impacts of Applied Microbiology (GIAM V) convened in Bangkok in November 1977, and attended by over 450 top researchers from around the world. There 17 papers were presented on tempeh, more than on any other single food. Those proceedings, edited by Steinkraus, were published in 1983 in *Handbook of Indigenous Fermented Foods*. Tempeh offers great hope for increasing local supplies of low-cost protein, while also providing local employment opportunities.

The rapid growth of tempeh in the U.S.A., Europe, and Japan is a promising sign for industrialized countries as well, caused in part by the growing interest in health, nutrition, and fitness, light and natural foods, and meatless diets with less cholesterol and saturated fats.

# Appendix B

# Tempeh Makers in the West

## U.S.A.

### ARKANSAS

**Fayetteville 72702**—Crystal Moon, PO Box 3308. Ph. 502-521-4932. Judy Foster & George Dewey.

**Fayetteville 72701**—Mountain Ark Trading Co, 120 South East St. Ph. 501-442-7191. Frank Head.

### CALIFORNIA

**Arcata 95521**—Arcata Soyfoods, PO Box 467. Ph. 707-822-7409. Tom Nawrocki.

**City Of Industry 91746**—Toko Baru, 16006 Amar Rd. Ph. 213-333-6220. Rudy Kohler.

**Escondido 92025**—Indocon, 1651 S Juniper #75. Ph. 714-434-4131. Max S T Tan & Bob I K Sih.

**La Crescenta 91214**—Tamarinda Foods, 3939 Santa Carlotta St. Ph. 213-248-1173. Winnifred H Dreeuws.

**Los Angeles 90046**—Namaste Foods, 7733 Hampton Ave #6. Ph. 213-851-6816. David Blomquist.

**Oakland 94608**—Amrit/Siddha Ashram, 1107 Stanford Ave. Ph. 415-655-8677. Tofu & Tempeh Maker.

**Richmond 94805**—Ottens Indonesian Foods, 322 Key Blvd. Ph. 415-620-9407. Irene & Mary Otten.

**S San Francisco 94080**—Pacific Tempeh, 161 Beacon St. Ph. 415-873-4444. Jim Miller.

**S San Francisco 94080**—Quong Hop & Co, 161 Beacon St. Ph. 415-873-4444. Travis Burgeson & Jim Miller.

**San Diego 92112**—Hi Pro Tempeh, Box 12516. Ph. 619-295-1864. T Michael Morearty.

**San Leandro 94577**—Soyfoods Unlimited, 14668 Doolittle Dr. Ph. 415-352-1320. John Robertson & Loni Stromnes.

**Santa Cruz 95062**—Western Soy Complements, 1560 Mansfield Dr #B. Ph. 408-479-0596. Jeremiah Ridenour.

**Santa Cruz 95062**—Western Soy Compliments, 3350 Coffee Ln. Ph. 408-479-0596. Jeremiah Ridenour.

**Santa Rosa 95402**—Heaven & Earth Soyfoods, PO Box 258. Ph. 707-525-9068. Linda Eucalyptus.

**Santa Rosa 95405**—Heaven & Earth Tempeh, 1521 Parker Dr. Ph. 707-525-9068.

**Sun Valley 91352**—Country Store Hlth Foods, 8720 Sunland Blvd. Ph. 213-768-6373. Joan Harriman.

**Valinda 91744**—Now Foods Enterprise, 15870 Alwood St. Ph. 213-918-5312. Al Bleeker, Dutch.

**Vallejo 94590**—Calindo Co, 617 Nebraska St. Ph. 707-644-8866. Ed Tirta.

### COLORADO

**Boulder 80301**—White Wave, 1990 N 57th Ct. Ph. 303-443-3470. Chip Mcintosh.

**Ft Collins 80526**—Nupro Foods, 1819 W Prospect Rd. Ph. 303-493-0138. Carol & John Hargadine.

### GEORGIA

**Atlanta 30317**—The Soy Shop, 1863 Memorial Dr SE. Ph. 404-377-8433. Sara & Steve Yurman.

**Decatur 30030**—Swan Gardens, 218 Laredo Dr. Ph. 404-378-1947. Dick Mcintyre.

### HAWAII

**Kaneohe/Oahu 96744**—Bali Hai Tempeh Co, PO Box 1342. Ph. 808-235-4658. Bonnie & Larry Fish.

**Kilauea Kauai 96754**—It's Natural, PO Box 711. Ph. 312-491-1144. Brian Schaefer.

### IDAHO

**Sandpoint 83864**—Star Soyfoods, Rte 2 Box 337. Ph. 208-265-4720. Penny Billiter.

### ILLINOIS

**Champaign 61820**—Midwest Soy Products, 608 S Belmont Ave. Ph. 217-398-5756. Anthony & Patricia Kao.

### MAINE

**Canaan 04924**—Soy Beings, Van Eeghen Rfd 2 Box 4196. Ph. 207-872-8790. Richard Tory.

**Rockland 04841**—Down East Soy, PO Box 1246. Ph. 207-596-0023. Rob Lovell.

### MARYLAND

**Baltimore 21230**—Bud Inc Soyfoods. Raleigh Industrial Ct, 1100 Wicomico St. Ph. 301-837-4034. Aaron Liu & Wang Chen.

### MASSACHUSETTS

**Brookline 02146**—Open Sesame Restaurant, 48 Boylston St. Ph. 617-277-9241.

**Greenfield 01301**—Tempeh Works, PO Box 870. Ph. 413-772-0991. Michael Cohen.

**Jamaica Plain 02130**—21st Century Foods Inc, 30A Germania St. Ph. 617-522-7595. Lucio Armellin & Norbert Belanger.

**Lenox 01240**—Kirpalu Yoga Retreat, West St Box 793. Ph. 717-754-3051. Chris Yorsten.

### MICHIGAN

**Ann Arbor 48104**—Soy Plant, 771 Airport Blvd #1. Ph. 313-663-8638. a Collective.

**Rochester 48063**—Harvest Earth Foods, 2789 Steamboat Springs. Ph. 313-254-2211. Jerry Skrocki.

### MISSOURI

**Jamestown 65046**—Imagine Foods Inc, RR 1 Box 11. Ph. 816-849-2583. David Carlson.

**Springfield 48063**—Hometown Tempeh, 300 N Waverly. Ph. 417-866-1337. Jim Hawkins.

### NEBRASKA

**Palmyra 68418**—Indonesian Tempeh Co, Rte 1 Box 146. Ph. 402-780-5634. Mr Gale Randall.

### NEW YORK

**Bronx 10474**—Panda Food Products, B249 Nyc Terminal Mkt. Ph. 212-441-6494. Saul Willner.

**Franklin Square 11010**—Appropriate Foods, 137 New Hyde Park Rd. Ph. 516-352-7222. Robert Werz.

**Ithaca 14850**—Thompson Tempeh, 371 Snyder Hill Rd. Ph. 607-273-2362. Bruce & Katie Thompson.

**Rochester 14607**—Northern Soy, 30 Somerton St. Ph. 716-442-1213. Andy Schecter & Norman Holland.

### NORTH CAROLINA

**Boone 28607**—Bean Mountain Soy Dairy, 121 W Howard St. Ph. 704-264-0890. James Stoertz.

### OHIO

**Cleveland 44121**—North Coast Tempeh Corp, 3555 Randolph Rd. Ph. 216-382-7494. Jeff Narten & Lynn Mcalister.

## OREGON

**Ashland 97520**—Ashland Soy Works, 280 Helman St. Ph. 503-482-1865. James Muhs.

**Corvallis 97339**—Food-Ray Tempeh, Box 2038. Ph. 503-754-6896. Janet Foudray.

**Eugene 97402**—Callix E Miller, Surata Soyfoods Inc, 302 Blair Blvd. Ph. 503-485-6990. Callix Miller.

## PENNSYLVANIA

**Mertztown 19539**—Cricklewood Soyfoods, Rd #1 Box 161. Ph. 215-682-4109. Karl & Renate Krummenoehl.

**State College 16801**—Spring Mills Foods, 304 S Pugh. Ph. 814-238-2082. Don Smith & Walt Tomasch.

## TENNESSEE

**Summertown 38483**—Farm Tempeh Shop, 156 Drakes Ln. Ph. 615-964-3574. Cynthia Bates, Makes Tempeh Primarily for the Farm Comm.

## VIRGINIA

**Charlottesville 22901**—Kingdom Foods, 1717-A Allied Ln. Ph. 804-979-2773. Eileen Judge.

## WASHINGTON

**Husum 98623**—Turtle Island Soy Dairy, PO Box 218. Ph. 509-493-2004. Seth Tibbot.

**Vashon 98070**—Island Spring Inc, PO Box 747. Ph. 206-622-6448. Luke Lukoskie.

## WISCONSIN

**Madison 53704**—Higher Ground Inc, 2049 Atwood Ave. Ph. 608-835-3793. Peter Ziegler.

**River Falls 54022**—Creative Soyfoods Inc, 526 N Clark St. Ph. 715-425-0467. David Nackerud.

# FOREIGN

**AUSTRALIA. Adelaide SA**—Light Wave Wholefood, 21 Gilbert St Newton 5074. Ph. 08-337-5500. Dominic Anfiteatro.

**AUSTRALIA. Leura/NSW 2781**—Blue Mountain Soyfoods, 3 Tennyson Ave. Gary Hepworth.

**AUSTRALIA. Melbourne**—Nectar Tempeh, PO Box 969 Geelong. Ph. 05-221-4458.

**AUSTRALIA VIC. Mt Waverley 3149**—Earth Angel Soyfoods, 53 Stanley Ave.

**AUSTRALIA. Queensland**—Mighty Bean Soyfoods. M.S. 1117 Cooloolabin, Via Yandina 4561. Ph. 07-146-7342. Michael & Julie-Anne Joyce.

**AUSTRALIA. Waverley 2024**—Earth Foods, 308 Bronti Rd. Ph. 99-4505. Swami Veetdharma.

**AUSTRALIA. 2480 Channon NSW**—Bodhi Farm, John Seed.

**AUSTRALIA. 6284 W Aust**—Frank Barker Tempeh, Post Office Cowarumup. Frank Barker.

**AUSTRIA. A-1160 Vienna**—P Zacharowicz & Stuart. Tempeh, Hasnerstr 159/18. Ph. NL. P Zacharowicz & P Stuart.

**AUSTRIA. A-5020 Salzburg**—Soyvita Austria, Herrengasse 30. Ph. 066-243-0985. Norbert Brunthaler.

**AUSTRIA. 2700 Wr Neustadt**—Naturkostladen, Deutschgasse 9. Ph. NL. Isabella Lorenz.

**BELGIUM. B9980 St Laurein**—De Hobbit, Waterstraat 4. Ph. 09-179-9622. Eric Dewilde.

**CANADA J9H4A4. Aylmer/QUE**—La Soyafriche, 38 Saratoga. Ph. 819-685-0815. Moruis Adams.

**CANADA V9L4T6. Duncan BC**—Thistledown Soyfoods, RR 5 Church Rd (5855). Ph. 604-748-9514. Jean & Jan Norris.

**CANADA. Edmonton ALTA**—Max Pierolie Tempeh, 2032 47th St. Ph. 403-462-1097. Max Pierolie.

**CANADA H4A2N9. Montreal PQ**—Emile Lefort Tempeh, 3877 Draper Ave. Ph. 514-489-6052. Emile Lefort.

**ENGLAND. London EC1**—Community Health Tempeh, 188 Old St. Ph. 01-251-4076. Andrew Leech & Jon Sandler.

**ENGLAND. London N6 4NA**—Paul's Tofu & Tempeh, 155 Archway Rd Highgate.

**FRANCE. 94400 Ivry**—Traditions Du Grain, 16 Ave Jean Jaures. Ph. 6-671-8988. Jean Luc Alonso, Anita Dupuy.

**INDONESIA. Bandung W Java**—Sjarim Oeben. Jalan Kopo, Gang Parsadi 44b/196b. Mr Oeben.

**INDONESIA. Bogor W Jave**—Manan B Samun, Cibuluh Rt 4/111.

**INDONESIA. Malang W Java**—Malang Tempeh Co-Op, Sanan.

**INDONESIA. Yogyakarta**—Tempe Giling Murni, Ketnggungan Ng Vii/51a. Pedro Sudjono.

**ITALY. 20122 Milan**—Giovanna Mazzieri Soyfds, Via S Tecla 3. Ph. NL. Giovanna Mazzieri.

**JAPAN 812. Fukuoka**—Torigoe Flour Milling Co. Higashi-Ku, 6-5-2 Hakozaki-Futo. Ph. 092-651-3269. Mr Kazuhiro Takamine.

**JAPAN 860. Kumamoto City**—Marukin Shokuhin Kogyo, 380 Yoyasu-Machi. Ph. 096-325-3232. Itsuo Kira, Pres: Moto-O Kira, Tempeh.

**JAPAN 444-21. Okazaki-Shi**—Marusan-Ai Tempeh, Aza-Arashita I Jingi-Cho. Ph. 056-445-3112. Naoki Kawai.

**JAPAN. Tokyo 131**—Takashin Shokuhin, Tachibana 1-29-2 Sumida-Ku. Ph. 03-613-5311.

**NETHERLANDS. Bleiswijk**—Consuma Bv, Postbus 175 2265 Zk. Ph. 01-892-7116. H P M Righter.

**NETHERLANDS. Dr Augustijnlaan**—Vanka-Kawat, Dr August Ynin 40. Ph. 07-098-4632. G L Van Kasteren.

**NETHERLANDS. Heerewaar-den**—Yakso, Voorne 13 6624 Kl. Ph. 08-877-2189. Tomas Nelissen & Peter Dekker.

**NETHERLANDS. S-Gravenhage**—Firma Sri Murti, Teylerstraat 84 2562 Rx. Ph. 07-065-5765. S Ghisa & Eddie Felix.

**NETHERLANDS. 1021 Amsterdam**—Manna, Meeuwenlaan 70. Ph. 02-032-3977. Hans Den Hoed, Klaus Hillberg.

**NETHERLANDS. 1065AB Amsterdam**—Firma E S Lembekker, Corn V Alkemadestraat 61. Ph. 02-015-1480.

**NETHERLANDS. 2288 Eg Rijswijk**—Haagse Tempe Fabriek, Treubstraat 9. Ph. 07-095-1974. U a Soffner, Tempe Chips, Sambal Goreng Tempe.

**NETHERLANDS. 6468-EK Kerkrade**—Tempeh Production Inc. Dappern Bv, Tunnelweg 107. Ph. 04-545-5803. Robert Van Dappern, Largest Tempeh Plant in Netherlands.

**NEW ZEALAND. Auckland 1**—Bean Supreme Soyfoods, Box 78-084 Grey Lynn. Ph. 0-976-4988. Trevor Johnston.

**SWEDEN S19063. 19063 Orsundsbro**—Aros Sojaprodukter, Bergsvagen 1. Ph. 01-716-0456. Ted Nordquist & Tim Ohlund.

**W GERMANY. 6454 Bruchkoebel**—Atlantis Tofurei, Insterburgerstr 7. Ph. 061-817-1438. J J Lilienthal.

**W GERMANY. 8000 Munich 2**—Byodo Naturkost, Thalkirchnerstr 57. Ph. 08-976-5241.

**W GERMANY. 8000 Munich 40**—Lukas Kelterborn Tempeh, Lerchenauerstr 11. Ph. 089-308-8624. Lukas Kelterborn.

**W GERMANY. 8210 Prien-Chiem**—Aueland Tofu/Sojaprod, Hub 2. Ph. 08-051-3416. Lukas Kelterborn.

# Weights, Measures, and Equivalents

## TEMPERATURE

C. = Celsius or Centigrade; F. = Fahrenheit
C. = 5/9 (F-32)
F. = (1.8 × C.) + 32
Freezing point of water = 32° F. = 0° C.
Boiling point of water = 212° F. = 100° C.
Body temperature = 98.6° F. = 37.0° C.

## WEIGHT (Mass)

1 ounce (avoirdupois) = 28.3495 grams = 16 drams = 437.5 grains
1 pound (avoirdupois) = 16 ounces = 453.59 grams = 0.45359 kg
1 (short or U.S.) ton = 2,000 pounds = 0.907 metric tons = 33.33 bushels of soybeans (used chiefly in the U.S. and Canada)
1 metric ton (MT or tonne) = 2,204.6 pounds = 1.1025 short tons= 36.75 bushels of soybeans
1 long ton = 2,240 pounds = 1.12 short tons = 1.016 metric tons = 37.33 bushes of soybeans (used chiefly in England)
100 grams = 3.527 ounces
1 kilogram = 1,000 grams = 2.2046 pounds
1 quintal (qt) = 100 kg = 220.46 pounds

## CAPACITY (U.S. Liquid Measure)

1 teaspoon = 60 drops
1 tablespoon = 3 teaspoons = 14.75 cc
1 fluid ounce = 2 tablespoons = 29.57 cc = 0.0296 liters = 1.8047 cu. in.
1 cup (U.S.) = 16 tablespoons = 8 fl. oz. = 236 cc
1 quart = 2 pints = 4 cups = 32 fl. oz. = 0.946 liters
1 gallon (U.S.) = 4 quarts = 231 cu. in. = 0.8333 imperial gallon = 3.785 liters
1 gallon (imperial) = 1.20 gallons (U.S.) = 4.456 liters
1 liter = 1,000 cc = 1.0567 liquid quarts = 4.237 cups = 61.03 cu. in. = 33.814 fluid ounces = 0.264 gallons = 0.220 Imperial gallons
1 kiloliter = 1,000 liters = 264.18 gallons (U.S.) = 35.315 cu. ft. = 2200.6 lbs. water

## CAPACITY (U.S. Dry Measure)

1 quart = 2 pints = 67.20 cu. in. = 1.1012 liters
1 peck = 537.605 cu. in. = 9.309 quarts = 8.809 liters
1 bushel = 4 pecks (ca. 8 gallons) = 2,150.42 cu. in. = 35.2390 liters

## CAPACITY (Japanese)

1 *go* (Japanese) = 10 *shaku* = 180 cc = 0.763 cups (U.S.)
1 *sho* = 10 *go* = 1.903 quarts = 0.476 gallons = 1.800 liters
1 *koku* = 10 *to* = 100 *sho* = 47.6 gallons = 180 liters
1 tablespoon = 15 cc = 3 teaspoons
1 cup = 200 cc = 0.847 U.S. cups = 13.5 U.S. tablespoons

## LENGTH (Linear Measure)

1 inch = 2.540 cm = 1,000 mils
1 foot = 12 inches = 30.48 cm
1 yard = 3 feet = 91.44 cm
1 fathom = 6 feet = 1.829 m
1 rod = 16.5 feet = 5.029 meters
1 mile = 5,280 feet = 320 rods = 1.609 km
1 nautical mile = 6,028 feet
1 league = 3 nautical miles
1 mm = 0.03937 inches
1 cm = 0.3937 inches
1 meter = 39.37 inches = 3.2808 feet = 1.094 yards
1 km = 1,000 m = 0.621 miles = 3,280.8 feet

## AREA (Square Measure)

1 sq. in. = 6.452 sq. cm
1 sq. ft. = 144 sq. in. = 929.03 sq. cm = 0.0929 sq. m
1 sq. yd. = 1296 sq. in. = 0.836 sq. m
1 sq. mile = 640 acres = 2.593 sq. km
1 acre = 43,560 sq. ft. (208.7 ft. on a side) = 4,840 sq. yds. = 4047 sq. m = 0.405 hectares
1 sq. cm = 0.1549 sq. in.
1 sq. meter = 10.764 sq. ft. = 1.196 sq. yd.
1 sq. km = 0.3856 sq. miles = 247.1 acres = 100 hectares
1 hectare = 10,000 sq. m (100 m on a side) = 2.471 acres

## VOLUME (Cubic Measure)

1 cu. in. = 16.387 cc
1 cu. ft. = 1728 cu. in. = 0.028 cu. m = 7.48 gallons
1 cu. yd. = 46,656 cu. in. = 0.765 cu. m
1 cc = 0.061 cu. in.
1 cu. m = 35.315 cu. ft. = 1.308 cu. yd. = 1 kiloliter (see Capacity)

## NATURAL EQUIVALENTS

Atmospheric pressure = 14.7 lb./sq. in.= 1.0 kg/sq. cm
1 gallon (U.S.) of water weighs 8.33 lbs. = 3.78 kg
1 cup of soybeans weighs 6.5 ounces = 182 grams
1 quart of soybeans weighs 1.62 lbs. = 0.736 kg
1 gallon of soybeans weighs 6.47 lbs. = 2.94 kg
1 bushel of soybeans weighs 60 lbs. = 27.24 kg;
it yields 10.7 lbs. (17.8%) crude soy oil plus 47.5 lbs. (79.2%) defatted soybean meal plus 1.8 lbs. (3%) manufacturing loss
1 metric ton of soybeans contains 36.75 bushels; it yields 400 lbs. of oil and 1850 lbs. of meal
1 *sho* of dry soybeans weighs 1.36 kilos or 3.0 pounds; it makes about 1.75 gallons of fresh soy puree (gô)

## YIELDS

1 bu./acre = 67.25 kg/ha (of soybeans)
1 MT/ha = 10 qt/ha = 1000 kg/ha = 14.87 bu./acre = 891 lb./acre (of soybeans)
1 g/sq. m/day = 8.9 lb./acre/day = 1.62 U.S. tons/acre/year = 10 kg/ha/day = 3.60 MT/ha/year

## ENERGY, WORK & PRESSURE

A BTU (British Thermal Unit) is the quantity of heat required to raise the temperature of 1 pound of water one degree Farenheit (near 39.2°F)
A calorie is the amount of heat required to raise the temperature of 1 kilogram of water 1 degree centigrade.
A watt is a unit of power equal to the rate of work represented by a current of 1 amp under a pressure of 1 volt.
1 horsepower = 746 watts = 0.746 kilowatts = 550 foot pounds per second
1 lb./sq. in. = 70.45 g/sq. cm. = 0.070 kg/sq. cm.
1 kg/sq. cm. = 14.19 lb./sq. in.

# Glossary

Most of the following foods are available (although often only in dried form) at Indonesian or Oriental food stores or import shops listed in the yellow pages of your phone directory. We have used the new Indonesian spellings as explained in the Preface. A lengthy and detailed version of this Glossary, with special emphasis on Indonesian fermented foods, is given in the professional edition of this book.

**AGAR:** A sea vegetable gelatin sold in Indonesia as dried yellowish whole plants. Sold in the West in the form of flakes, bars, powder, or strands

**AMARANTH, INDONESIAN** (*bayam*): Also called *Asian spinach* or *Chinese spinach*, the raw or cooked young leaves of *Amaranthus gangeticus* are highly nutritious and tasty.

**APEM:** Steamed rice bread. Made from white rice flour fermented overnight and steamed in cups; most popular in Bali.

**ARAK:** Distilled rice wine made from glutinous rice inoculated with ragi (see below).

**BANANAS** (*pisang*): Indonesia has at least seven popular types, all delicious.

**BREM:** Undistilled rice wine made from glutinous rice inoculated with ragi (see below).

**BUMBU:** A general term for spices, and more generally, seasonings and herbs.

**CANDLENUTS** (*kemiri*): This tan oily nut of the candlenut tree (*Aleurites moluccana*) resembles a macadamia nut; used to flavor and thicken cooked dishes. Substitute macadamia.

**CARAMBOLA** (*belimbing*): The fruit of the *Averrhoa carambola* tree has a pale yellow waxy skin and refreshing, juicy, slightly bittersweet flavor.

**CASSAVA** (*ketela pohon* or *singkong*): Also called *manioc* or *yuca*, the root of the *Manihot utilissima* is the source of the starch tapioca.

**CHAYOTE** (*labu siam* or *waluh jepang*): Also called *chajota, christophine, choko, vegetable pear,* or *Mexican chayote,* this pear-shaped, light-green tropical squash (*Sechium edule*) is 3 to 6 inches long and has a bumpy, corrugated surface. Substitute summer squash.

**CHILIES** (*chabé* or *lombok*): Also called *chili peppers, hot red peppers,* or *hot chilies,* there are hundreds of varieties but all are derived from two species of the Capsicum family (*C. annum* and *C. frutescens*). There are three basic types in Indonesia and all can be substantially "defused" by simply deseeding them as described in Preparatory Techniques. In general, the smaller the chili and the thinner the skin the hotter the flavor.

*Red chilies* (*chabé merah* or *lombok besar*): Also known in the West as Cayenne, the *Capsicum annum* (or *Capsicum frutescens longum*) is 3 to 4 inches long and quite hot.

*Green chilies* (*chabé* or *lombok hijau*): These partially ripened Capsicums are not as hot as red chilies, which they resemble in both size and shape.

*Fiery dwarf chilies* (*chabé rawit, lombok kechil, lombok rawit,* or *chengek*): The hottest of all chilies, *Capsicum frutescens* are also known as bird peppers, or bird's eye peppers. Only about 1 inch long, they are sun-dried and ground in the West to make chili powder.

**COCONUT** (*kelapa*, pronounced KLA-puh): The majestic coconut palm (*Cocos nucifera*) provides food, drink, and shelter for half the world.

*Coconut, grated* (*kelapa parut*): The best grated coconut is that prepared fresh as described in Preparatory Techniques. Dried shredded or grated coconut makes a fairly good substitute, and using the prepackaged product is quicker and easier. Avoid commercial brands that contain sugar. Substitute 1 cup dry shredded for 1 cup fresh grated coconut.

*Coconut milk and cream* (*santan* and *santan kanil*): Coconut milk (which should not be confused with coconut water, see below) is made by mixing the grated or pulverized meat of the coconut with an equal quantity of water and squeezing the liquid from the shreds as described in Preparatory Techniques. Less water is used to make cream. Both are key ingredients in creating the wonderful flavors of Indonesian cuisine. Canned and/or frozen coconut milks is now available throughout Hawaii and in a growing number of U.S. supermarkets.

*Coconut oil* (*minyak kelapa*): Indonesia's most popular all-purpose cooking oil. Substitute peanut oil or any other vegetable oil.

*Coconut water* (*air kelapa*): Also called coconut juice, this is the sweet clear liquid found in coconuts. Chilled, it makes a delicious, refreshing drink.

**CORIANDER** (*ketumbar*): The mild-flavored tan seeds of *Coriandrum sativum* are one of Indonesia's most popular spices. Fresh coriander leaves (*daun ketumbar*) are also known in the West as *Chinese parsley* or *cilantro*.

**CUMIN** (*jinten or jintan*): The tan ¼-inch-long seeds of the *Cuminum cyminum* are widely used together with coriander.

**DEGEH** (*dagé*): Dageh is prepared by a bacterial fermentation of various leguminous seeds and the remains, extracts, or waste products of various plants—cassava or potatoes peels, etc.

**DURIAN:** This football-sized fruit with its tough prickly hide, borne by the *Durio zibethinus* tree, is craved by some and hated by others.

**GINGERROOT** (*jahe*): The 4-inch-long knobby tan root of the *Zingiber officinale* is peeled, then grated or sliced. Now available at many food stores in the West. Two teaspons of powdered ginger may be substituted for 1 teaspoon of the fresh grated root; however, the flavor is quite different.

**JACKFRUIT** (*nangka*): This irregularly shaped fruit of the *Artocarpus integra* tree is closely related to the breadfruit and may weigh as much as 70 pounds. A fair substitute is pumpkin or Japanese *kabocha*.

**KANGKUNG LEAVES** (*kangkung*): Also called Chinese spinach, swamp cabbage, or (mistakenly) watercress, this water-grown plant (*Ipomoea reptans*) is cooked like spinach.

**KEMANGI LEAVES** (*daun kemangi*): These aromatic leaves of the *Ocimum canum* have a flavor and aroma resembling those of basil, savory, or mint. Substitute basil.

**KENCHUR ROOT** (*kencur*): Also called *aromatic ginger* or *lesser galangal*, this 1-inch-long rhizome of the *Kaempferia pandurata,* a member of the ginger family, has a mildly pungent, aromatic flavor. May be omitted if not available.

**KLUWAK:** This richly flavored nut of the *Pangium edule* tree comes in a 2-inch-wide, thick black shell and has a consistency like a soft chestnut. No substitute.

**KOJI:** Steamed rice, barley, or soybeans covered with a fragrant mycleium of white mold (usually *Aspergillus oryzae*). Used to make miso, taucho, shoyu, and soy sauce.

**LAOS ROOT** (*laos*): Also known as *lengkuas*, this 3- to 5-inch-long root comes from the greater galangal or galingale (*Alpina galanga*), a member of the ginger family, resembles gingerroot in appearance but with a milder and exotic, slightly pungent flavor. A ½-inch-thick slice of fresh laos equals ½ teaspoon of laos (or galangal) powder. There is no substitute, yet the flavor is so delicate it may generally be omitted, except in true Javanese cuisine.

**LEMONGRASS** (*sere or daun serai wangi*): Also known as *citronella*, *Cymbopogon citratus* is an aromatic grass with foot-long, sharp-edged blades and an ivory-colored stem tapering to a white bulbous base. It has a sweetish lemon flavor and delicate tang, popular in herbal teas and cookery. One teaspoon of lemongrass powder equals about one stalk of fresh lemongrass. Or substitute 2 strips of very thinly pared lemon rind.

**LIME LEAVES** (*daun jeruk purut*): These aromatic fresh or dried leaves of the Far Eastern wild lime tree *(Citrus hystrix)* are generally bruised and added to sauces, then removed before serving. Substitute Western lime or lemon leaves.

**MELINJO LEAVES** (*daun melinjo*): The leaf of the gnemon tree (*Gnetum gnemon*) is widely used in soups. Substitute spinach or kangkung.

**MISO:** *See* Taucho.

**MOCHI, INDONESIAN** (*uli*): Cakes of pounded glutinous rice.

**MUNG-BEAN SPROUTS** (*taogé* or *taugé*): Widely available in the West.

**OKARA** (*ampas tahu*): The residue of solids insoluble in water left over after the production of tofu or soymilk. Rich in dietary fiber and containing 3.5 percent protein, it is used in Indonesia to make tempeh and onchom.

**ONCHOM:** Like a growing number of scientists, we define onchom (now spelled *oncom* in Indonesia and formerly spelled *ontjom*) as a traditional West Javanese fermented food made from either peanut presscake or okara (see above) inoculated with *Neurospora* mold spores and sold in the form of mycelium-bound cakes, somewhat resembling tempeh except for their orange color and generally slightly larger size. In Indonesia, a number of closely related products that are inoculated, like tempeh, with *Rhizopus* mold spores have also traditionally been called onchom, thereby creating considerable confusion and ambiguity in terminology. Thus, when peanut presscake or okara is inoculated with *Rhizopus,* we call the resulting products *peanut-presscake tempeh* or *okara tempeh,* whereas in West Java they are both called *onchom.* For a lengthy discussion of onchom see *The Book of Tempeh,* professional edition.

**PALM SUGAR** (*gula merah*): The all-purpose, unrefined sweetener in Indonesia (called *jaggery* in India, Sri Lanka, and Burma), this natural and richly flavored dark-brown sugar, derived from the sap of the coconut or arenga palm, is sold in firm, slightly moist cakes. Substitute any natural dark-brown sugar, or honey.

**PANDANUS LEAF** (*daun pandan*): The dark-green leaf of the *Pandanus latifolia,* crushed or boiled, adds its distinctive vanilla-like flavor and natural green color to many confections, curries, and rice dishes.

**PEPPER** (*merica or lada*): Not to be confused with the more widely used chilies (see Chilies, above), both white and black pepper come from the berry of the tree *Piper nigrum,* which is native to Indonesia. To make black pepper (*merica hitam*), the berries are picked before they are ripe, sun-dried, then ground whole. To make white pepper, preferred by Europeans for its mellower flavor, the berries are allowed to ripen fully on the trees, then the outer hull is removed by buffing and the inner kernel bleached and ground.

**PETÉ BEANS** (*peté* or *petai*): This bitter, rather strong-flavored bean looks like a shiny lima and grows in foot-long pods on the large leguminous tree *Parkia speciosa.* Substitute lima beans for texture but *not* flavor.

**PETIS:** Shrimp, prawn, or fish paste. Substitute terasí, taucho, or miso.

**RAGI:** Indonesian yeast cakes. Broadly speaking, *ragi* refers to all types of starters, such as those used for bread, tempeh, or tapeh. Narrowly, it refers to the starter for tapeh, sold in the form of dry or crumbly discs each about 1¼ inches in diameter and ⅜ inch thick, made primarily of rice flour plus yeast and mold spores.

**SALAM LEAF** (*daun salam*): Also called *Indonesian bay* or *laurel leaf,* this aromatic green leaf of *Eugenia polyantha* is used like a

Western bay leaf. Sold dried. If not available, omit. A bay leaf makes only a poor substitute.

**SAYUR ASIN:** A salt-pickled and fermented vegetable preparation similar to sauerkraut.

**SHALLOTS** (*bawang merah*): Indonesian shallots (*Allium ascalonicum*) look like 1-inch-diameter cloves of garlic with a bright red to reddish-brown skin and a delicate flavor having mild garlic overtones. Use Western shallots or substitute 1 onion for 4 to 5 shallots.

**SHRIMP CRISPS** (*krupuk* or *kerupuk udang*): These look like fairly thick, giant salmon-pink potato chips, as long as a hand and half as wide. Made from tapioca, crisps come in many other flavors besides shrimp.

**SOURSOP** (*sirsak* or *nangka belanda*): The 6-inch-long, green-skinned fruit of the *Annona muricata* tree has white juicy flesh that is eaten raw, used in ice creams, or pureed in a blender with ice and a little sweetening to make "white mango juice," the most ambrosial drink we have ever tasted.

**SOY SAUCE, INDONESIAN** (*kechap* or *kecap*): Formerly spelled *ketjap*, this soy sauce, generally made from black soybeans, contains no wheat or grain and comes in three basic types: *Sweet Indonesian Soy Sauce* (*kecap manis*), which accounts for 90 percent of the nation's total soy sauce production, has a very thick consistency and a strong, sweet molasses flavor, since it may contain up to 50 percent palm sugar; numerous spices, such as star anise, help enrich the flavor. A recipe for making a similar product starting from Japanese soy sauce (shoyu) is given in Basic Preparatory Techniques. *Mellow Indonesian soy sauce* and *salty Indonesian soy sauce,* both called *kecap asin,* have a thinner consistency and contain less (or no) palm sugar.

**STAR FRUIT** (*belimbing wuluh*): The fruit of the *Averrhoa bilimbi* tree is the size of a finger and very sour, like a lemon — which makes a good substitute.

**TAMARIND** (*asam*): The tamarind is a huge tropical leguminous tree (*Tamarindus indica*) that bears edible leaves (*daun asam*) and elongated pods, the fruits. Tamarind paste (*asam*), the edible portion of the fruit, is a dark reddish-brown, sour, sticky, and rather fibrous pulp or paste. It is enclosed inside the bean-shaped, 4½-inch-long, brittle brown pod and surrounds several shiny black seeds. The pulp may be eaten raw but it is usually used to make a basic syrup or "tamarind water" (see Preparatory Techniques), which is used in Indonesian cookery much as we use vinegar or lemon juice to impart a tangy flavor. The latter make fairly good substitutes. Whole tamarinds are sold at some supermarkets in the West and the paste is sold in cakes at Indonesian, Chinese, Indian, Latin American, or gourmet specialty shops.

**TAPEH** (*tapi, tapai,* or *peuyeum*): Pronounced TAH-pay, this popular Indonesian fermented delicacy with a sweet and mildly alcoholic flavor is made from either cooked cassava or glutinous rice. Cassava tapeh (*tapé ketela*) and glutinous rice tapeh (*tapé ketan*) are especially popular as snacks and desserts. Foods of major worldwide interest, both are inoculated with ragi (see above).

**TAUCHO:** Formerly spelled *tao-tjo* (and now spelled *tauco* in West Java and *taoco* in Central and East Java), this is a fermented soybean paste (or chunky sauce) related to Japanese soybean miso. Produced and consumed mainly in West Java, it comes in four different flavors and consistencies, all of which are dark brown. *Sweet soft taucho* (*tauco cianjur*), the most popular variety, has the consistency of porridge interspersed with prominent soybean chunks, and contains 25 percent palm sugar. *Salty liquid taucho, firm dried taucho,* and *smoked dried taucho* are used in small quantities. For additional information see our *Book of Miso.* Substitute miso.

**TERASI:** Indonesian fermented shrimp or fish paste. Substitute petis, shrimp or anchovy paste, or salted overripe tempeh.

**TOFU** (*tahu*): Soybean curd. This is the Chinese-style product sold in firm white cakes, each about 3½ inches square and 1 inch thick. Prepared daily at over 10,900 small shops, tofu, like tempeh, is a key source of high-quality low-cost protein in the daily diet of the Indonesian people. For additional information see our *Book of Tofu.*

**TUAK:** Indonesian palm wine.

**TURMERIC** (*kunyit* or *kunir*): This is the starchy 2-inch-long, ½-inch-diameter orange-yellow rhizome or underground stem of the *Curcuma longa,* a plant related to ginger. In the West, it is sold mainly in its dried, powdered form.

**WINGED BEANS** (*kecipir*): This "soybean of the tropics" is a remarkable plant in that the entire plant (root, leaves, flowers, pods, and beans) is edible and tasty. The beans have as much high-quality protein as soybeans.

**NOTE:** Monosodium glutamate, a flavor intensifier generally known by its Chinese or Japanese brand name Vetsin (a mixture also containing lactose and salt), Accent or Aji-no-moto, is a highly refined, white crystalline powder that differs in structure from natural glutamic acid. When used in more than very small quantities, it is well known to produce in some people the "Chinese restaurant syndrome" characterized by headaches, burning sensations, a feeling of pressure in the chest, and other discomforting symptoms. It is produced by hydrolysis of molasses or glucose from tapioca, cornstarch, potato starch, etc. A committee of scientists selected by the U.S. Food and Drug Administration advises that MSG should not be given to infants under 12 months of age. U.S. baby-food manufacturers no longer use MSG in their products. We and many other people interested in healthy natural foods strictly avoid use of MSG.

# Bibliography

Adams, Ruth. 1979. How you can use soybeans. Today's Living. July. pp. 22-23, 38, 40.

Agricultural Research. 1966. From traditional Indonesian tempeh . . . Simple, uniform process produces new high-protein foods. August. pp. 8-9.

Agricultural Research. 1969. Tempeh: protein-rich food may increase disease resistance. 18(4):5.

Amar, D. and Grevenstuk, A. 1935. Bijdrage tot de kennis der bongkrekvergiftigingen. (Contributions to the knowledge of bongkrek poisoning.) Geneeskundig Tijdschrift voor Nederlandsch Indie 75: 104-06, 366-82.

Andersson, R. E. et al. 1977. Volatile compounds in tempeh. Presented at Symposium on Indigenous Fermented Foods, Bangkok, Thailand. Summarized in Steinkraus 1983.

Aramaki, Nancy. 1978. Acceptance evaluation of tempeh made from soybeans, bulgar, millet and aduki beans. San Luis Obispo, CA: California Polytechnic State University. Unpublished B.S. thesis in dietetics and food administration. 34 pp.

Arbianto, Purwo. 1971. Studies of Bongkrek Acid: Taxonomy of the Producing Bacterium, its Production and its Physiological Function. University of Wisconsin. Ph.D. thesis in Bacteriology. 194 pp.

Arbianto, Purwo. 1977. The bongkrek food poisoning in Java. Symposium on Indigenous Fermented Foods, Bangkok, Thailand. Summarized in Steinkraus 1983.

ASEAN Sub-Committee on Protein. 1980. Report of the Second ASEAN Workshop on Solid Substrate Fermentation. Kuala Lumpur, Malaysia, 27-29 Nov. 420 pp.

Autret, M. and van Veen, A. G. 1955. Possible sources of protein for child feeding in underdeveloped countries. American Journal of Clinical Nutrition 3(3):234-43.

Baars, J. K. and van Veen, A. G. 1937. The constitution of toxoflavin. Koninklijke Akademie van Wetenschappen, Amsterdam 40(6).

Bahi El-Din, Magboul, I., and Lein, H. T. 1977. Tempeh, a fermented soybean food. Sudan Journal of Food Science and Technol. 9:24-26.

Bai, R. G., et al. 1975. Processing and nutritional evaluation of tempeh from a mixture of soybean and groundnut. Journal of Food Science and Technol. 12:135-38.

Bai, R. G. et al. 1979. Studies on tempeh. II. Nutritive value of tempeh and its supplementary value to rice diets. Indian Food Packer 33(6):26-33.

Barret, C. 1977. Tempeh. Vegetarian Times No. 22. Nov.-Dec. p. 55.

Bates, C. et al. 1976. Beatnik tempeh making. Summertown, TN: The Farm. 20 pp. Mimeograph. Summarized in Steinkraus 1983 as "Utilization of tempeh in North America."

Bates, C. 1976. Tempe. Summertown, TN: The Farm. 2 p. Revised as Tempeh in 1977, for distribution with their commercial tempeh.

Batra, L. R. and Millner, P. D. 1976. Asian fermented foods and beverages. In Developments in Industrial Microbiology 17. pp. 117-28.

Beuchat, L. R. 1976. Fungal fermentation of peanut press cake. Economic Botany 30:227-34.

Bhavani Shankar, T. N. 1978. Studies on tempeh made from groundnut and soybean mixture. In Proceedings of the First Indian Convention of Food Scientists and Technologists No. 9.2. p. 95.

Bhumiratana, A. 1976. Small-scale processing of soybeans for food in Thailand. In Goodman 1976. pp. 143-46.

Boedijn, K. B. 1958. Notes on the Mucorales of Indonesia. Sydowia (Annales Mycologici) 12:321-62.

Boorsma, P. A. 1900. Scheikundig onderzoek van in Ned.-Indie inheemsche voedingsmiddelen. De sojaboon. (Chemical analysis of some indigenous foodstuffs in the Netherlands Indies. The Soybean.) Geneeskundig Tijdschrift voor Nederlandsch-Indië 40:247-59.

Bortz, Brenda. 1976. The joys of soy. II. Tofu and tempeh. Organic Gardening and Farming. March. pp. 128-31.

Brackman, Agnes de Keizer. 1974. The Art of Indonesian Cooking: The ABC's. Singapore: Asia Pacific Press.

Burkill, I. H. 1935. A Dictionary of the Economic Products of the Malay Peninsula. London: Crown Agents. pp. 1080-86. (The work is 2 volumes, 2400 pp.)

Cadwallader, S. 1982. Vegetarian soy tempeh. San Francisco Chronicle. Sun. Nov. 10. p. 43.

Calloway, D. H., Hickey, C. A. and Murphy, E. L. 1971. Reduction of intestinal gas-forming properties of legumes by traditional and experimental food processing methods. Journal of Food Science 36:251-55.

Charles, M. and Gavin, J. R. 1977. Engineering studies of solid substrate fermentations. I. Basic considerations and the tempeh fermentation. Presented at Symposium on Indigenous Fermented Foods, Bangkok, Thailand. Also in Steinkraus 1983. pp. 74–87.

Chen, L. H. et al. 1969. The potential of tempeh to serve as an antioxidant in lipids and in tissue. Federation of Am. Soc. for Exp. Biology Proceedings 28:306.

Chen, L. H. et al. 1972. Tissue antioxidant effect on ocean hake fish and fermented soybean (tempeh) as protein sources in rats. Journal of Nutrition 102:181–85.

Cohen, R. L. 1981. Bay Area companies excited about tempeh. San Francisco Business Journal. July 27. pp. 8-9.

Chen, S. 1980. Soy Foods. Taipei, Taiwan: American Soybean Assoc. 248 pp.

Chen, S. and Wang, E. 1981. Handbook of Soy Oil and Soy Foods. Taipei, Taiwan: American Soybean Assoc. 200 pp. (In Chinese, with English Table of Contents.)

Clute, Robin, and Andersen, Juel. 1983. Juel Andersen's Tempeh Primer: A Beginner's Book of Tempeh Cookery. Berkeley, CA: Creative Arts. 56 pp.

Compas, Le. 1982. Le Tempeh. "Usine" à proteines et vitamines. En provence d'Indonesie (Tempeh: A protein and vitamin factory from Indonesia.). No. 21. pp. 23-29. Spring.

Curtis, Paul R., Cullen, R. E. and Steinkraus, K. H. 1977. Identity of a bacterium producing vitamin B-12 activity in tempeh. Presented at Symposium on Indigenous Fermented Foods, Bangkok, Thailand. Summarized in Steinkraus 1983.

Cusamano, Camille. 1984. Tofu, Tempeh, & Other Soy Delights. Emmaus, PA: Rodale Press. 261 pp.

David, I. M. and Verma, J. 1981. Modification of tempeh with addition of bakla (Vicia faba). Journal of Food Technol. (India). 16:39-50.

de Bruyn, G. F., van Dulst, J. and van Veen, A. G. 1947. Biochemie in een Interneringskamp (Biochemistry in a concentration camp). Voeding 8:81-

Diet & Exercise. 1982. Spectacular soy food . . . tempeh. Summer. pp. 62-63, 74, 76.

Diokno-Palo, N. and Palo, A. M. 1968. Two Philippine species of Phycomycetes in tempeh production from soybean. Philippine Journal of Science 97:1-16.

Direktorat GIZI. 1967. Daftar Komposisi Bahan Makanan (Indonesian Food Composition Tables). Jakarta: Bhratara, Direktorat GIZI Departemen Kesehatan R.I.

Djurtoft, R. and Jensen, J. S. 1977. Tempeh produced from broad beans, cowpeas, barley, wheat, or from mixtures thereof. Presented at Symposium on Indigenous Fermented Foods, Bangkok, Thailand. Summarized in Steinkraus 1983.

Dorsett, P. H. and Morse, W. J. 1928–31. Log of the Dorsett-Morse Expedition to East Asia. 17 volumes of typewritten manuscript plus handwritten notebooks. Washington, DC: USDA Div. of Plant Exploration and Introduction. Only original and 2 microfilm copies at American Soybean Assoc., St. Louis, MO. One photocopy at Soyfoods Center. 6,000 pp.

Dosti, Rose. 1980. Tempeh: An old food moves out of ethnic kitchens. Los Angeles Times Part VII, p. 1. Part VIII. pp. 16, 18. July 31.

Dupont, A. 1954. Essential amino acids in some Indonesian food constituents. Thesis FIPIA, Bandung.

Dwidjoseputra, Dakimah. 1970. Microbiological Studies of Indonesian Ragi. Nashville, TN: Ph.D. thesis in Biology. 125 pp.

Dwidjoseputra, D. and Wolf, Frederick T. 1970. Microbiological studies of Indonesian fermented foodstuffs. Mycopathologia et Mycologia Applicata (The Netherlands) 41:211-22.

Ebata, J. et al. 1972. Beta-Glucosidase involved in the antioxidant formation in tempeh, fermented soybeans. Noka 46(7):323-29. (In Japanese with English summary.)

Ebine, Hideo. 1984. Tenpe no seishitsu to riyo-jo no shomondai (Various problems on the character and utilization of tempeh). Shokuhin Kogyo. June (2):20-24.

Ehrenberg, C. G. 1820. De Mycetogenesi. Nova Acta Academy Leopold 10(1):198.

Farm Foods. 1977. Tempeh. Summertown, TN. 2 page flyer. Revised 1978.

Farm, The. 1975. Farm Vegetarian Cookbook. Summertown, TN: The Book Publishing Co. 128 pp. Revised edition by L. Hagler. 1978. 223 pp.

Farm, The. 1977. Fermentation Funnies. Summertown, TN. 2 pp. About tempeh.

Farm, The. 1977. How we make and eat tempeh down on the farm. Mother Earth News. No. 47. Sept-Oct. pp. 105-08.

Farm, The. 1978. Make your own soyburger. East West Journal. July:58-63.

Fiering, S. 1981. Low technology soybean dehuller. Soyfoods. Winter. p. 52.

Food Engineering. 1981. Food from a fermenter looks and tastes like meat. May:117-18.

Food Processing magazine. 1971. Specialty fermented foods: Tempeh. Winter. pp. F7-F9 (Foods of Tomorrow section).

Ford, R. 1981. Soy Foodery Cookbook. Santa Barbara, CA: Self published. 78 pp.

Fritschner, S. 1983. Soybeans with pizzazz. Washington Post. Wed., March 2. pp. E1, E3.

Fukakura, N. et al. 1980. Survey on the acceptability of tempeh. Bulletin of Teikoku Gakuen. No. 6. pp. 33-39.

Gandjar, I. and Slamet, D. S. 1972. Tempe gembus hasil fermentasi ampas tahu (Okara tempeh). Penelitian Gizi dan Makanan 2:70-79.

Gandjar, I. and Hermana. 1972. Some Indoensian fermented foods from waste products. In Stanton 1972. pp. 49-54.

Gandjar, I. 1977a. Fermentasi Biji Mucuna pruriens DC dan Pengaruhnya Terhadap Kwalitas Protein (Fermentation of Mucuna pruriens DC seeds and its effects on protein quality). Bandung, Indonesia: Institute Teknologi Bandung, dissertation.

Gandjar, I. 1977b. Tempe benguk, tempe gembus, and tempe menjes. Presented at Symposium on Indigenous Fermented Foods, Bangkok, Thailand. Summarized in Steinkraus 1983.

Gandjar, I. 1977c. Fermentation of winged bean seeds—Tempe kecipir. Presented at Symposium on Indigenous Fermented Foods, Bangkok, Thailand. Summarized in Steinkraus 1983.

Gandjar, I. et al. 1977d. Tempe from the solid waste of a hunkwe (mung bean starch) factory. Presented at Symposium on Indigenous Fermented Foods, Bangkok, Thailand. Summarized in Steinkraus 1983.

Gandjar, I. 1978. Fermentation of winged bean seeds. In The Winged Bean. Los Banos, Laguna, Philippines: Philippine Council for Agriculture and Resources Research (PCARR). pp. 330-34.

Gandjar, I. 1981. Soybean fermentation in Indonesia. Advances in Biotechnology 2:531-34.

Gericke, J. F. C. and Roorda, T. 1875. Javaansch-Nederduitsch Handwoordenboek (Javanese–Low German Concise Dictionary). Amsterdam: Johannes Mueller. p. 378. Also in 2nd edition, 1901.

Gerras, Charles. 1984. Rodale's Basic Natural Foods Cookbook. Emmaus, PA: Rodale Press. 900 pp.

Gery, M. E. 1980. Tempeh: A tempting soyfood with culture. New Roots. No. 12. Nov. pp. 50-53.

Goldbeck, Nikki, and Goldbeck, David. 1983. American Wholefoods Cuisine. New York: New American Library. 580 pp.

Gomez, M. I. and Kothary, M. 1979. Tempeh from Red Kidney Beans. League for International Food Education Newsletter. June. p. 1-2.

Goodman, R. M., ed. 1976. Expanding the Use of Soybeans. Proceedings of a Conference for Asia and Oceania, held in Chiang Mai, Thailand, Feb. 1976. Urbana, IL; INTSOY Series No. 10. 261 pp.

Grant, M. W. 1952. Deficiency diseases in Japanese prison camps. Nature 169:91-92.

Grant, T. C. 1981. Tempeh: Molds make it happen. Learning. Feb. pp. 45, 48.

Gray, William D. 1970. The Use of Fungi as Food and in Food Processing. Cleveland, OH: Chemical Rubber Co. Press. 113 pp.

György, Paul. 1961. The nutritive value of tempeh. In Meeting Protein Needs in Infants and Children. Publication 843. National Academy of Sciences, National Research Council, Washington, D.C. pp. 281-89.

György, P., Murata, K. and Ikehata, H. 1964. Antioxidants isolated from fermented soybeans (tempeh). Nature 203(4947):870-72. Aug. 22.

György, P., Murata, K., and Sugimoto, Y. 1974. Studies on antioxidant activity of tempeh oil. Journal of the American Oil Chem. Soc. 51(8): 337-79.

Hackler, L. R. et al. 1964. Studies on the utilization of tempeh protein by weanling rats. Journal of Nutrition 82(4):452-56.

Hagler, L. 1982. Soja Total. Hamburg: Papyrus Verlag. 200 pp.

Hamlin, S. 1982a. Tempeh: "New" food for the 80s. Daily News (New York). Wed., Jan. 6. pp. 1, 3, 7. Good Living section.

Hamlin, S. 1982b. Tempeh: It doesn't look like much, this tasty bargain. Philadelphia Inquirer. Sun., Jan. 17. p. 8-F.

Hamlin, S. 1982c. Tempeh's an ugly food, but it's rich in nutrients. Miami Herald. Thurs. April 29. p. 6F.

Hannigan, K. J. 1979. Tempeh . . . a super soy. Food Engineering. Nov. p. 11.

Hardjo, Suhadi. 1964. Pengolahan dan pengawetan kedelai untuk bahan makanan manusia (Preparing and conserving tempeh for human consumption). Seminar Kedelai, Bogor.

Harsono, Hardjohutomo. 1958. Oxalis corniculata bagi pembikinan bongkrek (O.c. in the preparation of bongkrek). Kongres Ilmu Pengetahuan Nasional 1 (Report at the First National Science Congress) Malang, Indonesia. Vol. C. p. 117.

Harsono, Hardjohutomo. 1970. Pengganti tempe bongkrek (Substitutes for tempeh). Jakarta: Penerbit Pradnja Paramita.

Hartadi, S. 1980. Inoculum preparation for tempe and soysauce fermentation. In ASEAN 1980. pp. 256-62.

Hedger, J. N. 1982. Production of tempe, an Indonesian fermented food. In Primrose and Wardlaw eds. Sourcebooks of Experiments for the Teaching of Microbiology. pp. 597-602.

Heltova, Olivia. 1980. Tempeh de soya (Soy tempeh). El Alimento de Indonesia. Natura(Mexico). Dec. pp. 58-71.

Hermana and Sutedja. 1970. Advances in the preparation of tempeh. Part I. New method of preparing tempeh. Journal of the Indonesian Nutrition Assoc. GIZI Indonesia 2(3):167-68.

Hermana, and Roedjito, S. W. 1971. Preparation of tempe mold inoculum and observation on its activity during storage. (In Indonesian.) Penelitian Gizi dan Makanan 1:52-60.

Hermana. 1972. Tempe—An Indonesian fermented soybean food. In Stanton 1972. pp. 55-62.

Hermana, Roedjito, S. W., and Karjadi, D. 1972. Preparation of tempeh mold inoculum and observation on its activity during storage. Presented at the Fourth International Fermentation Symposium, Kyoto, Japan.

Hermana, Roedjito, S. W., and Karjadi, D. 1973. Advances in the preparation of tempeh. Part II. Preparation of tempeh mold inoculum and observation on its activity during storage. Journal of Indonesian Nutrition Assoc., GIZI, Indonesia.

Hermana. 1974. Saving the protein waste from processing of legumes of Indonesia. ASEAN 7FA/Wrks. GL1/Wop-13. Bogor, Indonesia: Ministry of Agriculture, Nutrition Research Institute.

Hesseltine, C. W. 1962. Research at Northern Regional Research Laboratory on fermented foods. In USDA/NRRC 1962. pp. 74-82.

Hesseltine, C. W., Camargo, R. de., and Rackis, J. J. 1963a. A mold inhibitor in soybeans. Nature 200:1226. Dec. 21.

Hesseltine, C. W., Smith, M., Bradle, B., and Ko, S. D. 1963b. Investigations of tempeh, an Indonesian food. Developments in Industrial Microbiology 4:275-87.

Hesseltine, C. W. 1965. A millenium of fungi, food, and fermentation. Mycologia 57:149-97.

Hesseltine, C. W. and Martinelli, A. 1966. Tempeh production in perforated bags. U.S. Patent 3,228,773. Jan. 11.

Hesseltine, C. W. and Smith, M. 1966. Cereal-containing varieties of tempeh and process therefor. U.S. Patent 3,243,301.

Hesseltine, C. W. et al. 1966a. Aflatoxin formation by Aspergillus flavus. Bacteriological Reviews 30:795-805.

Hesseltine, C. W. and Wang, H. L. 1967. Traditional fermented foods. Biotechnology and Bioengineering 9:275-88.

Hesseltine, C. W. 1967. Fermented products—Miso, sufu, and tempeh. In USDA/ARS 1967. pp. 170-80.

Hesseltine, C. W., Smith M., and Wang, H. L. 1967. New fermented cereal products. Developments in Industrial Microbiology 8:179-86.

Hesseltine, C. W. and Wang, H. L. 1969. Oriental fermented foods made from soybeans. Proceedings of the Ninth Dry Bean Research Conference, USDA ARS-74-50. pp. 45-62.

Hesseltine, C. W. and Wang, H. L. 1972. Fermented soybean food products. In Smith and Circle 1972. pp. 389-419.

Hesseltine, C. W., Swain, E. W., and Wang, H. L. 1976. Production of fungal spores as inocula for Oriental fermented foods. Developments in Industrial Microbiology 17:101-15.

Hesseltine, C. W. and Wang, H. L. 1977. Contributions of the Western world to knowledge of indigenous fermented foods of the Orient. Presented at Fifth International Conference on Global Impacts of Applied Microbiology, Nov. 1977, Bangkok, Thailand. 32 pp. In Steinkraus 1983. pp. 607-22.

Hesseltine, C. W. and Wang, H. L. 1979. Fermented foods, Chemistry and Industry. June 16. pp. 393-99.

Hunter, Beatrice T. 1973. Fermented Foods and Beverages. New Canaan, CT: Keats Publ. Co. 116 pp.

Ikehata, H. et al. 1964. Hakko daizu tempeh ni kansuru kenkyu. II. Koyoketsu busshitsu no seishitsu (Research on tempeh: II. The nature of hemolytic substances). (Lecture.) Eiyo 17(1):35.

Ikehata, H. et al. 1965. Hakko daizu tempeh ni kansuru kenkyu. II. Koyoketsu-sei busshitsu bunriho no kento (Research on tempeh: II. Investigations on separating hemolytic substances). (Lecture.) Eiyo to Shokuryo 18(5):13.

Ikehata, H., Wakaizumi, M., and Murata, K. 1968. Antioxidant and antihemolytic activity of a new isoflavone "Factor 2" isolated from tempeh. Agricultural and Biological Chemistry 32(6):740-46.

Ikehata, H. 1976. Daizu no furabonoido (Soybean flavonoids). Shokuhin Kogyo 10(16):42-47.

Iljas, Nasruddin. 1969. Preservation and Shelf-Life Studies of Tempeh. Ohio State University. M.S. thesis.

Iljas, N., Peng, A. C., and Gould, W. A. 1970. Tempeh. Find ways to preserve Indonesian soy food. Ohio Report 55:22. Jan-Feb.

Iljas, N. 1972. Development and Quality Evaluation of Soybean-Based Food—Tempeh. Ohio State University. Ph.D. Thesis. 146 pp.

Iljas, N., Gould, W. A. and Peng, A. C. 1973. New soybean food made from tempeh. Ohio Reports 58(6):125-26.

Iljas, N., Peng, A. C. and Gould, W. A. 1973. Tempeh: An Indonesian fermented soybean food. Wooster, OH: Ohio Agricultural Research and Development Center, Dept. of Horticulture. Horticulture Series No. 394. 36 pp.

Indonesian Department of Agriculture. 1967. Mustika Rasa (Gems of Taste: Indonesian Cookery). Jakarta: IDA. 1,123 pp.

Inui, T., Takeda, Y., and Iizuka, H. 1965. Taxonomical studies on the genus Rhizopus. Journal of General and Applied Microbiology 11:1-121. Supplement.

Jansen, B. C. P. 1923. Over de behoefte van het dierlijk organisme aan anti-beri-beri-vitamine, en over het gehalte van verschillende voedingsmiddelen aan dit vitamine (On the need of the animal organism for anti-beri-beri-vitamins, and on the content of this vitamin in various foods). Mededeelingen van de Burgelijke Geneeskundige Dienst van Nederlandsch Indie 1:68-

Jansen, B. C. P. and Donath, W. P. 1924. Over stofwisselingsproeven met ratten en over de verteerbaarheid der eiwitten van enige voedingsmiddelen (On metabolic tests with rats and on the digestibility of the protein in various foods). Mededeelingen van de Burgelijke Geneeskundige Dienst van Nederlandsch Indie 1:26-

Jensen, J. S. and Djurtoft, R. 1976. Preparation of Tempeh, a Fermented Food from Whole Grain Legumes and Cereals. Lyngby, Denmark: Technical Dept. of Biochemistry and Nutrition, Technical University of Denmark. 120 pp.

Jurus, A. M. and Sundberg, W. J. 1976. Penetration of Rhizopus oligosporus into soybeans in tempeh. Applied and Environmental Microbiology 32(2):284-87.

Jutono, T. et al. 1979. The preparation of usar, a traditional tempe inoculum. ASEAN Project Report.

Kanasugi, G. 1983. Tenpe no hokoku to setsumei (Tempeh: Report and explanation). Zenkoku Shokuhin Shimbun 453. June 1. p. 2.

Kao, C. 1974. Fermented Foods from Chickpea, Horsebean and Soybean. Kansas State University. Ph.D. thesis. 143 pp.

Kasuya, R. et al. 1967. Natto to tempeh no B-rui bitamin (B-vitamins in natto and tempeh). Kaseigaku Zasshi 18:362-64.

Kawarai, Fumiko, et al. 19?? Tenpe hakko katei ni okeru seibun no henka (Changes in tempeh during fermentation).

Kendig, J. 1984. Tempeh. Nutrition Health Review. Winter. pp. 20-21.

Keough, C. 1980. Tempeh—Really good, really cheap food. Organic Gardening. March. pp. 122-28.

Khumaidi, M. 1976. Role of Soybeans in Patterns of Indonesian Diets. Unpublished M.S. thesis. 57 pp.

Kidby, D. K., McComb, J. R., Snowdon, R. L., Garcia-Webb, P., and Gladstones, J. S. 1977. Tempeh production from Lupinus angustifolius L. Presented at Symposium on Indigenous Fermented Foods, Bangkok, Thailand. Summarized in Steinkraus 1983.

Ko, S. D. and Hesseltine, C. W. 1961. Indonesian fermented foods. Soybean Digest. Nov. p. 14-15.

Ko, S. D. 1964. Tempe, a fermented food made from soybeans. Presented at the International Symposium on Oilseed Protein Foods, May 11-16, Tokyo, Japan. 17 pp.

Ko, S. D. 1965. Tindjuan Terhadap. Penelitian Fermented Foods Indonesia (Observations and Investigations on Indonesian Fermented Foods). In Research di Indonesia 1945–1965, vol. 2. (R. M. Soemantri, ed.) Jakarta: P. N. Balai,Pustaka. pp. 209-23.

Ko, S. D. 1965. Tempe. In Research di Indonesia: 1945–1965. (M. Makaiansar and R. M. Soemantri, eds.) Vol. 2. Bidang Teknologi dan Industri. Jakarta: Departemen Urusan Research Nasional Republik Indonesia, p. 312.

Ko, S. D. 1974. Self-protection of fermented foods against aflatoxin. Proceedings of the Fourth International Congress on Food Science and Technology 3:244-53.

Ko, S. D., Kelholt, A. J., and Kampelmacher, E. H. 1977. Inhibition of toxin production in tempe bongkrek. Presented at Symposium on Indigenous Fermented Foods, Bangkok, Thailand. Summarized in Steinkraus 1983.

Ko, S. D. and Hesseltine, C. W. 1979. Tempe and related foods. In Economic Microbiology. Vol. 4, Microbial Biomass. A. H. Rose, ed., pp. 115-40. Academic Press.

Ko, S. D. 1981. Fermented foods of Indonesia except those based on soybeans. Advances in Biotechnology 2:525-30.

Kronenberg, C. 1980. Tempeh microbiology: A technique for culture preservation. Unpublished manuscript. 7 pp.

Kronenberg, C. 1981. Synthesis of an Antibiotic by Rhizopus oligosporus. Unpublished manuscript. 14 pp.

Kronenberg, C. 1983. Engineering tempeh incubators. Soyfoods. Summer pp. 56-62.

Kuehn, M. 1981. Tempe—Rich in protein, vitamins and minerals. Bestways. March. p. 79.

Kushi, Aveline. 1981. My favorite tempeh recipes. East West Journal. August. pp. 62-65.

Latuasan, H. E. and Berends, W. 1961. On the origin of the toxicity of toxoflavin. Biochemica et Biophysica Acta 52:502-08.

Leung, W-T. W. et al. 1972. Food Composition Table for Use in East Asia. U.S. Dept. of Health, Education and Welfare, and FAO. 334 pp.

Leviton, R. 1982. Tofu, Tempeh, Miso and Other Soyfoods. New Canaan, CT: Keats Publishing Inc. 26 pp.

Leviton, R. et al. 1982. The coming tempeh boom. Soyfoods No. 6. Winter. pp. 26-34.

Leviton, R. 1984. Tempeh in America: Selling sizzle, not steak. Whole Foods. March. pp. 28-29.

Lie, G. H. et al. 1976. Nutritive value of various legumes used in the Indonesian diet. In ASEAN Grain Legumes. M. A. Rifai, ed. Bogor, Indonesia: Central Research of Agriculture, Dept. of Agriculture.

Liem, Irene T. H., Steinkraus, K. H., and Cronk, T. C. 1977. Production of vitamin $B_{12}$ in tempeh, a fermented soybean food. Applied and Environmental Microbiology 34(6):773-76.

Loegito, Mas. 1977. Aflatoxins in tempeh. Presented at Symposium on Indigenous Fermented Foods, Bangkok, Thailand.

Loegito, Mas, and Soeparmo. 1977. Studies on aflatoxin concentration of tempe menjes and tempe kapuk, fermented foods made of disposal material in Malang municipality. Malang, Indonesia: Unpublished manuscript. 4 pp.

Lockwood, L. B., Ward, G. E., and May, O. E. 1936. The physiology of Rhizopus oligosporus. Journal of Agricultural Research 53:849-57.

Lyon, Alexander. 1974. Tempeh Instructions. Summertown, TN: The Farm. 3 pp.

Mallory, M. 1983. Introducing tempeh. Oakland Tribune. Wed., May 4, pp. D1-D3.

Martinelli, A. F. and Hesseltine, C. W. 1964. Tempeh fermentation: Package and tray fermentation. Food Technology 18(5):167-71.

Marusan-Ai. 1984. Myonichi no shoku seikatsu o ninau. Atarashii daizu tanpaku shokuhin (A new soy protein food for tomorrow). Okazaki city, Aichi prefecture, Japan: Marusan-Ai. 27 pp.

McComb, John. 1977. A Study of the Use of Sweet Lupins in Tempeh, an Oriental Food Fermentation. University of Western Australia. Bachelor of Agricultural Science thesis. 127 pp.

Mertens, W. K. and van Veen, A. G. 1933. De bongkrek-vergiftigingen in Banjoemas (Bongkrek poisoning in Banyumas). I. Geneeskundig Tijdschrift voor Nederlandsch Indie 73:1223-54.

Moore, Karen. 1979. Tofu, a Far East import offers potential as meat, fish, cheese substitute. Food Product Development 13(5):24. May.

Muljokusomo. 1962. Tempe dan Oncom (Tempeh and Onchom). Bandung: Penerbit Tarate. 36 pp.

Murata, K. 1963. Detection and isolation of the antioxidants from fermented soya beans: tempeh. Sixth International Congress of Nutrition, Edinburgh. Aug. p. 83.

Murata, Kiku. 1965. Nutritional value of tempeh. Annual Report of Research Conducted Under Grants Authorized by PL-480. Grant No. FG-Ja-110; All-NH-1. 15 pp.

Murata, K. and Ikehata, H. 1966. Hemolysis preventing antioxidant activity of synthesized 6.7.4'—trihydroxyisoflavone and that isolated from tempeh. Proceedings of the Seventh International Congress of Nutrition 5:656-59.

Murata, K., Ikehata, H. and Miyamoto, T. 1967. Studies on the nutritional value of tempeh. Journal of Food Science 32(5):580-86.

Murata, K. et al. 1968. Biosynthesis of B vitamins with Rhizopus oligosporus. Journal of Vitaminology 14:191-97.

Murata, K. et al. 1970. Studies on the nutritional value of tempeh. Part III. Changes in biotin and folic acid contents during tempeh fermentation. Journal of Vitaminology 16(4):281-84.

Murata K. 1970. Nutritive value of tempeh. Osaka Shiritsu Daigaku Kaseigakubu Kiyo 18:19-33.

Murata, K. et al. 1971. Studies on the nutritional value of tempeh. Part II. Rat feeding test with tempeh, unfermented soybeans, and tempeh supplemented with amino acids. Agricultural and Biological Chemistry 35(2):233-41.

Murata, K. 1977. Antioxidants and vitamins in tempeh. Presented at Symposium on Indigenous Fermented Foods, Bangkok, Thailand. Summarized in Steinkraus 1983.

Nakano, Masahiro. 1959. FAO Asia chiiki shokuhin kako kaigi ni shusseki shite (Attending the FAO Asian food processing conference). Shokuhin Kogyo Gakkai-shi 6(6):292-302.

Nakano, Masahiro, ed. 1967. Hakko Shokuhin (Fermented Foods). Tokyo: Korin Shoin. 244 pp.

Nakano, Masahiro, 1979. Tezukuri no kenko shokuhin. Hakko riyo no subete (Handmade healthy fermented foods). Tokyo: Nosan Gyoson Bunka Kyokai. 227 pp.

Nakazawa, Ryoji and Takeda, Yoshito. 1928. Nanyo-san ontjom, tempeh o tsukuru shijokin ni tsuite (On the filamentous fungi used to make onchom and tempeh in the South Pacific). Taiwan Sotokufu Chuo Kenkyusho, Kogyo-bu Hokoku 4:252-63.

Nakazawa, Ryoji. 1950–64. Bibliography of Fermentation and Biological Chemistry (Hakko oyobi Seibutsu Kagaku Bunken-shu).

11 vols. Tokyo: Nihon Gakujitsu Shinkokai/Hirokawa Publ. Co. (In European languages and Romanized Japanese.)

Nichibei Daizu Chosa-kai. 1960. Temupe no seizo-ho (Tempeh production methods). Tokyo: NDC. No. 130. 4 pp.

Nikkei Sangyo Shimbun. 1983. Daizu hakko no shizen shokuhin (A natural fermented soyfood—tempeh). June 30.

Nikkei Sangyo Shimbun. 1984. Indonesia daizu shizen shokuhin. Nihon de tenpe kokusai kaigi (An Indonesian natural soyfood. International tempeh symposium in Japan). July 13.

Nofziger, M. 1981. Tempeh and soy yogurt. Vegetarian Times 47. pp. 60-63.

Noguchi, Kazuko. 1984. Tenpe no seijo to chori kako (Tempeh, characteristics and cooking). Shokuhin Kogyo. June (2):38-46.

Noor, Muhammad, and Wahab, Abdul. 1975. Proses Pembuatan Tempe (Tempeh Production). Unpublished thesis from Biology Dept. of ITB in Bandung.

Noparatnaraporn, N. et al. 1977. Factors affecting fermentation and vitamin $B_{12}$ content in tempeh and tempeh-like products. Presented at Symposium on Indigenous Fermented Foods, Bangkok, Thailand. 10 pp. Summarized in Steinkraus 1983.

Noranizam, H. M. L. 1979. Studies of Rhizopus isolated from tempe and soy sauce. B.Sc. Honors Project. University of Malaysia.

Norton, R. and Wagner, M. 1980. The Soy of Cooking: A Tofu and Tempeh Recipe Book. Eugene, OR: White Crane, P.O. Box 3081, 97403. 24 pp. 1983 revised ed. 58 pp.

Noznick, P. P. and Luksas, A. J. 1970. Process for making tempa. U.S. Patent 3,489,570. Jan. 13. Assigned to Beatrice Food Co. 2 pp.

Nugteren, D. H., and Berends, W. 1957. Investigations on bongkrek acid, the toxine from Pseudomonas cocovenenans, Recueil des Travaux des Pays-Bas 76:13-27.

Ochse, J. J. 1931. Vegetables of the Dutch East Indies. Buitenzorg (Bogor), Java: Archipel Drukkerij. pp. 366, 372, 389-93, 398, 407-08, 732, 943-71.

Oda, Lorraine. 1983. Hawaii's tempeh pioneers. The Hawaii Herald. Fri., Oct. 7. p. 6.

Ohta, T., Ebine, H. and Nakano, M. 1964. Tenpe (Tempeh) ni kansuru kenkyu. 1. Indonesia-san tenpe funmatsu no hinshitsu to seijo ni tsuite (Research on tempeh: I. On the quality and characteristics of Indonesian-made tempeh). Shokuryo Kenkyujo, Kenkyu Hokoku 18:67-68.

Ohta, T., Ebine, H. and Nakano, M. 1964. Study on tempeh. Part 1. On the property of tempeh powder made in Indonesia. Report of the Food Research Institute, Tokyo 18:69.

Ohta, T. 1965. Tenpe. Nihon Jozo Kyokai Zasshi 60(9):778-83.

Ohta, T. 1971. Tenpe. In Watanabe et al. 1971. pp. 208-17.

Ohta, T. 1984. Seijinbyo o fusegu to Amerika de hyoban no Jawa Natto, tenpe (Tempeh, Java natto, prevents geriatric diseases and is popular in America). Watashi no Kenko. April. pp. 67-69.

Olszewski, N. 1978. Tofu Madness. Vashon, WA: Island Spring. 64 pp.

O'Neill, K. 1980. Tempe—a traditional food for tomorrow. Indonesian Circle No. 21, March. pp. 54-59.

O'Neill, K. 1981. America, God willing, becomes a tempe nation. Tempo. June 13. pp. 49-50.

Organic Gardening. 1982. Faster tempeh. March. pp. 102-07.

Packett, L. V. et al. 1971. Antioxidant potential of tempeh as compared with tocopherol. Journal of Food Science 36:798-99.

Podems, Marc. 1976. Comments on Organic Gardening and Farming Tempeh Questionnaire results. Emmaus, PA: Interoffice memo. July 21. 6 pp.

Prawiranegara, Dradjat D. 1960. Food and utilization of food resources. Proceedings, Fourth Pan Indian Ocean Science Congress. Section G. Human Ecology p. 55.

Pride, Colleen. 1984. Tempeh Cookery. Summertown, TN: The Book Publishing Co. 127 pp.

Prinsen Geerligs, H. C. 1895. Eenige Chineesche voedingsmiddeln uit Sojaboonen bereid (Some Chinese foods made with soybeans). Pharmaceutisch Weekblad Voor Nederland 32(33):1-2. Dec. 14.

Prinsen Geerligs, H. C. 1896. Einige chinesische Sojabohnenpräparate (Some Chinese soybean preparations). Chemiker-Zeitung 20(9):67-69 (Jan. 29).

Prinsen Geerligs, H. C. 1917. Ueber die anwendung von enzymwirkungen in der Ostasiatischen hausindustrie (On the application of enzymes in East Asian cottage industries). Zeitschrift für Angewandte Chemie, Wirtschaftlicher Teil 30(37):256-57.

Raintree, J. B. 1980. Leucaena tempe. League for International Food Education Newsletter. Oct. pp. 1-2.

Rathbun, Bonnie L. 1982. Mass and heat transfer effects in tempeh, a solid substrate fermentation. Ithaca, NY: Cornell University. M.S. thesis.

Rathbun, B. L. and Shuler, M. L. 1983. Heat and mass transfer effects in static solid substrate fermentations: Design of fermentation chambers. Biotechnology and Bioengineering 25:929-38.

Rivai, Abdul. 1980. Optimization of tempeh manufacture from eight soybean varieties and from okara. St. Paul, MN: University of Minn. Dept. of Food Science and Nutrition. M.S. Thesis. 63 pp.

Robeau, Alec. 1970. Cooking the Indonesian Way. New South Wales: A. H. & A. W. Reed.

Robinson, R. J. and Kao, C. 1974. Fermented foods from chickpea, horse bean, and soybean. Cereal Science Today 19:397. Abstract.

Robinson, R. J. and Kao, C. 1977. Tempeh and miso from chickpea, horse bean, and soybean. Cereal Chemistry 54:1192-97.

Rodale, Robert. 1977. Tempeh, a new health food opportunity. Prevention. July. p. 25-32.

Roelofsen, P. A. 1946. Tempeh-bereiding in Krijgsgevangenschap. Vakblad voor Biologen 26:114-16.

Roelofsen, P. A. and Thalens, A. 1964. Changes in some B vitamins during molding of soybeans by Rhizopus oryzae in the production of tempeh kedelee. Journal of Food Science 29(2):224-26.

Rothert, Y. 1981. Use of soybean well known but here comes tempeh. The Oregonian. June 24.

Rumphius, Georgius Everhardus. 1747. Herbarium Amboinese. Amstelaedami Vol. 5. p. 388.

Rusmin, S. and Ko, S. D. 1974. Rice-grown Rhizopus oligosporus inoculum for tempeh fermentation. Applied Microbiology 28(3):347-50.

Ruttle, J. 1977. Tempeh keeps 'em coming for more soybeans. Organic Gardening and Farming. Jan. pp. 103-10.

Saito, K. 1905. Rhizopus oligosporus, ein neuer technischer Pilz Chinas (R. oligosporus, a new industrial fungus from China). Zentralblatt für Bakteriologie Abt. II. 14:623-27.

Saito, Takamichi. 1909. Yuyo Hakkokin (Useful Fermentation Microorganisms). Tokyo: Hakubunkan-zo Han. pp. 14-17, 108-09, 122-25, 144-47, 162-64.

Sanke, Y., Miyamoto, T., and Murata, K. 1971. Studies on the nutritional value of tempeh: IV. Biosynthesis of folate compounds with Rhizopus oligosporus. Journal of Vitaminology 17(2):96-100.

Saono, S., Gandjar, I., Basuki, T., and Karsono, H. 1974. Microflora of ragi and some other traditional fermented foods of Indonesia. Annales Bogorienses 5(4):187-204. Feb.

Saono, S. et al. 1976 a. Microbiological studies of tempe, kecap and taoco. I. The microbial content and its amylolytic, proteolytic, and lypolytic activities. Progress Report Subproject III. b. ASEAN Project for Soybean and Low-Cost High Protein Foods. Jan-Dec. 1976.

Saono, S. et al. 1976 b. Microbiological studies of tempe, kecap, and taoco. I. Quantitative estimation and isolation of microorganisms from some products from West Java. In ASEAN Project on Soybean and Protein Rich Foods, Progress Report on Research Activities, Jan.—May, 1976. Appendix 7.

Saono, S. 1977. Microbiological studies of tempe, kecap and taoco. II. The amylolytic and proteolytic activities of the isolates. Progress Report Subproject III. b. ASEAN Project for Soybean and Low-cost High Protein Foods. Jan—Dec. 1977.

Saono, S. and Basuki, T. 1978. The amylolytic, lipolytic and proteolytic activities of yeasts and mycelial molds from ragi and some Indonesian traditional fermented foods. Annales Bogorienses 6(4):207-19.

Sastroamijoyo, M. S. A. 1971. An answer to the world food crisis? Australian National University Reporter 2(19). Nov. 26.

Seguin, Clare. 1982. Cooking with Tempeh. Higher Ground Press. P.O. Box 3128, Madison, WI 53704. 63 pp.

Serat Centini. 1846. Codex Orientalis 1814 of the Leiden University Library, Vol. I, p. 295. Reprinted in the Verhandelingen of the Bataviaasch Genootschap (Batavia, 1912-15), Vol. I-II, p. 82, Canto 31, stanza 212.

Shallenberger, R. S. 1967. Changes in sucrose, raffinose, and stachyose during tempeh fermentation. Proc. 8th Res. Conf. on Dry Beans, Belaire, MI. Aug. 1966. USDA Rept. ARS-74-41. pp. 68-71.

Shurtleff, W. and Aoyagi, A. 1975. The Book of Tofu. Brookline, MA: Autumn Press. 336 pp. Revised ed. 1983. Ten Speed Press, Berkeley, CA.

Shurtleff, W. and Aoyagi, A. 1976. Tempeh. Mother Earth News #39. May. p. 43.

Shurtleff, W. and Aoyagi, A. 1977. What is Tempeh? Lafayette, CA: New-Age Foods Study Center. 14 pp.

Shurtleff, W. and Aoyagi, A. 1977. Favorite tempeh recipes. Organic Gardening. June. pp. 112-14.

Shurtleff, W. R. 1978. Household preparation of winged bean tempeh, tofu, milk, and sprouts. In The Winged Bean. Los Banos, Laguna, Phillippines: Phillippine Council for Agriculture and Resources Research (PCARR). pp. 335-39.

Shurtleff, W. and Aoyagi, A. 1979. How to make tempeh at home. Vegetarian Times. Sept/Oct. pp. 26-29.

Shurtleff, W. and Aoyagi, A. 1979. The Book of Tempeh: A Super Soyfood from Indonesia. New York: Harper & Row. Paperback 160 pp. Professional hardcover 248 pp.

Shurtleff, W. and Aoyagi, A. 1979. American style tempeh: Our favorite recipes. Vegetarian Times. Nov/Dec. pp. 34-35.

Shurtleff, W. and Aoyagi, A. 1980. Tasty homemade tempeh. East West Journal. March. pp. 62-67.

Shurtleff, W. and Aoyagi, A. 1980. Tempeh Production. Lafayette, CA: The Soyfoods Center. 176 pp.

Shurtleff, W. and Aoyagi, A. 1981. La Soya y Sus Derivados: Tofu, Miso, Tempeh. Quadernos de Natura (Mexico). No. 20. 87 pp. Dec.

Shurtleff, W. and Aoyagi, A. 1981. Tempeh production in the community. Communities, Journal of Cooperative Living. No. 49. pp. 3-10.

Shurtleff, W. and Aoyagi, A. 1982. Teriffic tempeh. Ziriuz (Australia). March. pp. 7-8.

Shurtleff, W. et al. 1982. Dealing with tempeh contamination. Soyfoods. Winter. pp. 29-32.

Shurtleff, W. and Aoyagi, A. 1982. Using Tofu, Tempeh and Other Soyfoods in Restaurants, Delis & Cafeterias. Lafayette, CA: The Soyfoods Center. 116 pp. 2nd ed. 1984, 181 pp.

Shurtleff, W. and Aoyagi, A., eds. 1982. Soyfoods Labels, Posters and Other Graphics. Lafayette, CA: The Soyfoods Center. 185 pp. 2nd ed. 1984, 6 volumes, 685 pp. total.

Shurtleff, W. and Aoyagi, A. 1984. Soyfoods Industry and Market: Directory and Databook. Lafayette, CA: The Soyfoods Center. 203 pp. 4th ed.

Shurtleff, W. and Aoyagi, A. 1984. History of Tempeh. Lafayette, CA: The Soyfoods Center, 102 pp.

Shurtleff, W. and Aoygai, A. 1984. Tempeh, Tempeh Production in USA, and Tempeh Production in Indonesia: 3 color slide shows with 75, 77, and 72 slides respectively. Lafayette, CA: The Soyfoods Center. Narration with each set.

Shurtleff, W. and Aoyagi, A. 1985. Das Tempeh Buch (The Book of Tempeh). D-8201 Pittenhart. W. Germany: Ahorn Verlag. In Press.

Shyeh, B. J., Rodda, E. D., and Nelson, A. I. 1980. Evaluation of new soybean dehuller. Transactions of the ASAE 23(2):523-28.

Slamet, D. and Tarwotjo, I. G. 1971. The nutrient content of ontjom. Penelitian Gizi dan Makanan 1:49-52.

Smith, A. K. et al. 1963. Processing losses and nutritive value of tempeh, a fermented soybean food. Cereal Science Today 8:129.

Smith, A. K. et al. 1964. Tempeh: Nutritive value in relation to processing. Cereal Chemistry 41:173-81.

Smith, Dean A. and Woodruff, M. F. A. 1951. Deficiency diseases in Japanese prison camps. Privy Council. Medical Research Council Report Series. London. No. 274. p. 192.

Soetan, Sanif. 1956. Kedelai (Soybeans). Jakarta: Dinas Penerbitan Pustaka.

Sorensen, W. G. and Hesseltine, C. W. 1966. Carbon and nitrogen utilization by Rhizopus oligosporus. Mycologia 58:681-89.

Souser, M. L. and Miller, L. 1977. Characteristics of the lipase produced by Rhizopus oligosporus, the tempeh fungus. Abstracts of the Annual Meeting of the American Society for Microbiology 77:258.

Soyanews. 1981-83. Articles about tempeh in Sri Lanka. Colombo, Sri Lanka. P.O. Box 1024. Jan., May, June, July, 1981. Oct. 1983. June 1984.

Soybean Digest. 1965. Expect "entirely new foods" from Asian fermentations. May. pp. 26-27.

Soybean Digest. 1967. Cereals, soybeans combined in USDA food process. pp. 40-41.

Soybean Digest. 1967. USDA develops wheat-soy tempeh. Nov. p. 4.

Soybean Digest. 1975. Process for simulating meat. Dec. p. 31. Describes a sesame-soy tempeh developed by James Liggett of Foundation Foods, Inc.

Stahel, G. 1946a. Foods from fermented soybeans as prepared in the Netherlands Indies. II. Tempe, a tropical staple. Journal of the New York Botanical Garden 47(564):285-96.

Stahel, G. 1946b. Soy foods in New Guinea. Soybean Digest. Nov. pp. 14-15.

Stanton, W. R., Brook, E. J., and Wallbridge, A. J. 1968. Fermentation methods for protein enrichment of cassava. Presented at the Third International Fermentation Symposium, New Brunswick. Sept.

Stanton, W. R. and Wallbridge, A. J. 1969. Fermented food processes. Process Biochemistry 4(4):45-51.

Stanton, W. R. ed. 1972. Waste Recovery by Microorganisms: Selected Papers for the UNESCO/ICRO Work Study Held at the University of Malaya. Kuala Lumpur: Ministry of Education, Malaysia. 221 pp.

Steinkraus, K. H., Yap, B. H., Van Buren, J. P., Provvidenti, M. I., and Hand, D. B. 1960. Studies on tempeh—An Indonesian fermented soybean food. Food Research 25(6):777-88.

Steinkraus, K. H. et al. 1961. Studies on tempeh—An Indonesian fermented soybean food. In National Academy of Sciences, National Research Council Publication 843. pp. 275-79.

Steinkraus, K. H., Hand, D. B., Van Buren, J. P., and Hackler, L. R. 1962. Pilot plant studies on tempeh. In USDA NRRC 1962. pp. 83-92.

Steinkraus, K. H. 1964. Research on tempeh technology in the United States. Presented at the International Symposium on Oilseed Protein Foods, Tokyo. May 10-16. 9 pp.

Steinkraus, K. H., Van Buren, J. P., Hackler, L. R., and Hand, D. B. 1965. A pilot plant process for production of dehydrated tempeh. Food Technology 19(1):63-68.

Steinkraus, K. H., Lee, C. Y. and Buck, P. A. 1965. Soybean fermentation by the ontjom mold Neurospora. Food Technology 19(8):119-20.

Steinkraus, K. H. and van Veen, A. G. 1971. Biochemical, nutritional and organoleptic changes occurring during production of traditional fermented foods. In Global Impacts of Applied Microbiology, GIAM 3. (Y. M. Freitas and F. Fernandez, eds.). University of Bombay. pp. 444-50.

Steinkraus, K. H. 1974. Research on traditional Oriental and Indian fermented foods. Current Science and Technology, Special Report No. 16. April. pp. 10-13 (Cornell Univ.)

Steinkraus, K. H. 1975. Exotic fermented foods for Americans. New York's Food and Life Sciences Quarterly 8(1):8-11.

Steinkraus, K. H. 1977a. Contributions of Asia to Western food science. In Global Impacts of Applied Microbiology. GIAM: State of the Art 1976 and Its Impact on Developing Countries. W. R. Stanton and E. J. DaSilva, eds. Universiti Malaya Press.

Steinkraus, K. H. 1977b. Enhancement of food quality by fermentation. New York State Agr. Exp. Station Special Report No. 26.

pp. 21-25.

Steinkraus, K. H. 1978. Tempeh—An Asian example of appropriate/intermediate food technology. Food Technology 32(4):79-80. April.

Steinkraus, K. H. 1979. Transfer of tempe technology. League for International Food Education Newsletter. December. pp. 1-2.

Steinkraus, K. H. 1982. Fermented foods and beverages: The role of mixed cultures. In Microbial Interactions and Communities, Vol. 1: (A. T. Bull and J. H. Slater, eds.) Academic Press. pp. 407-42.

Steinkraus, K. H. ed. 1983. Handbook of Indigenous Fermented Foods. New York: Marcel Dekker. 671 pp.

Steinkraus, K. H. 1983. Fermented foods, feeds and beverages. Biotechnology Advances 1:31-46.

Steinkraus, K. H. 1983. Industrial applications of Oriental fungal fermentations. In The Filamentous Fungi. Vol. IV. Fungal Technology. (J. E. Smith, D. R. Berry and B. Kristiansen, eds.) pp. 171-89.

Steinkraus, K. H. 1984. Solid-state (solid-substrate) food/beverage fermentations. Acta Biotechnologica 4(2):83-88.

Stillings, B. R. and Hackler, J. P. 1965. Amino acid studies on the effect of fermentation time and heat-processing of tempeh. Journal of Food Science 30:1043-1048.

Sudarmadji, S. 1975. Certain Chemical and Nutritional Aspects of Soybean Tempe. Michigan State University, Ph.D. thesis. 155 pp.

Sudarmadji, S. and Markakis, P. 1977. The phytate and phytase of soybean tempeh. Journal of the Science of Food and Agriculture 28:381-83.

Sudarmadji, S. and Markakis, P. 1978. Lipid and other changes occurring during the fermentation and frying of tempeh. Food Chemistry 3(3):165-70.

Sugimoto, Y. and Murata, K. 1972. Studies on the antioxidative factors in tempeh. Osaka Shiritsu Daigaku Kaseigakubu Kiyo 20:13-19. (In Japanese with English summary.)

Suhadi, D. et al. 1979. Studies on the performance of selected Rhizopus strains in tempe fermentation. ASEAN Project Report.

Sunset Magazine. Nov. 1983. Indonesian tofu? It is versatile, cheese-like tempeh. p. 256.

Takamine, Kazuhiro. 1984. Tenpe no seizoho to riyo o meguru gijutsu-teki hatten (Technical advances in tempeh production and utilization). Shokuhin Kogyo. June. (2):25-33.

Tammes, P. M. L. 1950. De bereiding van tempe (The preparation of tempeh). Landbouw 22:267:70.

Tanuwidjaja, Lindayati. 1977. Utilization of defatted soybean flour in tempeh fermentation. Presented at Symposium on Indigenous Fermented Foods, Bangkok, Thailand. Summarized in Steinkraus 1983.

Thio, G. L. 1972. Introduction of soybeans for human nutrition, Republic of Zambia. Amsterdam, Netherlands: Royal Tropical Institute, Dept. of Agric. Research. 48 pp.

Thio, G. L. 1975. Small Scale and Home Processing of Soya Beans with Applications and Recipes. Amsterdam, Netherlands: Royal Tropical Institute; Dept. of Agric. Research Communication 64. 51 pp. Revised ed. 1978. No. 64a. 59 pp.

Today's Living. 1974. How can you use soybeans? Let us count the ways. Dec. pp. 20-23, 46, 48.

USDA NRRC. 1962. Proceedings of Conference on Soybean Products for Protein in Human Foods. Peoria, IL: USDA Northern Regional Research Laboratory. 242 pp. Conference was Sept. 13-15, 1961.

USDA ARS. 1967. Proceedings of International Conference on Soybean Protein Foods. Peoria, IL: USDA Agricultural Research Service. ARS-71-35. 285 pp. Conference was Oct. 17-19, 1966.

Vaidehi, M. P. 1981. A few soyabean products requiring better attention. The Lal-Baugh Journal 26(2). April-June.

Vaidehi, M. P. and Vijayakumari, J. 1981. Soya Delights: Recipes for the Use of Soybean. Hebbal, Bangalore 560-024, India: University of Agricultural Sciences. 102 pp.

van Buren, J. P., Hackler, L. R., and Steinkraus, K. H. 1972. Solubilization of soybean tempeh constituents during fermentation.

Cereal Chemistry 49(2):208-11.

Van Damme, P. A., Johannes, A. G., Cox, H. C. and Berends, W. 1960. On toxoflavin, the yellow poison of pseudomnas co-covenenans. Recueil des Travaux Chimiques des Pays-Bas 79:255-67.

van Veen, A. G. 1932. Over het B₂ vitamine gehalte van verschillende Indische voedingsmiddelen (On the vitamin B-2 content of various Indonesian foods). Geneeskundig Tijdschrift voor Nederlandsch Indie 72:1377-99.

van Veen, A. G. and Mertens, W. K. 1933a. De bongkrek vergiftigingen in Banjoemas. II (Bongkrek poisoning in Banyumas). Geneeskundig Tijdschrift voor Nederlandsch Indie 73:1309-34.

van Veen, A. G. and Mertens, W. K. 1933b. On the isolation of a toxic bacterial pigment. Proceedings of the Acad. Sci. Amsterdam. 36:666-70.

van Veen, A. G. and Mertens, W. K. 1934a. Die giftstoffe der sogenannten bongkrek-vergiftungen auf Java (The poison in the so-called bongkrek poisoning in Java). Recueil des Travaux Chimiques des Pays-Bas 53:257-66. and Chemical Abstracts 28:2426.

van Veen, A. G. and Mertens, W. K. 1934b. Das toxoflavin, der gelbe Giftstoff der Bongkrek (Toxoflavin, the yellow poison in bongkrek). Recueil des Travaux Chimiques des Pays-Bas 53:398-404.

van Veen, A. G. and Mertens, W. K. 1935. Over het Bacterium cocovenenans en zijn verspreiding (On the Bacterium co-covenenans and its propagation). Publ. Geneeskundig Lab. Dienst Volksgez. Nederlandsch Indie.

van Veen, A. G. 1935. Het B₁ gehalte van voedingsmiddelen (The vitamin B-1 content of foods). Geneeskundig Tijdschrift voor Nederlandsch Indie 75:2050-64.

van Veen, A. G. and Mertens, W. K. 1935. Die bongkreksäure, ein blutzucker-senkender stoff (Bongkrek acid, a substance that lowers blood sugar). Recueil des Travaux Chimiques des Pays-Bas 54:373-80. and Chemical Abstracts 29:4094-5.

van Veen, A. G. and Mertens, W. K. 1936. Der Einfluss der bongkrek-säure auf den kohlehydratstoffwechsel (The influence of bongkrek acid on carbohydrate metabolism). Archives Neerlandaises de Physiologie de l'Homme et des Animaux 21:73-103.

van Veen, A. G. and Baars, J. K. 1938. Ueber das Toxoflavin, ein isomeres von I-methyl-xanthin (On toxoflavin, an isomer of I-methyl-xanthin). Recueil des Travaux Chimiques des Pays-Bas 57:248-64.

van Veen, A. G. 1946. De voeding in de Japansche interneerings-kampen in Nederlandsch Indie (The food in Japanese concentration camps in the Netherlands Indies). Voeding 7:173-86.

van Veen, A. G. 1950. Bongkrek acid, a new antibiotic. Documenta Neerlandica et Indonesica de Morbix Tropicis 2:185-88.

van Veen, A. G. and Schaefer, G. 1950. The influence of the tempeh fungus on the soya bean. Documenta Neerlandica et Indonesica de Morbix Tropicis 2:270-81.

van Veen, A. G. 1962. Panel discussion on problems involved in increasing world-wide use of soybean products as foods—Possible contribution of FAO. In USDA NRRC 1962. pp. 210-13.

van Veen, A. G. 1967. The bongkrek toxins. In Biochemistry of Some Foodborne Microbial Toxins. R. I. Mateles and G. N. Wogan, eds. Cambridge, MA: MIT Press. pp. 43-50.

van Veen, A. G., Graham, D. C. W., and Steinkraus, K. H. 1968. Fermented peanut press cake. Cereal Science Today 13(3):96, 98-99.

van Veen , A. G. and Steinkraus, K. H. 1970. Nutritive value and wholesomeness of fermented foods. Journal of Agricultural and Food Chemistry 18(4):576-78.

van Veen, A. G. 1973. Toxic properties of certain unusual foods. In Toxicants Occurring Naturally in Foods, 2nd ed. Washington, DC: National Academy of Sciences, pp. 179-82.

Vorderman, A. G. 1893. Analecta op bromatologisch gebied I (Writings on mold-fermented foods. I). Geneeskundig Tijdschrift voor Nederlandsch Indie 33:359-60.

Vorderman, A. G. 1902. Analecta op bromatologisch gebied. IV (Writings on mold-fermented foods. IV). Geneeskundig Tijdschrift voor Nederlandsch Indie 42:395, 411-431.

Wagenknecht, A. C. et al. 1961. Changes in soybean lipids during tempeh fermentation. Journal of Food Science 26(4):373-76.

Wagner, M. 1980. Tempeh: Food for the future. New Age. July. pp. 68-71.

Walker, R. 1978. How to make tempeh. Alive. #18. p. 13.

Wang, H. L. and Hesseltine, C. W. 1965. Studies on the extracellular proteolytic enzymes of Rhizopus oligosporus. Canadian Journal of Microbiology 11:727-32.

Wang, H. L. and Hesseltine, C. W. 1966. Wheat tempeh. Cereal Chemistry 43(5):563-70.

Wang, H. L., Ruttle, D. I., and Hesseltine, C. W. 1968. Protein quality of wheat and soybeans after Rhizopus oligosporus fermentation. Journal of Nutrition 96 (1):109-14.

Wang, H. L., Ruttle, D. I., and Hesseltine, C. W. 1969. Antibacterial compound from a soybean product fermented by Rhizopus oligosporus. Proceedings of the Society for Experimental Biology and Medicine 131:579-83.

Wang, H. L. and Hesseltine, C. W. 1970. Multiple forms of Rhizopus oligosporus protease. Archives of Biochemistry and Biophysics 140:459-63.

Wang, H. L., Vespa, J. B., and Hesseltine, C. W. 1972. Release of bound trypsin inhibitors in soybeans by Rhizopus oligosporus. Journal of Nutrition 120(11):1495-1500.

Wang, H. L., Ellis, J. J., and Hesseltine, C. W. 1972. Antibacterial activity produced by molds commonly used in Oriental food preparations. Mycologia 64 (1):218:21.

Wang, H. L., Vespa, J. B., and Hesseltine, C. W. 1974. Acid protease production by fungi used in soybean food fermentation. Applied Microbiology 27:906-11.

Wang, H. L. et al. 1975a. Free fatty acids identified as antitryptic factor in soybeans fermented by Rhizopus oligosporus. Journal of Nutrition 105(10):1351-55.

Wang, H. L., Swain, E. W., and Hesseltine, C. W. 1975b. Mass production of Rhizopus oligosporus spores and their application in tempeh fermentation. Journal of Food Science 40:168-70.

Wang, H. L. et al. 1976. Directions for tempeh fermentation. Peoria, IL: USDA NRRC. 1 page. Sent to those ordering tempeh starter.

Wang, H. L. et al. 1977. An Inventory of Information on the Utilization of Unprocessed and Simply Processed Soybeans as Human Food. Peoria, IL: USDA Northern Regional Research Center. AID AG/TAB-225-12-76. 197 pp.

Wang, H. L., Swain, E. W., and Hesseltine, C. W. 1977. Calling all tempeh lovers. Organic Gardening and Farming. June. pp. 108-11.

Wang, H. L. and Hesseltine, C. W. 1979. Mold-modified foods. In Microbial Technology, 2nd ed. Vol. II, H. J. Peppler ed. New York: Academic Press. pp. 95-129.

Wang, H. L. and Hesseltine, C. W. 1981. Use of microbial cultures: Legume and cereal products. Food Technology. Jan. pp. 79-83.

Wang, H. L. 1984. Tofu and tempeh as potential protein sources in the Western diet. Journal of the American Oil Chem. Soc. 61(3):528-34.

Watanabe, Tadao. 1984. Tenpe to kosan kasei (Tempeh and antioxidative activity). Shokuhin Kogyo. June (2):35-37.

Watanabe, Tokuji, Ebine Hideo, and Ohta Teruo. 1971. Daizu Shokuhin (Soyfoods). Tokyo: Korin Shoin. 270 pp.

Wehmer, C. 1900. Die 'Chinesische Hefe' und der zogennante Amylomyces (= Mucor rouxii) (The Chinese yeast and the so-called Amylomyces). Centralblatt für Bakteriologie. Abt. II. 6:353-

Wehmer, C. 1900. Der javanische Ragi und seine Pilze (Javanese ragi and its fungi). Centralblatt für Bakteriologie. Abt. II. 6:610-19; and 1901 7:313-27.

Went, F. A. F. C. and Prinsen Geerligs, H. C. 1895. On Rhizopus Oryzae? Verhandling Koninklijk Acad. Wetenschap. 2 sections, 4 parts, Nos. 2, 3.

Went, F. A. F. C. and Prinsen Geerligs, H. C. 1895. Beobactungen ueber die Hefearten und zuckerbildenden Pilze der Arakfabrikation (Observations on the yeast varieties and saccharifying fungi used in making arak, rice brandy). Centralblatt für Bakteriologie. Abt. II. 1:501-

Went, F. A. F. C. and Prinsen Geerligs, H. C. 1896. Ueber die Hefearten und zuckerbildenden Pilze der Arakfabrikation (On the yeast varieties and saccharifying fungi used in making arak, rice brandy). Verhandelingen Koninkl. Akad. Wetenschap 4.3.

Went, F. A. F. C. 1901. Monilia sitophila (Mont.) Sacc., ein technischer Pilz Javas (Monilia sitophila, and industrial fungus from Java). Centralblatt für Bakteriologie. Series 2. 7:544-50; and 1902 8:313-

Winarno, F. G., Fardiaz, S. and Daulay, D. 1973. Indonesian fermented foods. Presented to Regional Graduate Nutrition Course, SEAMEO, Jan. Bogor Agricultural University. 25 pp.

Winarno, F. G. et al. 1976. The Present Status of Soybean in Indonesia. Bogor, Indonesia: FATEMETA, Bogor Agricultural University. 128 pp.

Winarno, F. G. and Karyadi, D. 1976. Nutrition and processing of soybeans. In Goodman 1976. pp. 137-42.

Winarno, F. G. 1979. Fermented vegetable protein and related foods of Southeast Asia with special reference to Indonesia. Journal of the American Oil Chemists' Soc. 56:363-66.

Wolf, Ray. 1981. Home Soyfood Equipment. Emmaus, PA: Rodale Press. 84 pp.

Yap, Bwee Hwa. 1960. Nutritional and Chemical Studies on Tempeh, an Indonesian Soybean Product. Report on Special Problem. Ithaca, NY: Cornell University. M.S. thesis. 50 pp.

Yeoh, Q. L. and Mercian, Z. 1977. Malaysian tempeh. Presented at Symposium on Indigenous Fermented Foods, Bangkok, Thailand. Summarized in Steinkraus 1983.

Yomiuri Shimbun. 1983. Niowanai natto (Natto that doesn't smell: Tempeh). May 20. Evening ed. p. 14.

Yueh, M. H. et al. 1979. Process for producing a fried snack containing tempeh. U.S. Patent 4,151,307. April 24. Assigned to General Mills Inc. 3 pp.

Zamora, R. G. and Veum, J. L. 1979. The nutritive value of dehulled soybeans fermented with Aspergillus oryzae or Rhizopus oligosporus as evaluated by rats. Journal of Nutrition 109:1333-39.

Zycha, H., Siepmann, R., and Linneman, G. 1969. Mucorales. D-3301, Germany: Verlag von J. Kramer.

# Index

Dan Rosenstrauch

William Shurtleff and Akiko Aoyagi Shurtleff spent their formative years on opposite sides of the Pacific. Born in California on 28 April 1941, Bill received degrees in engineering, honors humanities, and education from Stanford University. He taught physics for two years in Nigeria in the Peace Corps and has lived and traveled extensively in East Asia and Third World countries. He speaks seven languages, four fluently, including Japanese.

Akiko Aoyagi Shurtleff, born in Tokyo on 24 January 1950, received her education there from the Quaker-run Friends' School and the Women's College of Arts. She has worked as an illustrator and designer in Japan's modern fashion industry and America's emerging soyfoods industry.

Since 1972 Bill and Akiko have been working together, doing research and writing books about soyfoods. They have lived for six years in East Asia, mainly in Japan, studying with top soyfoods researchers, manufacturers, nutritionists, historians, and cooks. Over 500,000 copies of their eleven books on soyfoods are now in print. The titles and publishers are listed on the title page at the beginning of this book.

In 1976 Bill and Akiko founded The Soyfoods Center, and since that time they have worked to introduce soyfoods, especially traditional low-technology soyfoods, to the Western world. They feel that soyfoods can play a key role in helping to solve the world food crisis

while providing high-quality low-cost protein and healthier diets for people everywhere. Their work has led to the establishment of hundreds of soyfoods businesses making tempeh, tofu, soymilk, miso, and other soyfoods, and to the publication by others of more than 25 books about these foods. Their nationwide tours and many lectures, demonstrations, and media appearances have drawn widespread acclaim.

Their global view and uniquely holistic, interdisciplinary approach are aimed at presenting the best of both traditional lore and modern scientific knowledge about soyfoods in a language accessible to both laymen and professionals.

By constantly addressing the problems of world hunger, the suffering of human beings and animals, and the perennial longing for good health and liberation, they hope to make their work relevant everywhere and a force for planetary renaissance.

If you would like to help in the larger work related to soyfoods and world hunger, if you have questions or suggestions related to this book, or if you would like to receive a free copy of their Soyfoods Center Catalog, the authors invite you to contact them.

THE SOYFOODS CENTER
P.O. Box 234
Lafayette, CA 94549 USA
(Phone: 415-283-2991)

# TEMPEH PRODUCTION

### The Book of Tempeh, Volume II

The authors have prepared a technical manual, *Tempeh Production,* containing detailed information on the commercial preparation of tempeh on a large or small scale in developed or developing countries. Instructions are also included for making tempeh starter. Available from The Soyfoods Center.

# About the Soyfoods Center

The Soyfoods Center, founded in 1976 by William Shurtleff and Akiko Aoyagi, has offices in California. Our basic goals and activities are related to soyfoods and world hunger.

**Soyfoods:** Our center is, above all, a source of information about soyfoods, especially tofu, soymilk, tempeh, and miso, about which we have done extensive research and written books and recipe pamphlets. Like a growing number of people, we feel that soybeans will be one of the key protein sources of the future on planet Earth, and that both traditional and modern soyfoods from East and West will serve as important sources of delicious, high-quality, low-cost protein in the diets of people everywhere, regardless of their income. We are interested in each of the following soyfoods, listed here in what we consider to be their approximate order of potential worldwide importance: tofu (soybean curd), soy flour, soymilk, tempeh, shoyu (natural soy sauce), textured soy protein (TVP), miso, whole dry soybeans, soy protein isolates and concentrates, roasted soybeans or soynuts, fresh green soybeans, roasted whole soy flour (kinako), soy sprouts, yuba, and natto. We have developed hundreds of tasty and nutritious Western-style recipes for the use of these foods and compiled extensive, up-to-date information on their nutritional value, history, and production.

**World Hunger:** Presently more than 15,000,000 people die each year of starvation and malnutrition-caused diseases; three fourths of these are children. We constantly relate our work to this urgent problem of world hunger by studying and developing creative, low-cost, village-level methods for soyfood production using appropriate technology, by traveling and speaking in less developed countries, and by sending complementary copies of our publications to and communicating with key soyfoods researchers and producers in these countries.

**Meatless Diets:** Over half of all agricultural land in the United States is now used to grow crops (such as corn, soybeans, oats, and wheat) that are fed to animals. The affluent American diet is emerging as a major cause of world hunger as well as of degenerative diseases such as heart disease and cancer. Soyfoods, which are low in cost, high in protein, low in saturated fats, free of cholesterol, and relatively low in calories, can be used as delicious replacements for meats and dairy products as part of meatless or vegetarian diets. We encourage the adoption of such diets which help to make best use of the planets' precious food resources, are conducive to the development of a healthy body and clear mind, kind to animals, economical, and ecologically sound.

**Commercial Soyfood Production:** We encourage and aid people throughout the world in starting community or commercial production of soyfoods by providing technical manuals, technical advice, materials, and equipment. We have helped to establish the *Soyfoods Association of North America* (SANA) and its international publication *Soyfoods,* to found *Bean Machines, Inc.* (a company selling tofu and soymilk equipment), and to develop catalogs of large- and small-scale equipment. We have compiled various technical manuals and presently serve as consultants for a wide variety of companies.

**Lecture Demonstrations:** We have done more than one hundred programs relating to soyfoods for natural food groups, research scientists, food technologists, nutritionists, commercial producers, university audiences, international symposia, home economists, and cooking schools. We have also done numerous television and radio programs and cooking classes throughout the world. We welcome invitations.

**Soyfoods Library and Information Center:** We have the world's largest library of documents on soyfood and the soyfoods industry—8,000 documents from 1100 B.C. to the 1980s. Over 6,500 of these have computerized records, quickly accessible by 180 keywords.

**International Soyfoods Center Network:** Our growing network of Soyfoods Centers is helping to introduce soyfoods to countries around the world. Contact us if you'd like to start a branch in your country.

**New Lifestyles:** Our work is deeply involved in the development of lifestyles conducive to the welfare and survival of all beings on planet Earth. Thus we encourage voluntary simplicity, self-sufficiency (particularly food self-sufficiency on personal, regional, and national levels), right livelihood, a deeper understanding of selfless service, and of daily life and work as a spiritual practice, ecological awareness, wholistic health, appropriate technology, the rapid development and adoption of solar energy, and the phasing out of nuclear energy.

**Publications and Catalog:** Our Center has published a number of full-sized specialty books on soyfoods including *Tofu & Soymilk Production; Miso Production; Tempeh Production; Soyfoods Industry and Market: Directory and Databook;* and *History of Soybeans and Soyfoods.* We also provide a free catalog listing our other widely distributed books on tofu, miso, and tempeh, materials such as pamphlets, tofu kits, and slide shows related to soyfoods and a list of soyfoods manufacturers in North America and Europe.

**Your Financial Support and Help:** Our work, now reaching people throughout the world, is not supported by government or corporate funds. We do, however, welcome contributions of any size from individuals and private foundations to aid us in furthering the soyfoods revolution and helping to put an end to world hunger. We have established *Friends of The Center* for supporters willing to contribute $35.00 or more; smaller contributions are also welcomed. If you would like to contribute your time and energy to our work, please contact us:

Bill Shurtleff & Akiko Aoyagi Shurtleff
The Soyfoods Center
P.O. Box 234
Lafayette, CA 94549 USA
Tel. 415-283-2991